LONDON MATHEMATICAL SOCIETY LECTURE NOTE SERIES

Managing Editor:
Professor N.J. Hitchin,
Mathematical Institute, 24–29 St. Giles, Oxford OX1 3DP, UK

All the titles listed below can be obtained from good booksellers or from Cambridge University Press. For a complete series listing visit
http://publishing.cambridge.org/stm/mathematics/lmsn/

London Mathematical Society Lecture Note Series. 321

Surveys in Modern Mathematics

Edited by

VICTOR PRASOLOV
YULIJ ILYASHENKO
Independent University of Moscow

CAMBRIDGE
UNIVERSITY PRESS

CAMBRIDGE
UNIVERSITY PRESS

University Printing House, Cambridge CB2 8BS, United Kingdom

One Liberty Plaza, 20th Floor, New York, NY 10006, USA

477 Williamstown Road, Port Melbourne, VIC 3207, Australia

314-321, 3rd Floor, Plot 3, Splendor Forum, Jasola District Centre, New Delhi - 110025, India

103 Penang Road, #05-06/07, Visioncrest Commercial, Singapore 238467

Cambridge University Press is part of the University of Cambridge.

It furthers the University's mission by disseminating knowledge in the pursuit of education, learning and research at the highest international levels of excellence.

www.cambridge.org
Information on this title: www.cambridge.org/9780521547932

First published 2005

A catalogue record for this publication is available from the British Library

ISBN 978-0-521-54793-2 Paperback

Contents

The Independent University of Moscow and Student Sessions at the IUM

Student Sessions at the Independent University of Moscow is a new tradition in the mathematical life of Moscow. The Independent University (briefly the IUM) itself is a young child of the new times in Russia. In this introduction, a brief description of the IUM is presented.

The history of the IUM begins with a meeting held in the summer of 1991 at Moscow High School No. 57. The meeting was initiated by N. Konstantinov. According to his suggestion, the future team of the Independent University simply decided to start teaching university courses in mathematics, beginning in September 1991. The subsequent history of this meeting characterizes the historical period when it occurred. Had it taken place in Stalin's time, all the participants of the meeting would have been immediately arrested. Had it opened in Brezhnev's time, nothing would have resulted from the meeting. Since it happened in Gorbachev's time, it turned out to be the beginning of the history of the Independent University of Moscow. The founders of the Independent University instituted a small fund from which the IUM was supported during the first period of its work.

The founders were organized into the Scientific Council of the IUM, presided over by V. I. Arnold and consisting of A. A. Beilinson, the late R. L. Dobrushin, L. D. Faddeev, B. M. Feigin, Yu. S. Ilyashenko, A. G. Khovanskii, A. A. Kirillov, S. P. Novikov, A. N. Rudakov, M. A. Shubin, Ya. G. Sinai, and V. M. Tikhomirov. Professors P. Deligne and R. MacPherson, both of whom have actively supported the IUM since its foundation, are Honorary Members of the Scientific Council.

In the first years, the administration was carried out by N. Konstantinov with his students and friends working as assistants: S. Komarov was responsible for economic and financial matters, V. Imaikin prepared the lecture notes, M. Vyalyi organized the teaching process. During the first year, the IUM worked in the School of Informational Technologies near Moscow State University. During the next four academic years, Moscow High School No. 2 kindly invited the IUM to have classes in its building in the evenings. We are especially grateful to the director of the school, P. V. Khmilinskii, for his hospitality.

In 1994, the Prefect of the Central District of Moscow, A. Muzykantskii, proposed that we organize a new institution, related both to high school and university mathematics, to which a building might be officially presented by the authorities. The bureaucratic work needed for the functioning of this new institution and for solving numerous administrative problems related to getting the new building was enormous. We began to look for an executive director for this new institution who would be able to carry out this work. As I said to one of my older friends and colleagues, we needed a person who would be a professional in the administrative world and would understand our university ideals. "Don't bother," my friend answered, "such a person simply does not exist." But we were lucky to find two people of the kind we were dreaming about: I. Yashchenko and V. Furin, both alumni of the Moscow State University. At that time, both had successful enterprises; in parallel, I. Yashchenko continued his mathematical research work.

Producing all the necessary documentation was a full time job, and in half a year it resulted in a gift from the Moscow government: in June 1995, the Major of Moscow, Yu. Luzhkov, signed the ordinance giving a new institution, the Moscow Center of Continuous Mathematical Education, an unfinished building in the historical center of Moscow. The IUM was required to find, by its own efforts, $1 000 000 needed to finish the construction of the building.

At that time, it was a brick four-story house without a roof, with unfinished staircases and floors covered by crushed bricks, like after a bombing. We then declared that we would find the necessary sum, having no concrete sources whatever in mind, only hoping that, for such a good enterprise, the money would eventually be found. Indeed, in August 1995, the Moscow government granted $1 500 000 for finishing the construction of the building and furnishing it, and in a year it was concluded, according to a project presented by the IUM team. On September 26, 1996, the inauguration ceremony of the new building took place, and two closely related institutions, the IUM and the MCCME, began to work in it.

Besides the support of the IUM, the MCCME carries on a lot of activities related to high school education: various mathematical olympiads, lectures for high school teachers, conferences dedicated to educational problems, and so on.

During the last twelve years, when the Moscow Mathematical Society, and later the MCCME, became directly involved in the organization of the famous Moscow Mathematical Olympiads, it regained and exceeded its former popularity. Last year, three thousand high school students participated in the Olympiad, and the number of awards equalled the total number of participants of the Moscow Olympiad of 1992.

Other activities of the MCCME include a conference on educational problems, organized in 2000. Beginning in 2001, the MCCME organizes an annual Summer School, which brings together high school and university students with lecturers of the highest level, academicians Anosov, Arnold, and the late Bolibrukh included.

The present status of the Independent University is as follows. The first President of the Independent University was M. Polivanov, a mathematical physicist and philosopher, who passed away a year after the beginning of his Presidency. The IUM has two colleges, the Higher College of Mathematics and the Higher College of Mathematical Physics. The former was first headed by A. Rudakov, and now it is headed by Yu. Ilyashenko; the latter was headed by O. Zavialov, now by A. I. Kirillov. We have about 100 students in both colleges and about 40 freshmen each year. The graduate school of the IUM was founded in 1993 as a result of the initiative of A. Beilinson, B. Feigin, and V. Ginzburg. Twenty seven people have graduated from this school and passed their Ph.D. theses as of now.

At present, most of our male students study in parallel at two universities, say Moscow State and the IUM, in order to have military draft exemption. Therefore, our classes take place in the evenings.

The IUM gives a chance to create their own mathematical schools to mathematicians not involved in the teaching process at Moscow State University. The seminars of B. Feigin, S. Natanzon, O. Sheinman, O. Shvartsman, M. Tsfasman, and V. Vassiliev have been continuing for several years at the IUM.

Lecture courses at the IUM were given by D. V. Anosov, V. I. Arnold, A. A. Kirillov, S. P. Novikov, Ya. G. Sinai, V. A. Vassiliev, A. A. Belavin, V. K. Beloshapka, B. M. Feigin, S. M. Gusein-Zade, Yu. S. Ilyashenko, A. G. Khovanskii, I. M. Krichever, A. N. Rudakov, A. G. Sergeev, V. M. Tikhomirov, M. A. Tsfasman, and many others. The courses of Arnold (PDE), Vassiliev (Topology), and Anosov (Dynamical Systems) were published as books later.

The IUM provides teaching possibilities to professors who have full time positions in the West now. They are realized in the form of crash courses, usually one month long but so intensive that they are equivalent to semester courses. Such courses were given by A. A. Kirillov, A. Khovanski, I. Krichever, A. Katok (who is a Foreign Member of the IUM faculty), P. Cartier, and D. Anosov. In 1995–96, A. Khovanski gave a regular course in honors calculus; he got permission to be on leave from Toronto University, where he had a full position at the time.

The IUM tries to be a place to which Russian mathematicians can return after their work abroad, if they will. At present, we have seven young faculty members who obtained their Ph.D. abroad but are now teaching at the IUM.

Beginning in 2001, the IUM launched a new periodical, the *Moscow Mathematical Journal*. Among the authors of the papers already published and presented are V. I. Arnold, P. Deligne, G. Faltings, V. Ginzburg, A. Givental, A. J. de Jong, A. and S. Katok, C. Kenig, A. Khovanski, A. A. Kirillov, Ya. Sinai, M. Tsfasman, A. Varchenko, D. Zagier, and many others.

In the spring of 2001, the IUM organized a Study Abroad Program, called Math in Moscow (MIM), for foreign students. They are invited to the IUM for one semester to take mathematical and nonmathematical courses and to plunge into Russian cultural life. The credits for these courses are transferable to North American and Canadian universities. Up to now, the MIM program was attended by students from Berkeley, Cornell, Harvard, MIT, McHill, universities of Montreal and Toronto, Penn State, and many others.

In order to support young researchers, the Möbius Competition for the best research work of undergraduate or graduate students was organized in 1997 and sponsored by V. Balikoev and A. Kokin, both alumni of the Moscow Institute of Mathematics and Electronics. The winners were A. Kuznetsov (1997), V. Timorin (1998), A. Bufetov (1999) (all from the IUM), S. Shadrin and A. Melnikhov (2000) (MSU), A. Ershler (2001) (St. Petersburg University), V. Kleptsyn and L. Rybnikov (2002), S. Chulkov (2003, first place), and S. Oblezin and S. Shadrin (2004, second place). Recently, thanks to the initiative of V. Kaloshin (Caltech) who raised extra funds, the number of stipends was increased from one to three, and the duration was extended from one to two years.

Last but not least, the IUM has organized Student Sessions, which were held beginning in 1997. The first lecture was delivered by Arnold, one of our Founding Fathers, President of the Scientific Council of the IUM. The lectures given in 1998–2000 are presented to the reader. The lectures were intended for a large audience, from students to professional researchers. They contained no proofs or technical details. The objective was to give panoramas of whole research areas and describe new ideas.

Beginning in 2001, the Sessions were transformed into a regular mathematics research seminar, called Globus. This seminar brings together mathematicians from all sides of Moscow. It is in a sense parallel to the sessions of the Moscow Mathematical Society and intended for a similar audience. The lectures are taped and collected into volumes. Two volumes of these lectures will appear in Russian soon.

Of course, numerically the IUM plays a negligible role in Russian cultural life, but its influence, in my opinion, is far from negligible. It may be characterized by a quotation from the Gospel:

The Kingdom of Heaven is like unto leaven, that a women took and hit into three measures of meal till the whole was leavened. (Mt, 13:33)

The lectures at the Student Sessions and later at the Globus seminars were tape recorded. Then these records were decoded and edited by Professor V. Prasolov, translated into English, and sent to the authors to make final corrections. It is a hard job to transform speech into written text. This volume, as well as the subsequent ones prepared for publication in Russian, would never have appeared without the energy and devotion of V. Prasolov. The organizers of the Student Sessions, as well as of the Globus seminars, are cordially grateful to him.

V. I. Arnold

Mysterious mathematical trinities

Lecture on May 21, 1997

I shall try to tell about some phenomena in mathematics that make me surprised. In most cases, they are not formalized. They cannot even be formulated as conjectures. A conjecture differs in that it can be disproved; it is either true or false.

We shall consider certain observations that lead to numerous theorems and conjectures, which can be proved or disproved. But these observations are most interesting when considered from a general point of view.

I shall explain this general point of view for a simple example from linear algebra.

The theory of linear operators is described in modern mathematics as the theory of Lie algebras of series A_n, i.e., $\mathfrak{sl}(n+1)$, and formulated in terms of root systems. A root system can be assigned to any Coxeter group, that is, a finite group generated by reflections (at least, to any crystallographic group). If we take a statement of linear algebra which refers to this special case of the group A_n and remove all the content from its formulation, so as to banish all mentions of eigenvalues and eigenvectors and retain only roots, we will obtain something that can be applied to the other series, B_n, C_n, and D_n, including the exceptional ones E_6, E_7, E_8, F_4, and G_2 (and, sometimes, even to all the Coxeter systems, including the noncrystallographic symmetry groups of polygons, of the icosahedron, and of the hypericosahedron, which lives is four-dimensional space).

From this point of view, the geometries of other series $(B, C, ...)$ are not geometries of vector spaces with additional structures, such as Euclidean, symplectic, etc. (although formally, they, of course, are); they are not daughters of A-geometry but its sisters enjoying equal rights.

The above classification of simple Lie algebras, which is due to Killing (and, hence, attributed to Cartan), has an infinite-dimensional analogue – in analysis. The algebraic problem solved by Killing, Cartan, and Coxeter has an infinite-dimensional analogue in the theory of Lie algebras of diffeomorphism groups. Given a manifold M, the group $\mathrm{Diff}(M)$ of all diffeomorphisms of M naturally arises. This group (more precisely, the connected component of the identity element in this group) is algebraically simple, i.e., it has no normal divisors. There exist other similar "simple" theories, which resemble the geometry of

1

manifolds but differ from it. They were also classified by Cartan at one time.[1]
Having imposed a few fairly natural constraints, he discovered that there exist
six series of such groups:

Diff(M);

SDiff(M), the group of diffeomorphisms preserving a given form of volume;

SpDiff(M, ω^2), the group of symplectomorphisms.

Next, there are complex manifolds and groups of holomorphic diffeomorphisms.

There is also the very important contact group, the group of contactomor-
phisms.

Finally, there are conformal versions of some of these theories. I shall not
describe them in detail.

The idea which I mentioned is that, in these theories, there is something
similar to the passage from theorems of linear algebra, i.e., from the A_n root
system, to other root systems. In other words, in the whole of mathematics
(at least, of the geometry of manifolds), there are higher-level operations (e.g.,
symplectization) that assign analogues from the theory of manifolds with volume elements or of symplectic manifolds to each definition and each theorem of
manifold theory. This is by no means a rigorous statement; such an operation
is not a true functor.

For example, an element of the Lie algebra of the diffeomorphism group
is a vector field. The symplectization of a vector field is a Hamiltonian field
determining the Hamilton equation

$$\frac{dq}{dt} = \frac{\partial H}{\partial p}, \quad \frac{dp}{dt} = -\frac{\partial H}{\partial q}.$$

Other situations are more involved. It is difficult to understand what becomes of notions of linear algebra under the passage to other geometries. But
even dealing with the symplectomorphism group, we take only the vector fields
that are determined by a single-valued Hamiltonian function rather than the
entire Lie algebra of this group. These vector fields form the commutator of the
Lie algebra, which does not coincide with the entire Lie algebra of the symplectomorphism group. Still, when we are able to find regular analogues for some
notions of some geometry in another geometry, the reward is very significant.

Consider two examples.

1. Symplectization. So-called Arnold's conjectures (1965) about fixed points
of symplectomorphisms were stated in an attempt to symplectize the Poincaré–
Euler theorem that the sum of indices of the singular points of a vector field

---

[1] See, in particular, E. Cartan. *Selected Works* (Moscow: MTsMNO, 1998) [in Russian].
(*Editor's note*)

Figure 1

on a manifold is equal to the Euler characteristic. They estimate the number of closed trajectories for Hamiltonian vector fields by means of the Morse inequalities (i.e., in terms of the number of critical points of a function on the manifold).[2]

We start with formulating the following simpler assertion. It was stated by Poincaré as a conjecture and proved by Birkhoff.

Theorem 1. *Suppose that a self-diffeomorphism of a circular annulus preserves area and moves the points of each of the boundary circles in the same direction and the points of different circles in opposite directions (Fig. 1). Then this diffeomorphism has at least two fixed points.*

This assertion follows from a slightly more general theorem about fixed points of diffeomorphisms of the torus.

Theorem 2. *Let F be a diffeomorphism of the torus $T^2 = \mathbb{R}^2/\mathbb{Z}^2$ defined by $x \mapsto x + f(x)$ in the standard coordinate system. Suppose that F preserves area and "preserves the center of gravity," i.e., the mean value of the function f (considered as a function on the torus with standard metric) is zero. Then F has at least four fixed points.*

This is related to the fact that the sum of Betti numbers for the torus is equal to 4.

The first proof of this theorem was obtained by Y. Eliashberg. But nobody had verified this proof. A surely correct proof was published in 1983 by Conley and Zehnder, and this proof initiated a whole large theory – symplectic topology (developed by Chaperon, Laudenbach, Sikorav, Chekanov, Gromov, Floer, Hofer, Givental, and many other authors).[3] In recent months, there have been

---

[2] V. I. Arnold. On one topological property of globally canonical mappings of classical mechanics. In V. I. Arnold. *Selected Works–60* (Moscow: Fazis, 1997) [in Russian], pp. 81–86. (*Editor's note*)

[3] On symplectic topology see, e.g., V. I. Arnold. The first steps of symplectic topology. In V. I. Arnold. *Selected Works–60* (Moscow: Fazis, 1997) [in Russian], pp. 365–389. See also the references cited on p. XL of this book. (*Editor's note*)

communications that the initial conjectures (that the number of fixed points of an exact symplectomorphism is not smaller than the minimum number of critical points of a function on the manifold, at least for symplectomorphisms and generic functions) had been proved at last (by several independent groups in different countries).

2. Another example: the passage from $\mathbb{R}$ to $\mathbb{C}$. Using this example, it is, possibly, easier to explain the essence of the matter. We shall consider the passage from the real case to the complex one. There are real geometry and complex geometry. How can we pass from real geometry to complex geometry? For example, in real geometry, there is the notion of manifold with boundary, on which the notions of homology and homotopy are based. In general, the whole of topology essentially uses the notion of boundary.

We may ask: What becomes of the notion of boundary under complexification?

If we admit that all mathematics can be complexified, then, in particular, we must admit that various notions of mathematics can be complexified. Let us compose a table of transformations of various mathematical notions under complexification.

The complexification of the real numbers is, obviously, the complex numbers. Here the matter is very simple.

In the real case, there is Morse theory. Functions have critical points and critical values. Morse theory describes how level sets change when passing through critical values. What shall we obtain if we try to complexify Morse theory?

The complexification of real functions is holomorphic (complex analytic) functions. Their level sets have complex codimension 1, i.e., real codimension 2. In particular, they do not split the ambient manifold; the complement to a level set is by no means disconnected.

In the real case, the set of critical values of a function does split the real line. Therefore, generally, passing through a critical value affects the topology of a level set. For the complex analytic functions, this is not so. Their sets of critical values do not split the plane of the complex variable. Therefore, in the complex case, the level sets of a function that correspond to different noncritical values have the same topological structure. But in going around a critical value, a monodromy arises. This is a self-mapping of the set level (determined up to isotopy).

In the real case, the complement to a critical value consists of two components; thus, its homotopy group π_0 is $\mathbb{Z}_2$. In the complex case, the complement to a critical value is connected and has fundamental group $\mathbb{Z}$. Therefore, it is natural to regard π_1 as the complexification of π_0 and the group $\mathbb{Z}$ as the complexification of the group $\mathbb{Z}_2$.

Going further, we see that this approach turns out to be quite consistent. The complexification of the Morse surgeries (which refer to elements of the group π_0 of the set of noncritical values of a real function) is a monodromy (a representation of the group π_1 of the set of noncritical values of a complex function).

Monodromies are described by the Picard–Lefschetz theory, which is a theory of branching integrals.[4] In this sense, as the complexification of Morse theory we can regard the Picard–Lefschetz theory. As the complexification of a Morse surgery (attachment of a handle to a level set) we take the monodromy in a neighborhood of a nondegenerate critical point of a holomorphic function. This operation is the so-called Seifert transformation, that is, twisting a cycle on a level set. It consists in twisting a cylinder in such a way that one base of the cylinder remains fixed and the other base makes a full turn. In both cases, real and complex, the simplest operations correspond to singular points determined by sums of squares.

We can go even further. In the real theory, there are Stiefel–Whitney classes with values in $\mathbb{Z}_2$. Under complexification, they become Chern classes with values in $\mathbb{Z}$. Everything is consistent: the complexification of $\mathbb{Z}_2$ is indeed $\mathbb{Z}$.

The complexification of the projective line $\mathbb{R}P^1 = S^1$ is the complex projective line $\mathbb{C}P^1 = S^2$ (the Riemann sphere). Thus, the Riemann sphere is the complexification of the circle. It contains a circle (the equator). On this sphere, there is the theory of Fourier series defined on this circle and the theory of Laurent series which have two poles (at the poles of the sphere).

Let us find out what the complexification of the boundary of a real manifold is. First, we must algebraize the notion of boundary. A manifold with boundary is specified by an inequality of the form $f(x) \geqslant 0$. The correct complexification of this inequality is the equation $f(x) = y^2$. This equation specifies a hypersurface in the (x, y)-space, the standard projection of which on the x-space determines a double branched covering with branching along the boundary. Thus, the complexification of a manifold with boundary is a double covering with branching over the complex boundary.

This approach proved very fruitful. I invented the trick with a covering in 1970, when working on Hilbert's 16th problem about the arrangement of ovals of an algebraic curve of given degree n.[5] A polynomial of degree n in two variables determines a set of curves in the (real projective) plane. Hilbert's

---

[4] See V. A. Vasil'ev. *Branching integrals* (Moscow: MTsNMO, 2000) [in Russian]. (*Editor's note*)

[5] References to the literature on Hilbert's 16th problem are cited in D. Hilbert. *Selected Works* (Moscow: Factorial, 1998) [in Russian], vol. 2, p. 584. (*Editor's note*)

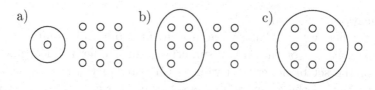

Figure 2

problem is to find all possible arrangements for a given n. In the cases of $n = 1$ and $n = 2$, this problem has been solved by ancient Greeks. The cases of $n = 3$ and 4 have been studied by Descartes and Newton. The case of $n = 5$ has been handled too, and the case of $n = 6$ resisted until the second half of this century, and it was included in Hilbert's problem. This part of the problem was solved by D. A. Gudkov.

For the degree 8, the question of how these ovals can be arranged is open, even for generic curves.

For $n = 6$, the answer to this question was given by Gudkov. Namely, the maximum possible number of ovals is 11, and the ovals can be arranged in the projective plane in three ways: ten ovals have pairwise disjoint interiors, and the eleventh oval contains one, five, or nine of them inside (Fig. 2, a–c). (On the projective plane, the complement to a disk is a Möbius band. Therefore, the notion "inside" is well-defined.)

Gudkov noticed that, for all arrangements which he managed to construct, the number of ovals satisfied certain congruences modulo 8. For example, the numbers 1, 5, and 9 differ by 4, and the Euler characteristics of the positivity sets of the polynomials that specify the curves, by 8. In all other examples, similar congruences hold. This suggests that there are four-dimensional manifolds somewhere about, because, as is well known, congruences modulo 8 and modulo 16 play a crucial role in the topology of 4-manifolds.

Thus, we have to find a 4-manifold related to a real algebraic curve. After several-month efforts to construct a suitable 4-manifold from an algebraic curve, I had guessed at last that what should be taken is precisely the double covering of the complement to the algebraic curve. Applying the powerful methods of four-dimensional topology to this 4-manifold, Rokhlin[6] (1972) and I[7] (1971) succeeded in proving that there exist no real algebraic curves not satisfying

---

[6] The works of V. A. Rokhlin on real algebraic geometry are collected in V. A. Rokhlin. *Selected Works* (Moscow: MTsNMO, 1999) [in Russian]. (*Editor's note*)

[7] V. I. Arnold. On the arrangement of the ovals of real plane algebraic curves, involutions of four-dimensional smooth manifolds, and the arithmetic of integer quadratic forms. In V. I. Arnold. *Selected Works–60* (Moscow: Fazis, 1997) [in Russian], pp. 175–187. (*Editor's note*)

Gudkov's congruences. The new field of mathematics, real algebraic geometry, that emerged in this way was later developed very far by Viro, Kharlamov, Nikulin, Shustin, Khovanskii, and many other mathematicians.

For $n = 8$, the proved constraints leave about 90 possible arrangements of 22 ovals. And about 80 arrangements have been constructed.

The same construction works in the theory of singularities. First, singularities for Coxeter schemes with simple constraints, i.e., for the series A, D, and E, were constructed. There remain B, C, F, and G, for which angles between vectors different from 90° and 120° occur. These cases resisted all the attempts to describe them by means of singularities, until I guessed that the same construction of a double branched covering should be applied. This construction makes it possible to construct singularities of caustics and wave fronts, which correspond to the remaining Weyl groups (with the only exception of G_2).

The correctness of a complexification can be confirmed only by results. *Per se*, it has no *a priori* definition. The situation is similar to what happened in the past century, when it was discovered that the theory of integral equations is parallel to the theory of linear operators. And even earlier, the same situation occurred in projective duality. Projective duality remained mysterious until its meaning was clarified. Similarly, until functional analysis was developed, the parallelism between theorems about integral equations and analogous theorems of linear algebra, including the Fredholm theorems, remained mysterious.

Nowadays, a similar situation occurs when physicists use the operation of summation over a continuous index, and mathematicians do not want to understand what it means.

What is more, mathematicians were using real numbers for thousands of years, but they knew nothing rigorous about these numbers except the ancient Greek discovery that not all of them exist: no rational number raised to the second power can equal two. All numbers in Greek mathematics were rational; thus, on the one hand, real numbers were used, but on the other hand, nobody knew what they were, because there was no rigorous definition.

Approximately the same happens to what I am going to talk about – the ternarity phenomenon.

The Ternarity Phenomenon

The ternarity phenomenon is that, to all of the pairs considered above, a third term can be added. In mathematics, objects very often occur in triplets. In many cases, these triplets form commutative diagrams.

The simplest triplet is obtained by adding the quaternions $\mathbb{H}$ to the real and complex numbers. But there are other triplets too.

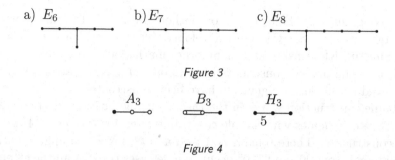

Figure 3

Figure 4

Example 1. The simple Lie algebras E_6, E_7, and E_8. They correspond to the Dynkin diagram shown in Fig. 3, a–c, respectively. Each vertex of the graph corresponds to a vector. If vertices are joined by an edge, then the angle between the corresponding vectors is equal to $120°$, and if they are not joined, then this angle is equal to $90°$. The group generated by the reflections with respect to the orthogonal complements to these basis vectors in Euclidean space is precisely the Weyl group.

Example 2. In stationery stores, triangles of three types are sold: equilateral triangles, right-angled triangles with angles of $45°$, and right-angled triangles with angles of $30°$. It turns out that these triangles correspond to the diagrams E_6, E_7, and E_8; together with these diagrams, they form a 3×2 commutative diagram.

In this mystic theory of ternarities, many nontrivial theorems and similar commutative diagrams have already been obtained. Proving everything in this lecture is out of the question. First, I shall write out examples, and then, explain the relations between them.

Example 3. The tetrahedron, octahedron, and icosahedron. Their symmetry groups are the Weyl groups A_3, B_3, and H_3, respectively. Their Dynkin diagrams (which were introduced earlier by Witt and Coxeter) are shown in Fig. 4.

The numbers of edges in the corresponding polyhedra are 6, 12, and 30, respectively. These numbers have the forms $2 \cdot 3$, $3 \cdot 4$, and $5 \cdot 6$. As we might expect, subtracting 1 from the first factors in these products, we obtain 1, 2, and 4, i.e., the real dimensions of $\mathbb{R}$, $\mathbb{C}$, and $\mathbb{H}$.

Let us describe the symmetry group of the tetrahedron. It is generated by the reflections with respect to its symmetry planes, the number of which is equal to the number of edges, i.e., to 6. These planes split space into 24 parts, called the Weyl chambers. Let us describe these chambers. All the planes pass through the center. Therefore, they can be represented by projective lines in

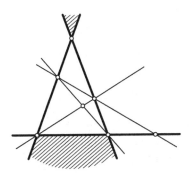

Figure 5

$\mathbb{R}P^2$. These lines divide the projective plane into 12 parts (see Fig. 5). Take one of these parts. It corresponds to a trihedral angle, which is represented by a triangle in the projective plane (this triangle is hatched in Fig. 5). Let us extend the sides of the triangle (the extensions are shown by thick lines in Fig. 5). In the 3-space model, these three lines correspond to three planes passing through the origin. They divide space into eight parts. In the projective plane, these parts correspond to four parts, being four large triangles composed of 12 small triangles. The large triangles contain 1, 3, 3, and 5 small triangles, respectively (in Fig. 5, only one of the large triangles looks like a real triangle; this is the one consisting of five small triangles). Thus, we obtain the formula

$$24 = 2(1 + 3 + 3 + 5).$$

The number of chambers – 24 – is, of course, the order of the symmetry group of the tetrahedron. Similarly, the order of the symmetry group of the octahedron equals

$$48 = 2(1 + 5 + 7 + 11),$$

and for the icosahedron, we have

$$120 = 2(1 + 11 + 19 + 29).$$

Note that, increasing the terms in these three formulas by 1, we obtain systems of weights (powers of basis invariants) for A_4, B_4, and H_4, respectively. *A propos*, these weights are precisely the numbers of vertices, faces, and edges in the tetrahedron, octahedron, and icosahedron (possibly, two is the number of three-dimensional faces?).

The polyhedra corresponding to the above Weyl groups determine the polynomials

$E_6 = x^3 + y^4 + z^2,$	$E_7 = x^3 + xy^3 + z^2,$	$E_8 = x^3 + y^5 + z^2.$
$\widetilde{E}_6 = x^3 + y^3 + z^3,$	$\widetilde{E}_7 = x^4 + y^4 + z^2,$	$\widetilde{E}_8 = x^3 + y^6 + z^2.$

The corresponding singularities are called elliptic and parabolic (the latter are related to the so-called affine root systems). We shall briefly describe the passage from polyhedra to polynomials further on.

We continue the list of trinities.

The characteristic classes:

Stiefel–Whitney classes,	Chern classes,	Pontryagin classes.

Next is the row-trinity

$S^1 \to S^1 = \mathbb{R}\mathrm{P}^1,$	$S^3 \to S^2 = \mathbb{C}\mathrm{P}^1,$	$S^7 \to S^4 = \mathbb{H}\mathrm{P}^1.$

The first mapping is the double covering of the boundary of the Möbius band over its base S^1. The second is the Hopf bundle with fiber S^1. By the way, this demonstrates that complexification is a delicate operation. The point is that the complexification of S^1 is S^3 in one case and S^2 in the other. This is explained by that the two circles in the covering of the Möbius band are quite different. One circle is $\mathbb{R}\mathrm{P}^1$, and the other is $SO(2)$. Under complexification, $\mathbb{R}\mathrm{P}^1$ transforms into $\mathbb{C}\mathrm{P}^1$ and $SO(2)$ transforms into $SU(2) \cong S^3$.

The third mapping is the Hopf bundle with fiber S^3.

I shall write out several more trinities. I am more or less sure of the next two rows, but the meaning of the rows following them is not completely clear so far.

S^0-bundle,	S^1-bundle,	$SU(2)$-bundle?

The first object refers to a mere double covering, the second to a bundle into complex lines, and the third, apparently, to a bundle into quaternion lines (maybe, endowed with some "hyperconnection" complexifying the complex connection of the second row).

They correspond to

monodromy of a flat connection,	curvature of a connection (2-form),	4-form (hypercurvature?).

The degree to which the objects labeled by the question mark are twisted must be measured by some 4-forms leading to the Pontryagin characteristic classes. These forms must measure some hyper-Kähler incompatibility of complex structures.

Yet another trinity:

polynomials,	trigonometric polynomials,	modular polynomials.

The polynomials are the holomorphic self-mappings of the sphere under which infinity has precisely one preimage. The trigonometric polynomials are the Laurent polynomials from $\mathbb{C}[t, t^{-1}]$ for which one of the points has two preimages. The modular polynomials are the rational functions for which the preimage of one point consists of three points, 0, 1, and ∞.

The last trinity was suggested by Givental:

homology,	K-theory,	elliptic homology.

As mentioned, the polyhedron corresponding to a Weyl group is related to a polynomial in three variables. We shall describe this correspondence for the example of the icosahedron, whose symmetry group consists of 120 elements. The orientation-preserving symmetries form a subgroup of order 60 in $SO(3) \cong \mathbb{RP}^3$. Consider the double covering of the group $SO(3)$ by the group $\text{Spin}(3) = SU(2) \cong S^3$. Each element of $SO(3)$ corresponds to two points from S^3. The 60-point subgroup in $SO(3)$ under consideration corresponds to 120 points in S^3. It is these points that form the binary group of the icosahedron.

The binary group of the icosahedron is contained in $SU(2)$; therefore, it acts on $\mathbb{C}^2$. Let us denote the orbit space of this action by X. The orbit space X is a two-dimensional complex surface with singularities embedded in $\mathbb{C}^3$. It is described by the equation $x^3 + y^5 + z^2 = 0$. This is proved by means of the theory of invariants. Namely, we are looking for basis invariants. They are binary forms with zeros at the vertices, the midpoints of the edges, and the centers of the faces of the icosahedron, respectively. The degrees of these binary forms are equal to 12, 20, and 30. Then, we seek a relation between them; it is called a *syzygy*. Such a syzygy was found in the past century by Schwarz. Under an appropriate normalization of the basis invariants x, y, and z, it takes the form $x^3 + y^5 + z^2 = 0$.

Now, let us explain how to construct E_6, E_7, and E_8 from formulas (polynomials). Instead of equating the polynomials to zero, we equate them to some $\epsilon \neq 0$, i.e., consider the surface $x^3 + y^5 + z^2 = \epsilon \neq 0$, known as the Milnor fiber. For the Milnor fiber, we can consider homology in the middle dimension. The Milnor fiber is a complex 2-manifold; its real dimension equals 4, and therefore the middle dimension is 2. On the two-dimensional homology of the Milnor fiber, the monodromy group acts; it is described by paths in the base space of the versal deformation. When we go around the discriminant of the corresponding family of polynomials which has such a singular point at the worst parameter value, the Picard–Lefschetz theory arises, which describes self-mappings of the Milnor fiber. It turns out that the intersection form of the Milnor fiber is the very Euclidean structure in which the basis cycles are specified by the Dynkin diagram, and the monodromy corresponding to the simplest course around the discriminant (the Seifert transformation) is precisely the reflection in the corresponding mirror. Therefore, E_6, E_7, and E_8 are the monodromy groups corresponding to the singularities determined by the polynomials specified above.

Finally, let me explain how $\widetilde{E}_6$, $\widetilde{E}_7$, and $\widetilde{E}_8$ are obtained from polyhedra. Take a simplest standard representation of the corresponding symmetry group of a polyhedron in 2-space and consider the tensor product of some representation with the standard one. This tensor product decomposes into irreducible terms with some coefficients. The matrix of these coefficients is the "Cartan matrix," which describes the corresponding affine root system.

The table summarizing the trinities looks as follows:

$\mathbb{R}$	$\mathbb{C}$	$\mathbb{H}$
E_6	E_7	E_8
A_3	B_3	H_3
D_4	F_4	H_4
$x^3 + y^4 + z^2$	$x^3 + xy^3 + z^2$	$x^3 + y^5 + z^2$
$x^3 + y^3 + z^3$	$x^4 + y^4 + z^2$	$x^3 + y^6 + z^2$
$60°, 60°, 60°$	$45°, 45°, 90°$	$30°, 60°, 90°$
tetrahedron	octahedron	icosahedron
$6 = 2 \cdot 3$	$12 = 3 \cdot 4$	$30 = 5 \cdot 6$
quadratic forms	Hermitian forms	hyper-Hermitian forms
Stiefel–Whitney classes	Chern classes	Pontryagin classes
$S^1 \to S^1 = \mathbb{R}\mathrm{P}^1$	$S^3 \to S^2 = \mathbb{C}\mathrm{P}^1$	$S^7 \to S^4 = \mathbb{H}\mathrm{P}^1$
double coverings	S^1-bundles	$SU(2)$-bundles
monodromy of flat connection	curvature of connection (2-form)	4-form (hypercurvature?)
polynomials	trigonometric polynomials	modular polynomials
homology	K-theory	elliptic homology

V. I. Arnold

The principle of topological economy in algebraic geometry

Lecture on May 21, 1997

This part of the lecture is not related to the first part, so it can be understood independently. We start with an example.

Example 1. In $\mathbb{CP}^n$, consider two algebraic varieties X and Y of complementary dimensions. In general position, they intersect in finitely many points. Let $[X]$ and $[Y]$ be the homology classes realized by the varieties X and Y, and let $[X] \circ [Y]$ be the intersection index of these classes (which is an integer). It is equal to the number of "positive" intersection points of X with Y minus the number of "negative" intersection points. Thus, the number $\#(X \cap Y)$ of all intersection points is not smaller than the intersection index $[X] \circ [Y]$ (and has the same parity). The *Bézout Theorem* asserts that $\#(X \cap Y)$ is equal to the number $[X] \circ [Y]$, i.e., there is no inequality! The point is that the orientation of complex manifolds is such that each intersection makes a contribution of $+1$, not -1, in the total intersection index. Negative intersections are "expensive," they increase the number of intersection points of X with Y in comparison with the "topologically necessary" number. *A propos*, the same considerations imply that a polynomial of degree n has precisely n roots, not more.

This (well-known) and the following (newer) examples lead to a "principle of economy," which, in its turn, can be used to state further conjectures. These conjectures can be verified in particular cases; sometimes, they can be proved and become theorems. But in most cases, they remain conjectures, i.e., assertions which we may try to disprove, for a long time.

Example 2. The following assertion has long been a conjecture. Consider a Riemannian surface X of genus g specified by an irreducible polynomial of degree n in $\mathbb{CP}^2$. Its homology class $[X] \subset H_2(\mathbb{CP}^2, \mathbb{Z})$ is the generator $[\mathbb{CP}^1]$ of the group $H_2(\mathbb{CP}^2, \mathbb{Z}) = \mathbb{Z}$ taken n times; the number n is the degree of the polynomial. As is known, the genus g of the surface X can be calculated by the Riemann formula (though, some people say that Riemann did not know what the genus is). The formula is

$$g = \frac{(n-1)(n-2)}{2}.$$

The question arises: Is it possible to realize the homology class $[X]$ by another smooth surface, maybe real rather than complex, of smaller genus? In other words, does a complex algebraic variety realize the minimum genus for a given homology class $[X]$? The *Thom conjecture* is that a complex surface with $g = (n-1)(n-2)/2$ handles is indeed the most economical realization of the class $n \cdot [\mathbb{CP}^1] \subset H_2(\mathbb{CP}^2, \mathbb{Z})$. Any smooth real surface realizing this class has as many handles or more. This conjecture was proved only recently by Kronheimer and Mrowka, who employed "heavy artillery," namely, the theory of Donaldson, Gromov, Witten, etc., which originates from the ideas of quantum field theory.

Example 3 (Milnor's conjecture). This conjecture was also proved only recently by the same authors. Given an arbitrary knot, we can always transform it into an unknotted circle in several unknottings. An unknotting is a crossing change (passing above is changed for passing below) on a suitable diagram (that is, projection on the plane) of the given knot. Unknotting is made by a sword; Alexander the Great used the same method to untie the Gordian knot. The Gordian (unknotting) number of a knot is the minimum number of crossing changes necessary to transform the given knot into the trivial one.

Knots are related to singularities as follows. Consider an algebraic curve K^2 with a singularity at the origin and a small sphere $S^3 \subset \mathbb{C}^2$ centered at zero. The intersection $N^1 = K^2 \cap S^3$ is a knot (or link) in S^3. For example, the semicubic singularity $x^2 = y^3$ of a curve in $\mathbb{C}^2$ corresponds to the trefoil knot.[1] Milnor considered the question how to obtain the unknotting number of a knot N from the algebraic properties of the polynomial specifying the curve K. He suggested the following method for untying. Let us specify the curve K parametrically; in the case under consideration, its equations are

$$x = t^3, \quad y = t^2.$$

In the general case, the similar equations $x = t^n$ and $y = f(t)$, where $f(t)$ is a holomorphic function, are given by the theory of Puiseux series (which was discovered by Newton). We transfer the functions $x(t)$ and $y(t)$ into general position by adding terms of lower degrees, for example,

$$x = t^3 - \epsilon t, \quad y = t^2.$$

We obtain a curve K' without singular points (at least near the origin) except some number δ of self-intersection points. This number δ (the Milnor number) can be expressed in terms of certain algebraic invariants of the curve K. It is a

---

[1] This fact is proved in, e.g., J. Milnor. *Singular Points of Complex Hypersurfaces* (Princeton (USA): Princeton Univ. Press, 1968), Section 1.

"candidate" for the unknotting number. The point is that there exists a method for performing unknottings "corresponding" to double points; the number of unknottings equals the number of double points. The Milnor conjecture is that this "algebraic" method for untying the knot of a singularity is most economical; the knot cannot be made trivial in less than δ unknottings. This is yet another manifestation of the economy principle.

The recent proof of this conjecture uses the same "heavy artillery" (essentially originating from quantum physics) as the proof of the Thom conjecture.

Example 4 (the Möbius theorem). Possibly, Möbius came to the Möbius band in working on the following problem. Consider a projective line $\mathbb{RP}^1$ in the projective plane $\mathbb{RP}^2$. This is an infinitely degenerate curve: its curvature vanishes at each point. Under a small perturbation, a generic curve with a finite number of inflection points arises. What is the minimum number of inflection points?

Let us try to apply the economy principle rather than guess the answer by experimenting. According to the principle, we must consider the simplest algebraic model of the phenomenon we are interested in, namely, the appearance of inflection points under a perturbation of the line. Probably, this model is most economical, i.e., contains the minimum possible number of inflection points.

Thus, we must deform the line in the class of algebraic curves of degree as small as possible and count the number of emerging inflection points. According to the principle of topological economy, it is impossible to obtain a smaller number of inflection points. We cannot manage with first-degree curves, because these are straight lines. The second-degree curves are not suitable either, because they have a quite different topology: the line $\mathbb{RP}^1$ is not contractible in $\mathbb{RP}^2$, while the circle (as well as other quadrics) is.

The curves of the third degree are sufficient. For example, we can specify the line in affine coordinates x, y by the equation $y = 0$ and its perturbation by the equation $y \cdot f(x, y) = \epsilon$, where ϵ is a small number and $f(x, y)$ is a second-degree polynomial having no real roots. Examining curves of the third degree, we see that most economical is the curve $y(1 + x^2) = 1$, or $y = 1/(1 + x^2)$. It has three inflection points, two with abscissas $\pm 1/\sqrt{3}$ and one at infinity. Indeed, any curve having no inflection point at infinity lies on one side of the tangent at the infinite point in a neighborhood of this point, and in the affine chart, it lies on both sides from the asymptote, as the hyperbola. The curve under consideration approaches the asymptote $y = 0$ from above both times when going to infinity. Thus, at the infinite point, the curve intersects its tangent; this is possible only at an inflection point.

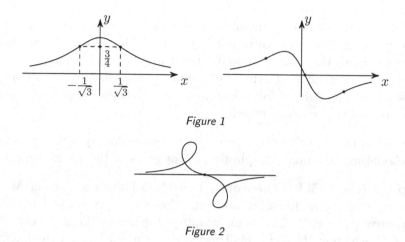

Figure 1

Figure 2

It is worth noting that the notion of an inflection point is purely projective. It is based only on the multiplicity of the intersection of the curve with its tangent at a given point and involves no metric. Indeed, the multiplicity of the intersection of two curves is preserved by any diffeomorphism, and projective transformations take projective lines to projective lines.

The curve shown in Fig. 1 on the left has two finite inflection points and one infinite inflection point; on the right, all of the three inflection points are finite. In any case, the total number of inflections points is odd, which is implied by the nonorientability of the Möbius band (a neighborhood of $\mathbb{RP}^1$ in $\mathbb{RP}^2$ is precisely the Möbius band).

The Möbius theorem asserts that the *number of inflection points is an odd number not smaller than three, i.e., it cannot equal 1.*

In this connection, I shall formulate one more theorem and yet another conjecture. *If a line is deformed strongly, the number of inflection points is still no less than three as long as the line remains embedded.* Continuing to deform the line, we can obtain an immersed (not embedded) curve (Fig. 2) having only one inflection point. The conjecture is that *obtaining a (regular homotopy) curve with one inflection point from an embedded curve with three inflection points requires changing the Legendre type of the knot; i.e., there must occur a moment when the curve is tangent to itself and the orientation of two branches at the point of tangency coincide.*

At this moment, a self-intersection of the Legendre knot corresponding to the curve under consideration in the space of contact elements of the plane occurs. Thus, this conjecture (it is similar to the Chekanov conjecture on quasi-functions, which generalizes the conjectures on fixed points and on Lagrangian intersections that I stated at the beginning of the preceding lecture; see p. 2) asserts that the number of inflection points cannot become equal to 1 until the Legendre type of the corresponding knot is changed.

At present, it is proved (by D. Panov) that, to remove two out of the three inflection points (i.e., to leave only one inflection point) while leaving the Legendre knot intact, it is necessary to create at least nine inflection points on some curve in the course of homotopy.

Example 5. For small deformations of the line, the Möbius theorem is a special case of the Sturm theorem, which gives an estimate for the number of zeros of trigonometric Fourier series. Let f be a 2π-periodic function whose Fourier expansion begins with harmonics of the nth order:

$$f(x) = \sum_{k \geqslant n}(a_k \cos kx + b_k \sin kx).$$

What minimum number of zeros can f have on one period? According to the general principle, we must consider the simplest case, i.e., the series consisting of only one harmonic, $f(x) = \cos kx$ (or $f(x) = \sin kx$, which is the same thing). This function has $2k$ zeros. The Sturm theorem (proved by Hurwitz) asserts that the *number of zeros of the Fourier series is not smaller than that of its first nonzero harmonic.*

A similar assertion for the Fourier integral is not yet proved. Let

$$F(x) = \int f(k)e^{-ikx}\, dk,$$

where $f(-k) = \bar{f}(k)$ (thus, F is a real function). Suppose that $f(k) = 0$ for $|k| \leqslant \omega$. The Grinevich conjecture is that the *average number of zeros of $F(x)$ in a long interval is not smaller than that for the "first harmonic"* $\sin \omega x$, i.e., than π/ω. This assertion, which illustrates the general economy principle, is very likely to be true, but it has not been proved so far.

Let me explain why the Möbius theorem is implied by the Sturm theorem (for small deformations). Consider the projective plane with an embedded deformed line. On the sphere (which doubly covers $\mathbb{RP}^2$), we have the deformation of the equator described by an odd function $f(\phi)$ ($f(\phi + \pi) = -f(\phi)$), where ϕ is the longitude and $f(\phi)$ is the perturbation of the latitude at the point ϕ. Calculations show that the inflection points of the perturbed equator correspond to the values of ϕ for which

$$f''(\phi) + f(\phi) = 0. \tag{$*$}$$

Actually, this is true only in the first approximation; but the question about the number of inflection points still does reduce to the question about the number of the roots of an equation that has form $(*)$ but involves a different function $\tilde{f}$ instead of f: $\tilde{f}''(\phi) + \tilde{f}(\phi) = 0$. From the oddness of f, we can derive that the function $\tilde{f}$ is also odd, i.e., its Fourier expansion contains only harmonics with odd numbers: the first, third, fifth, etc. Thus, the Fourier series of the function $L\tilde{f} = \tilde{f}'' + \tilde{f}$ starts with at least third harmonic, because $(\sin\phi)'' + \sin\phi = 0$ and $(\cos\phi)'' + \cos\phi = 0$: the differential operator

$$L = \frac{d^2}{d\phi^2} + 1$$

kills $\sin\phi$ and $\cos\phi$. Therefore, the equation $L\tilde{f} = 0$ has at least six roots in the interval $[0, 2\pi]$, and the number of inflection points of the perturbed line in $\mathbb{RP}^2$ is two times less; thus, it is at least three.

As well as the Möbius theorem, the Sturm theorem (which generalizes it) must have (and indeed has) generalizations to the case of large deformations, where variations of some curves in multidimensional spaces are considered instead of functions. This gives rise to a theory of convex curves in the projective space, which contains many interesting problems (see, e.g., V. I. Arnold. Topological problem of wave propagation theory. *Usp. Math. Nauk*, **51** (1) (1996), 3–50 and the references cited therein).

Example 6. Let us try to find generalizations of the Möbius theorem to surfaces and, accordingly, of the Sturm theorem to functions of two variables. Consider the projective plane $\mathbb{RP}^2$ embedded in projective space $\mathbb{RP}^3$. Take a small deformation of the embedding. How many "inflection points" does the obtained surface have?

First, we must explain what "inflection" means. The points of a surface are divided into elliptic, parabolic, and hyperbolic points, depending on the second quadratic form. At an elliptic point, the surface is convex, at a hyperbolic point, the two principal curvatures have different signs, and at a parabolic point, one of the two principal curvatures vanishes. Here we can manage without the second quadratic form, metric, etc.; elliptic, hyperbolic, and parabolic points are defined projectively invariantly. It is sufficient to consider the pair formed by the surface and its tangent plane at a given point. The distance from a close point of the surface to this tangent plane determines a quadratic form whose degeneracy (nondegeneracy) and signature do not change under projective transformations.

Thus, the parabolic points (which play the role of inflection points in the case under consideration) form quite definite curves on surfaces in projective space. How many domains of ellipticity (bounded by parabolic curves) do appear under infinitesimal deformations of the plane? This problem is not solved. However, our method immediately yields the conjecture that this number is never smaller than in the case of a simplest algebraic deformation. Such an algebraic deformation was studied by B. Segre in 1942. It turned out that, on each cubic surface in $\mathbb{RP}^3$ being a deformation of $\mathbb{RP}^2$ (i.e., having no handles; surfaces with handles were also examined by Segre), there are four curves of parabolic points. F. Aicardi from Trieste conjectured that the number of parabolic curves is always at least four and performed many computer experiments confirming this conjecture. The conjecture of Aicardi is that the situation is similar to that of the Sturm theorem: given a lower harmonic with $2k$ zeros (a cubic deformation with four inflection lines), it is impossible to reduce the number of zeros (inflection curves) by adding harmonics (perturbations) of higher degree, even with large coefficients.

This conjecture has numerous special cases. Some of them were proved, but the conjecture itself resists all efforts; even approaches to proving it are not seen. There exist fairly many different proofs of the Möbius and Sturm theorems, but all of them have obscure origination. It is also unclear how to generalize these proofs to the multidimensional case. The conjecture stated above is, probably, the simplest multidimensional generalization of the assertions of these theorems.

Remark 1. A generic surface has even more special, "inflection," points than parabolic points. At each hyperbolic point, there are two asymptotic directions; straight lines with these directions have an abnormal ($\geqslant 2$) order of tangency with the given surface. They are also tangent to the intersection curves of the surface with the tangent plane at the given hyperbolic point. Thus, in a hyperbolic domain, there is a field of crosses. As a byproduct, the question about global topological constraints on these fields of crosses arises: What fields of crosses can be realized by surfaces? The same question can be asked about affine surfaces and even about surfaces being the graphs of functions $z = f(x, y)$. This problem is studied very poorly. I shall tell about one related invariant.

Approaching a parabolic curve, the asymptotic directions move toward one another, and the field of crosses in a hyperbolic domain transforms into a field of (asymptotic) directions on the parabolic curve. At some points, it is tangent to the curve itself. These points are even more special: they are vertices of swallow tails of the dual surface. Segre established that, on a cubic surface, there are precisely six such points. This suggests the second conjecture of Aicardi that, under arbitrary (not necessarily cubic) deformations of the projective plane, at least six tangency points of an asymptotic direction with the parabolic curve emerge.[2]

---

[2] In the beginning of September 1997, D. Panov communicated that he disproved both conjectures of Aicardi by constructing a surface with only one parabolic curve without special points. Apparently, the dual surface (front) is fairly complex, so there still remains a possibility to generalize the Möbius theorem by proving that it is separated from the cubic surfaces by

Remark 2. Consider the surface M in the space of all lines tangent to the initial surface in projective space formed by the asymptotic directions of this initial surface. Topologically, it is the double of the hyperbolic domain N. On this surface, there is a smooth single-valued field of directions (with singularities only at the special points where the asymptotic direction is tangent to the parabolic curve), which is projected to the field of crosses on the hyperbolic domain. Every point inside N has two preimages in M, at which the projection $M \to N$ is regular, and every point on the boundary of N has one preimage in M, which is a fold point. The surface M has as many handles as many ellipticity domains are contained inside N. On M, there is a smooth dynamical system whose integral curves (they are called asymptotic lines) are tangent to the given field of directions; the projection $M \to N$ maps the (smooth) asymptotic lines to curves with semicubic singular points on the parabolic curve ∂N. Very little is known about the properties of this dynamical system. It is known that (in the case of a cubic surface) it has 27 periodic solutions (in the complex domain), that is, straight lines on the cubic surface which are, obviously, its asymptotic lines. This dynamical system suggests many interesting questions. What properties does it have? For example, is it integrable? What will studying this system by the usual means of perturbation theory near periodic solutions give? What are the properties of the Poincaré first-return map, which moves the curve of parabolic points along integral curves? Is this system generic? Is there chaos?

To conclude, I shall show how simpler problems can be obtained from the problem under consideration. We need a method for deforming a plane in space in such a way that we could control the curvature of the obtained surface and keep track of the parabolic points. There are several such methods; each of them gives an algorithm that determines the set of parabolic curves and, sometimes, the points of tangency of a parabolic curve with an asymptotic direction from some combinatorial data.

One of the methods is as follows: we take a triangulation of the initial plane and slightly move each vertex away from the plane, upward or downward. We obtain a polyhedral surface, which is a deformation of the plane. Certainly, it must be smoothed in some way that would allow us to trace the emergence of parabolic points. It is easy to apply this method to the Möbius theorem, because it is clear where the inflection edges of the polygonal line approximating our curve are. For polyhedra, we can define parabolic curves too; we can define them on the very polyhedron or on (some "standard") smoothing of this polyhedron. Next, for a given perturbation of the triangulation, we define (combinatorially!)

---

some suitable discriminant, in addition to the obviously necessary discriminant of the birth of swallow tails on the dual surface.

a certain number, which must be no smaller than 4 if the Aicardi conjecture is true.

Another method is based on considering partial differential equations and leads to the following conjecture (I omit the details). Consider the odd functions on the sphere S^2 in $\mathbb{R}^3$; they have the property $f(-x) = -f(x)$, where $-x$ is the point diametrically opposite to x. It is these functions that describe the deformations of projective plane. Now, consider the spherical functions, i.e., the functions on the sphere that are eigenfunctions for the spherical Laplace operator. Thus, x, y, and z are eigenfunctions with eigenvalue -2, and the equation $\Delta f + 2f = 0$ has no other solutions – its solution space is three-dimensional. So, the spherical functions of degree 1 are precisely all linear homogeneous functions of x, y, and z; all of them are odd. However, it turns out that there exists a different theory of spherical functions, which has not been considered by the classics for some reason. We can consider spherical functions with singularities, e.g., with poles of multiplicity not exceeding a certain number. Then the equation $\Delta f + 2f = 0$ has generalized solutions, in addition to the usual spherical functions. To be more precise, the right-hand side of the equation is then equal to some linear combination of δ-functions (and their derivatives) supported at the points where the function f has poles (rather than vanishes). It turns out that almost all points of a surface for which the deformation is determined by such a function are hyperbolic, and only smoothing around the poles yields curves of parabolic points bounding domains of ellipticity. In addition, at some particular points, the second quadratic form of such a surface vanishes completely. As a rule, near such a point, the surface has the form of a monkey saddle. This approach leads to the conjecture that *the number of singular points + the number of monkey saddles* $\geqslant 8$. Here $8 = 2 \cdot 4$, because we consider the sphere S^2 instead of $\mathbb{RP}^2$. We take into account monkey saddles because there may be points at which the function f is too well approximated by a linear function, i.e., points where the second differential $d^2 f$ is not merely degenerate but vanishes.

Vanishing of $d^2 f$ usually occurs at isolated points. A simplest example is $f(x, y) = x^3 - 3xy^2$; it is easy to show that all points of this surface (except zero) are hyperbolic; it has no elliptic points, and the form $d^2 f$ vanishes only at the point $(0, 0)$.

In general, we can consider (in a fixed local chart) the mapping

$$(x, y) \longmapsto \begin{pmatrix} A & B \\ B & C \end{pmatrix}, \qquad (**)$$

where

$$A = \frac{\partial^2 f}{\partial x^2}, \quad B = \frac{\partial^2 f}{\partial x\,\partial y} = \frac{\partial^2 f}{\partial y\,\partial x}, \quad C = \frac{\partial^2 f}{\partial y^2}.$$

Thus, to a function $f(x,y)$, we can relate a surface on the three-dimensional space with coordinates A, B, C that is the symmetric square of the cotangent space. We have obtained, so to speak, the second Gaussian mapping.

In this space, there is a cone of degenerate forms; it is specified by the equation $AC = B^2$. Inside the cone are positive and negative definite forms, and outside it are hyperbolic forms. The Laplace equation $A + C = 0$ (as well as some other equations) means that the entire surface-image of the mapping under consideration lies in the hyperbolic domain completed by the point 0 (for the Laplace equation, in the plane $A + C = 0$, which intersects the cone only at the origin). Hence there are no elliptic points on the surface-preimage; only vanishing of $d^2 f$ at some points is possible. If we slightly move the function f, then the image of the corresponding surface will also slightly move, and parabolic curves (corresponding to the intersections of the surface-image with the cone) will spring up. Therefore, the monkey saddles must be taken into account as well as the poles, because, under a small deformation, they also give birth to parabolic curves. This is where the last conjecture comes from.

Finally, I shall state one more conjecture (by M. Herman), which fits into the framework of the same theory of extremal algebraic objects. Although, in this case, the role of algebraic object is played by versal deformations. But it is known that, in the theory of singularities, versal deformations "behave like algebraic objects." For some reason, Herman stated this conjecture for caustics of the Lagrangian, or gradient, mapping. But the same question is, probably, open not only for the gradient mapping but also for an arbitrary mapping from $\mathbb{R}^n$ to $\mathbb{R}^n$.

Thus, consider a generic mapping $f \colon M^n \to M^n$ or a projection $f \colon L \to \mathbb{R}^n$ of a Lagrangian submanifold $L \subset \mathbb{R}^{2n}$ of the symplectic space. A typical example is $y = \operatorname{grad} f(x)$, where $x \in \mathbb{R}^n$ and $f \colon \mathbb{R}^n \to \mathbb{R}$. (To such a form, any Lagrangian mapping can be reduced by a suitable choice of coordinates.)

Consider the critical points of this mapping, i.e., the points where $\det(\partial y/\partial x) = 0$, and the corresponding critical values of y; in the symplectic case, the set of critical values forms a caustic.

The *degree of a mapping at a point* y is defined as follows. We take a regular value y' close to y and count (as usual, with orientation taken into account) the preimages of y' – not all of them but only those close to x.

The degree thus defined does not depend on the choice of the point y'. Suppose that it vanishes. The conjecture asserts that, *in this case, the point y' can be chosen so that it has no preimages at all (in a neighborhood of x).* If the degree equals k, then the conjecture asserts that *there exists a point having precisely k preimages, not more.*

Further comments on these problems had been published in:

[1] V. I. Arnold. Remarks on parabolic curves on surfaces and on higher-dimensional Möbius–Sturm theory. *Funct. Anal. Appl.*, **31** (4) (1997), 3–18.

[2] V. I. Arnold. Topological problems of the theory of asymptotic curves. *Proc. Steklov Inst. Math.*, **225** (1999), 11–20.

[3] V. I. Arnold. Towards the Legendrian Sturm theory of space curves. *Funct. Anal. Appl.*, **32** (2) (1998), 75–80.

[4] V. I. Arnold. Symplectic geometry and topology. *J. Math. Phys.*, **41** (6) (2000), 3307–3343.

[5] V. I. Arnold. On the problem of realisation of a given Gaussian curvature function. *Topol. Methods Nonlinear Anal.*, **11** (2) (1998), 199–206.

[6] V. I. Arnold, The longest algebraic and trigonometric curves and pseudo periodic Harnack theorem. *Funct. Anal. Appl.*, **36** (3) (2002), 1–10.

[7] V. I. Arnold. Astroidal geometry of hypocycloids and Hessian topology of hyperbolic polynomials. *Russ. Math. Surveys*, **56** (6) (2001), 1019–1083.

[8] V. I. Arnold. *Wave Fronts and Topology of Curves* (Moscow: Phasis, 2002) [in Russian], esp. Chapter 4, pp. 81–111.

[9] D. A. Panov. Möbious theorem on three inflection points. *Funct. Anal. Appl.*, **32** (1) (1998), 29–39.

[10] D. A. Panov. Parabolic curves and gradient mappings. *Proc. Steklov Inst. Math.*, **212** (1998), 271–288.

[11] V. I. Arnold. *Partial Differential Equations* (Springer-Verlag, 2003).

Yu. I. Manin

Rational curves, elliptic curves, and the Painlevé equation

Lecture on October 1, 1997

I would like to tell you a story, which – as it seems – can be instructive in various aspects. On the one hand, it is an attempt to solve a very recent problem. On the other hand, it turns out that in the solution one has to use very classical results, which have been obtained during the first 30 years of the twentieth century, and then the interest in them has somehow disappeared. They were forgotten, then reconsidered for completely different reasons, and so on.

This is a story of only a partial success. Starting to work with pleasure, like a hound following a fresh track, you reach a certain place, but the problem wagging its tail disappears somewhere behind a corner. So there is something left to do. I hope very much that those who cannot or do not want to follow all the mathematics still will be involved, interested in other things. And among those wanting to follow the mathematics in detail, somebody will be interested enough to work on the problem itself.

Technically, the problem I am speaking about concerns the attempt to understand the structure of the potential of quantum cohomology of the projective plane. I am just going to write down explicitly what this means. The projective plane, say over the field of complex numbers, yields its three-dimensional cohomology space, which is spanned by the cohomology classes of the whole plane Δ_0, of a line Δ_1, and of a point Δ_2. This three-dimensional space has coordinates which I denote by x, y, z:

$$H^*(\mathbb{P}^2, \mathbb{C}) = \{x\Delta_0 + y\Delta_1 + z\Delta_2\}.$$

The theory of quantum cohomology yields a formal series (the quantum potential of this cohomology) that looks as follows:

$$\Phi(x, y, z) = \frac{1}{2}(xy^2 + x^2z) + \sum_{d=1}^{\infty} N(d)\frac{z^{3d-1}}{(3d-1)!}e^{dy},$$

$N(d)$ being the number of rational curves of degree d on the plane which pass through $3d-1$ points in a general position. Thus, $N(1)$ is the number of straight lines passing through two points, i.e., $N(1) = 1$. Then, $N(2)$ is the number of conics passing through 5 points; this is also 1. The reader can try to check that

24

$N(3) = 7$; this is somewhat more difficult. Of course, the fact that this problem has a finite number of solutions, is easily checked by counting constants: a curve is given by parameters, then one imposes linear conditions on these parameters (namely, the requirement to pass through given points). This being not very easy to do, however, a little of algebraic geometry enables one to expose this argument correctly.

It is striking that these numbers $N(d)$ satisfy absolutely nontrivial bilinear relations allowing us to express them consecutively, one through another.

Removing from $\Phi(x, y, z)$ the classical contribution $\frac{1}{2}(xy^2 + x^2 z)$ whose sense I explained yesterday in my lecture at the Moscow Mathematical Society, one gets the function

$$\phi(y, z) = \sum_{d=1}^{\infty} N(d) \frac{z^{3d-1}}{(3d-1)!} e^{dy}$$

that already does not depend on x. A theorem proved by Maxim Kontsevich a few years ago, which constitutes a motivation for our joint work on quantum cohomology, looks as follows.

Theorem 1 (Kontsevich). *The function ϕ satisfies the differential equation*

$$\phi_{zzz} = \phi_{yyz}^2 - \phi_{yyy}\phi_{yzz}.$$

This single equation is equivalent to all the associativity equations for the function $\Phi(x, y, z)$. (The associativity equations state that the third partial derivatives of the function $\Phi(x, y, z)$ are the structure constants of an associative algebra.)

This equation is also equivalent to the following recurrent formula:

$$N(d) = \sum_{k+l=d} N(k)N(l)k^2 l \left[l \binom{3d-4}{3k-2} - k \binom{3d-4}{3k-1} \right]$$

for $d \geqslant 2$, and $N(1) = 1$.

If you forget the above-mentioned interpretation of the numbers $N(d)$, then the equivalence of the three statements is easily checked by a straightforward computation. However, the proof of the fact that the above defined numbers $N(d)$ defined above satisfy this recurrence relation, is an algebraic geometry argument requiring considerable effort. However, if one is allowed some hand-waving, then it can be proved in a rather intuitive way.

The problem that I have in mind and that excites me until now is to understand the analytical behavior of the function ϕ. It is given by the formal series in a neighborhood of zero. It is not difficult to show that it has a nonzero convergence domain. And then everything becomes unclear: the function has singularities, but of an uncertain type. And, it is generally unclear what function it is. Maybe it is something known: a solution of the hypergeometric equation, or a zeta function, or a modular function. Most likely, it is neither of these three. The work I am going to tell you about, arose from attempts to understand the nature of this function.

Although the differential equation for the function ϕ is simple, it is of somewhat nonclassical kind. However, it was known that this differential equation can be reduced by rather cumbersome nonlinear changes of variables to one of classical equations which are called "Painlevé-six." These equations form a four-parameter family depending on the parameters $\alpha, \beta, \gamma, \delta$. There is a standard notation $\mathrm{PVI}_{\alpha,\beta,\gamma,\delta}$ for these equations which is almost a century old:

$$
\frac{d^2 X}{dt^2} = \frac{1}{2} \left(\frac{1}{X} + \frac{1}{X-1} + \frac{1}{X-t} \right) \left(\frac{dX}{dt} \right)^2 - \left(\frac{1}{t} + \frac{1}{t-1} + \frac{1}{X-t} \right) \frac{dX}{dt}
$$
$$
+ \frac{X(X-1)(X-t)}{t^2(t-1)^2} \left(\alpha + \beta \frac{t}{x^2} + \gamma \frac{t-1}{(X-1)^2} + \delta \frac{t(t-1)}{(X-t)^2} \right).
$$

Here I shall omit the explanation of how and why the initial equation for ϕ can be reduced to PVI, and devote the rest of the lecture to an attempt to understand what this implies for the initial mysterious function.

At the beginning of the century Painlevé studied the following classification problem. Consider a differential equation, to begin with, say, of the following kind:

$$
\frac{dy}{dt} = F(y, t),
$$

with the initial condition $y(t_0) = c$, the function F being complex analytic. You start to continue the solution analytically and look where you shall meet a singularity. The singularity can be of various kinds: it can be a usual pole, or a branching point, or an essential singularity. The position of the singularity, in general, depends on the value of the initial constant c. Here various situations can arise: some singularities can move and some can remain fixed. The question was the following. (Its motivation is not completely clear to me up to now, but maybe there was some inner logic of the development of analysis up to the time when this question was posed.) One wants to understand what can be differential equations with the property that only their poles can move and everything essential (essential singular points and branching points) does not depend on c. It turned out that for small orders of the equation (for the first and even for the second order) the answer to this question yields a classification: one can write explicitly a reasonably small family of equations to which everything is reduced by changes of variables. With the first order everything is simple, and the answer was known long ago. With the second order, Painlevé and his students tried to succeed for a long time. And when Painlevé wrote his final paper on the subject, he missed PVI. The calculations were huge, and he made a computational error. No more than a year latter the error was found. The first to write out PVI correctly was B. Gambier (1906). The second was Fuchs who did it independently of Gambier and from a completely different viewpoint.

To miss PVI is a great pity, because – in some sense – the beginning of the list consists of classical equations, and this is the first nonclassical one. In this list of nonclassical equations PVI is the most general, all the rest are obtained by specialization and passing to an appropriate limit. Painlevé missed, due to his error, not a special, but a general, even the most general equation. This can happen to everybody, and I generally request you not to be severe to errors in good papers. The very fact that a paper stated a right question, gave a half-right answer and stimulated further research, is much more important than this or that error of the author. We are all human, and I quite disapprove a widely spread custom to attribute a proof of a theorem to the person who has corrected the last mistake in somebody else's proof. This is unfair, and, as it seems to me, leads to a wrong appreciation of mathematics.

Thus, we have the Painlevé-six equation, written in fact by Gambier and Fuchs (the son of the famous Fuchs; his paper is in a sense devoted to papers of his father, that he mentions with great respect). What is more interesting in the Fuchs paper (which I read with great pleasure and which I should have read 20 years earlier, but simply did not know about its existence) is that he obtained this family of equations by a method completely different from that of Painlevé. Fuchs' method is much nearer to me. More precisely, Fuchs obtained these equations in two ways. The first one is via so called isomonodromic deformations of linear differential equations, but let us not stop at this point. The other way is related to the fact that he discovered a very nice and geometric way to write out these equations.[1]

When you look at the equation PVI, if you have ever worked with elliptic curves, you realize immediately that the curve

$$Y^2 = X(X - 1)(X - t)$$

should play some role here. Though it is not quite clear what this role is: the equation looks cumbersome. And Fuchs wrote this equation in a very remarkable way, which made me almost jump up as soon as I saw it.

Before I continue the history of PVI, I would like to recall what we have begun with. We have a four-parameter family of equations plus two more pa-

---

[1] On Painlevé equations and isomonodromic deformations, see K. Iwasaki, H. Kimusa, S. Shimomusa, and M. Yoshida M. *From Gauss to Painlevé* (Braunschweig: Vieweg Verlag, 1991). (*Editor's note*)

rameters coming from the initial conditions. Thus, we have a six-parameter family of functions, which contain the function describing the quantum potential of $\mathbb{P}^2$, with which I have started. What would we like to do in order to describe this function? We would like to calculate which constants and which initial conditions are related to the quantum potential. And then, after we calculate all this, we would like to see what the Painlevé equation says about the function with such constants and initial conditions. Unfortunately, *a priori* we do not expect but a moderate success. And this is one more interesting story related to the history of PVI. Many efforts were applied to a precise statement and proof of the theorem which Painlevé stated as follows: "Almost all solutions of this new system of equations are nonclassical functions." This is hard to formulate, hard to prove, and hard to believe that this statement is useful. Indeed, one must give a precise definition of what is a classical function. You define an iterative process: those functions that you already have can be taken as coefficients of a new linear differential equation; the solutions of this equation can be added. Then they can be taken as coefficients of an algebraic equation, and solutions to this equation can also be added. However, you never know whether you did not miss some important operation. For example, it seems to me that in the statement and proof of this theorem one very important operation was indeed missed, namely, taking the inverse function. But nevertheless, the general belief is that most of Painlevé transcendental functions are some new functions. However, it is not excluded *a priori* that the only function we are interested in is nevertheless classical. Because among the PVI solutions a lot of classical functions are known. We are interested in one point in a six-dimensional space. We want to find its analytic sense and do not know the answer. This is what I meant when I said that you run what you think to be the right track, but the problem wags its tail and disappears behind a corner.

Now I return to the theorem of Fuchs (1907). Consider the integral

$$\int_{\infty}^{(X,Y)} \frac{dx}{\sqrt{x(x-1)(x-t)}}.$$

It is defined only up to an integral over a closed contour on the elliptic curve corresponding to the given value of t, i.e., defined up to a period. Periods of an elliptic curve satisfy a linear differential equation, hence one can remove the indefiniteness of the integral by applying a differential operator. Let us do so and write down an equation

$$t(t-1)\left[t(t-1)\frac{d^2}{dt^2} + (1-2t)\frac{d}{dt} - \frac{1}{4}\right]\int_{\infty}^{(X,Y)} \frac{dx}{\sqrt{x(x-1)(x-t)}}$$

$$= \alpha Y + \beta\frac{tY}{x^2} + \gamma\frac{(t-1)Y}{(x-1)^2} + \left(\delta - \frac{1}{2}\right)\frac{t(t-1)Y}{(x-t)^2},$$

where $Y^2 = X(X-1)(X-t)$. This equation is equivalent to the previous form of PVI.

This new form is better, because the left-hand side has now a transparent algebraic geometry sense, and a very wide context is known in which such equations can be generalized, in contrast to the Painlevé equations, which are naturally embedded into an absolutely different context.

This new form takes the problem away to a completely different area, namely, to the realm of algebraic geometry, which is nice: the initial problem being algebro-geometric, we embedded it into some nonlinear differential equations, and then return again to algebraic geometry.

I prefer to look at this equation as at a nonhomogeneous Picard–Fuchs equation. By a Picard–Fuchs equation I mean an equation of the following type. You have a family of tori, or curves, or algebraic varieties, and you take their periods over closed cycles which depend on the parameters. These periods – as functions of a parameter – satisfy homogeneous linear differential equations whose right-hand side equals zero. It turns out that if in the right-hand side one writes not zero but what is written above, then one gets the Painlevé equation.

What I have done after that should have been done many, many years ago. I do not understand why this story did not develop further. If you have obtained such an elliptic curve, then it is quite obvious that you can calculate this equation in other geometric representations of the same object. And the most natural geometric representation of the same object is, of course, the following. You change the t plane to the upper half-plane with the coordinate τ, by interpreting a point of the upper half-plane as a generator of the period lattice of the torus. Then over the point τ you have the torus $\mathbb{C}/\langle 1, \tau \rangle$. The differential turns simply into the differential dz. The integral becomes just equal to z. Hence, without any computations, we know that the differential operator written above turns into $d^2/d\tau^2$. The reason is that the above operator was of the second order and annihilated two periods. We now have periods 1 and τ, hence the operator annihilating them is nothing but $d^2/d\tau^2$. Hence all the left-hand side turns into $d^2/d\tau^2$: you see how much shorter it becomes. The right-hand side is not quite trivial. To write the answer, one should use a series of classical formulas from the theory of elliptic functions. Now, let me write the answer, it is rather simple:

$$\frac{d^2 z}{d\tau^2} = \frac{1}{(2\pi i)^2} \sum_{j=0}^{3} \alpha_j \wp_z \left(z + \frac{T_j}{2}, \tau \right).$$

Here T_j are the periods $(0, 1, \tau, 1 + \tau)$; the constants α_i are the same as above but suitably renormalized: $(\alpha_j) = (\alpha, -\beta, \gamma, 1/2 - \delta)$; $\wp_z$ is the derivative over

z of the Weierstrass function

$$\wp(z,\tau) = \frac{1}{z^2} + \sum_{m,n}{}' \left(\frac{1}{(z+m\tau+n)^2} - \frac{1}{(m\tau+n)^2} \right).$$

The Weierstrass function is the simplest series which one can construct from z and τ so that it be double periodic with periods 1 and τ.

This form of the Painlevé equation is already simple enough; I myself can memorize it. Why this form of the Painlevé equation was not known before my paper is for me a full and absolute mystery. Or rather it is a proof of the fact that the interest to the Painlevé equation has disappeared for some reasons, and when it reappeared again, people did not look at the classical papers on PVI.

What is nice in this reformulation, is that one can immediately notice something. Let us remember that we are interested in solutions of PVI, and even if we believe that most of them are nonclassical, it may happen that our solution is classical. And, generally, let us look whether there are classical solutions. One of them is seen immediately; if all α_j are zero, then we get the equation $d^2z/d\tau^2 = 0$, whose all solutions are linear functions. By the way, in the initial form of the equation this solution is not seen. Actually this simple remark can be used in an absolutely nontrivial way, by the reasons which are hidden very deeply. I shall speak about this below.

Corollary 1. *For $(\alpha, \beta, \gamma, \delta) = (0, 0, 0, 1/2)$, all solutions are classical functions.*

Those of you who do not want to think geometrically should just imagine that passage from X, Y, t to z, τ is but a substitution, a nonlinear change of variables. Though, of course, it is much more efficient to look at this geometrically.

Before we move further, I shall state one corollary more. Classical Weierstrass functions satisfy many identities, and in particular the so called Landen identities. These identities relate the Weierstrass functions of the arguments z and τ and of the arguments z (with a shift on a semiperiod) and 2τ. It is rather clear that if one changes τ by 2τ, then one passes to functions which are periodic with respect to a smaller lattice (a sublattice of index 2). Averaging such functions over semiperiods, one can again obtain a function periodic with respect to the initial lattice. This leads to the Landen identities which I shall not write out explicitly. Now, if one looks whether the Painlevé equation can be transformed so as to pass to a sublattice of index 2, then we come to some unexpected symmetries.

Corollary 2. *The Painlevé equation with constants* $(\alpha_0, \alpha_1, \alpha_0, \alpha_1)$ *can be transformed according to the Landen transform, and one gets the formulas which calculate the relation between solutions with these constants and solutions of the equation with the constants* $(4\alpha_0, 4\alpha_1, 0, 0)$.

Having calculated the constants corresponding to the equation responsible for quantum cohomology, we obtain that it is of a magnificent kind:

$$\frac{d^2 z}{d\tau^2} = -\frac{1}{2\pi^2} \wp_z(z, \tau).$$

Those of you who worked in symplectic geometry and Hamiltonian mechanics, looking at this equation can easily see that it is Hamiltonian.

Corollary 3. *The* PVI *equation is Hamiltonian:*

$$\frac{dz}{d\tau} = \frac{\partial H}{\partial y}, \qquad \frac{dy}{d\tau} = -\frac{\partial H}{\partial z},$$

where

$$H = \frac{y^2}{2} - \frac{1}{(2\pi i)^2} \sum \alpha_j \wp\left(z + \frac{T_j}{2}, \tau\right).$$

The Hamiltonian essentially depends on τ (the parameter τ plays the role of complex time).

The fact that the equation PVI is Hamiltonian was known in classics. However, the formulas were tremendous, and their geometric sense was unclear. The new representation makes it clear that the equation is Hamiltonian and it also helps us to study its geometric sense. I shall omit precise statements here.

The main general result on the six-parameter family of solutions of the Painlevé equation is a completely mysterious symmetry; I am going to pass to its description.

All the hope on the symmetries is that we know a number of classical solutions, and the symmetry group enables us to construct new solutions out of known ones. It is surprising that here the group of symmetries is very large and absolutely nonevident. These symmetries were discovered and rediscovered many times. However, up to very recent times they could not be understood and written out in a comprehensible form. I would like to mention the names of Schlesinger and Okamoto, and also add the names of Arinkin and Lysenko. These are two students of Drinfeld who have written a very nice paper which partially clarifies what is going on here. Let me state the answer.

I must introduce new parameters $a_i^2 = 2\alpha_i$. On the space with the coordinates (a_i) one has an action of the group of symmetries W generated by the following transformations:

(a) $(a_i) \mapsto (\pm a_i)$;

(b) permutations of a_i;

(c) $(a_i) \mapsto (a_i + n_i)$, where $n_i \in \mathbb{Z}$ and $\sum_{i=0}^{3} n_i$ is even.

The main series of symmetries is c). This group of symmetries can be lifted to an action on the whole structure and, in particular, it transforms solutions of the Painlevé equation with given parameters to solutions with some other parameters.

The main discovery here is due to Schlesinger (1924). Schlesinger embedded this problem into a series of others which are now called Schlesinger equations or the theory of isomonodromic deformations of linear differential equations. He founded the basics of this theory, and discovered a great number of discrete transforms from one equations to others. A particular case of these transforms is the group of symmetries described above. Later on, there appeared physical papers where consequences of all these symmetries were described. Japanese school made a lot here. Okamoto discovered a subset on which these transforms form a group (Schlesinger transforms form only a semigroup and they mix equations with one another; this is a rather incomprehensible thing). And Okamoto discovered that on PVI the transforms form a group. Arinkin and Lysenko, whom I mentioned above, in a sense returned to Schlesinger's ideology for the particular case of an equation on $\mathbb{P}^1$ with four singular points to which the Painlevé equation is reduced. They showed that in a certain reasonable sense this construction gives the full group of birational automorphisms of a properly defined geometric object.

Corollary 4. *All the* PVI *with* $(a_i) \in \mathbb{Z}^4$ *with the even sum of* a_i *have completely classical solutions.*

Indeed, we know this for $a_i = 0$.

Recently, the Painlevé equation has been much studied by Hitchin. In one of his papers it is proven that for the constants $(0, 0, 0, 2)$ the solutions are completely classical, and explicit formulas are given. He did not notice that this point is just one element of an infinite orbit.

At this step, I became rather optimistic and decided that certainly the equation for the quantum potential is covered by this net. It is indeed almost covered: it lies in the middle between two classical solutions, namely, in our case one has the parameters $(a_i) = (0, 0, 0, 1)$ which lie at the middle point between $(0, 0, 0, 0)$ and $(0, 0, 0, 2)$ and for them it is known that the solutions are classical. However, we know nothing about the middle. For example, it can happen that these points possess not a two-parameter family of classical solutions but a one-parameter one.

Perhaps here I should mention the two remaining parameters. Four parameters give the place of the equation in the hierarchy, and two more parameters are the initial conditions for the concrete function of interest for me.

For me it turned out convenient to calculate the initial conditions for the most degenerate curve with complex multiplication: $\tau_0 = e^{2\pi i/3}$. It is interesting that the initial condition was the following: $z(\tau_0)$ is a point of third order on the elliptic curve.

We must stop here.[2]

Note (January 14, 1998). After this lecture, I have got by electronic mail a reference to a note by Painlevé dated 1906, in which he deduces the form of his equation with the Weierstrass $\wp$-function. I breathed with relief. So this formula was discovered at its proper time and simply forgotten, and good ideas, even being forgotten, necessarily reappear again.

Bibliography

[1] Yu. I. Manin. Sixth Painlevé equation, universal elliptic curve, and mirror of $\mathbf{P}^2$. In *Geometry of Differential Equations*, eds. A. Khovanskii, A. Varchenko, and V. Vassiliev, (Providence, RI: Amer. Math. Soc., 1998), pp. 131–151. Preprint alg–geom/9605010.

[2] K. Okamoto, Studies in the Painlevé equations I. Sixth Painlevé equation PVI. *Ann. Mat. Pure Appl.*, **146** (1987), 337–381.

[3] B. Dubrovin, Geometry of 2D topological field theories. *Springer LNM*, **1620** (1996), 120–348.

[4] D. Arinkin and S. Lysenko. On the moduli of $SL(2)$-bundles with connections on $P^1(\{x_1, ..., x_4\})$. *Int. Math. Res. Notices*, **19** (1997), 983–999.

---

[2] The interested reader is referred to Yu. I. Manin. *Frobenius Manifolds, Quantum Cohomology, and Moduli Spaces* (Providence, RI: Amer. Math. Soc., 1999). (*Editor's note*)

A. A. Kirillov

The orbit method and finite groups

Lectures on December 28, 1997 and January 3, 1998

1

Truly, I am somewhat confused by what I would call an overqualified audience, for I was promised second- and first-year students. But let us hope that at least the notes of this lecture will be delivered at the right door.

I shall start with a topic not related to the lecture, namely, what is the news in mathematics. This is a separate topic, which can take the whole two hours. But still, I would like to touch upon two things. I shall only briefly mention the first of them: this is the last paper of Maxim Kontsevich, which has closed the topic of deformation quantization and which is being agitated at all mathematical centers.[1] The second is a trivial proof of the so-called cosmological theorem of Conway. This topic is intended for first-year students. Conway is a quite extraordinary mathematician, who deals with quite unexpected things. For example, it came to his mind to investigate the so-called audioactive operator. Imagine that you are like a Chukchi man, who goes in the tundra and sings whatever he sees. Suppose, you see a number, say one. You see one one, and you write: "One one (11)." Thereby, the second term of our sequence is obtained. You see the second term of the sequence, two ones, and write: "Two ones (21)." This is the third term of the sequence. Reading it aloud, we obtain: "One two, one one (1211)." The next term is "One one, one two, two ones (111221)." As a result, we obtain the sequence

$$1, \ 11, \ 21, \ 1211, \ 111221, \ 312211, \ \ldots .$$

Conway became interested in the asymptotic properties of this sequence. It is a fairly easy exercise to prove that it contains no digits other than 1, 2, and 3. The next easy exercise is to prove that, if the first and second terms of such a sequence contain only 1, 2, and 3, then the other terms also contain only these digits.

It turns out that the pronunciation operator has precisely one fixed point, 22. All the other sequences begin to grow, but in a very irregular manner:

---

[1] M. Kontsevich. Formality conjecture. Deformation theory and symplectic geometry. In *Math. Phys. Stud.*, **20** (Dordrecht: Klüwer Acad. Publ., 1996), pp. 139–156. (*Editor's note*)

the length increases sometimes very strongly and sometimes not very strongly. Conway proved that, for all sequences consisting of digits 1, 2, and 3 (except the sequence at the fixed point 22), the ratio between the lengths of the words a_n and a_{n-1} tends to a certain limit, which equals $1.301577269\ldots$. This limit is an algebraic number, and the experts in dynamical systems, certainly, guess that this is the maximum eigenvalue of some operator, but I shall omit the details.

This is not related to the topic of the lecture, but I spent several minutes expressly so that everyone who wanted to come would come and sit down.

Now, I proceed to the lecture. I am going to briefly tell about some problems arising in relation to the application of the orbit method to finite groups. Since the lecture is intended for able first-year students, I shall sometimes say words which the listeners must not understand at the moment. Later, they can ask elder friends, or guess themselves, or read in a book what these words mean and, thereby, settle a local incomprehension. There should be no conceptual incomprehensions, because I am going to change the paradigm (as Yurii Ivanovich Manin says) every several minutes, so that, at any moment, you can forget everything said before and start to attend afresh.

On the whole, the lecture is about representation theory. I believe that every first-year student at Independent University has an idea of groups, linear spaces, linear operators, and linear representations of groups (i.e., in a more scientific language, homomorphisms of groups to groups of invertible linear operators) and understands that this science is useful; representation theory has diverse applications, which I shall not talk about now.

My contribution to representation theory is that I have suggested the orbit method. To a certain degree, this method is a combination of two things, symplectic geometry and representation theory. Or, at a higher level of abstraction, this is a combination of modern mathematical physics and mathematics. The point is that mathematical physics already gradually supplants mathematics in canonically steady mathematical fields and invents new approaches and new problems, solving old ones as a byproduct. About half of mathematical journals are gradually being filled by papers in which mathematical physics plays a role, sometimes decisive.

I shall spend several minutes explaining what symplectic geometry is, although I have been told that this subject is not quite unfamiliar here. I shall start with considering a smooth manifold M. A smooth manifold is something on which mathematical analysis can be developed. Namely, this is a set M which locally has the structure of Euclidean space. This means that the set M can be covered by open domains[2] U_i in such a way that each domain ad-

---

[2] When I talk about open domains, I imply that M is a topological space rather than a

mits a one-to-one mapping to a domain in Euclidean space: $U_i \xrightarrow{\phi} V_i \subset \mathbb{R}^n$. So everything that we know from mathematical analysis, namely, the rules for handling functions of many variables, differentiation, integration, substitutions, differential equations, and so on can be transferred to the manifold M.

The simplest example of a manifold is the circle S^1. It is impossible to introduce one coordinate on the circle. Again, this is a problem for first-year students: Prove that any continuous function on the circle takes one value at two different points and, therefore, cannot serve as a coordinate. On the other hand, it is easy to figure out how to specify two domains on the circle so that such a coordinate would exist on each of them. We can introduce a coordinate on the entire circle except at the top point by projecting the circle from the top onto the horizontal line. The top point itself has no image, but we can use another projection, from the bottom point, and map everything except this point to the same horizontal line. We obtain two charts; together they cover our manifold, the circle. We call these local coordinates x_+ and x_-. For a circle of radius 1, they are related as $x_+ x_- = 1$ ($x_+ x_- = r^2$ for a circle of radius r).

So far, our main object is a manifold, i.e., a set on which we can do mathematical analysis. I also said the word "group." In a group, elements can be multiplied. Usually, groups arise as groups of transformations of something. Combining the two structures, of a group and of a manifold, we obtain a Lie group. This is our main object of study.

A Lie group is an object being a manifold and a group simultaneously. This means that it is endowed simultaneously with local coordinates and with a multiplication law possessing the usual properties of associativity, existence of inverse elements, and the existence of an identity element.

The same circle is an excellent example of a group. The simplest way to introduce multiplication is as follows. Imagine that the circle has radius 1 and lies not simply in the real plane but in the complex plane. Then we can specify it by the equation $|z| = 1$ and introduce a group law in the form of the usual multiplication of complex numbers: $z_1 z_2$. If two numbers have modulus 1, then their product also has modulus 1.

I forgot to say that the two structures, of a group and of a manifold, must be related to each other in a natural way. Namely, on the manifold, the notion of a smooth function is defined. Defining a manifold, I intentionally forgot to mention that the correspondence $U_i \xrightarrow{\phi} V_i$ is arbitrary, but at the regions where two coordinate systems arise, we must require that the transition from one coordinate system to the other is implemented by means of smooth functions. Smoothness means the existence of several (one, two, three, and so on)

---

mere set, i.e., open and closed subsets of M are defined.

continuous derivatives. As a rule, it is convenient to assume that all functions are infinitely differentiable. So, the relation between the group axioms and the axioms of a manifold is very simple: the group law and the mapping which takes each element to its inverse must be smooth functions.

Exercise. Write down the group law for the circle in the coordinates x_+ and x_- and show that it is a continuous function.

In what follows, we shall deal with representation theory. I shall consider only unitary representations. This means that a group G is mapped to the group aut(H), where H is a Hilbert space. A Hilbert space is a generalization of a linear space in two directions. Namely, the dimension of the space is not bounded and can be infinite. Secondly, as a rule, the complex rather than real field is considered. We could consider real representations too, but they reduce to complex representations, and many facts and theorems over the complex field are much simpler. For a Hilbert space, there is a notion of the inner product of two vectors, which is inherited from finite-dimensional spaces. Thus, we can introduce norms of vectors and the notion of orthogonal vectors. By the automorphisms of H I understand the operators which preserve all the structures in H, namely, the structure of linear space and inner product. Such operators are called unitary.

Thus, to each element g of the group G we assign a unitary operator $U(g)$ in such a way that the functional equation $U(g_1 g_2) = U(g_1)U(g_2)$ holds: a product of elements of the group is mapped to a product of operators. There is a fairly well-developed science about how to deal with unitary representations. In particular, there are natural notions of equivalent representations and of decomposition of representations into sums. The representations that do not admit decompositions into direct sums are called indecomposable. They turn out to be irreducible, i.e., they have no nontrivial subrepresentations. The first problem that arises in considering any group is to describe the set of all irreducible unitary representations of this group up to equivalence. The set of equivalence classes of irreducible unitary representations of a group G is denoted by $\widehat{G}$. In the English mathematical-physical language, there is a very convenient abbreviation for this long term, namely, *unirrep* (UNItary IRREducible REPresentation). These unirreps are the main object of our study; it would be good if, for each group, we had a list of such unirreps and their properties: how they behave under restriction to a subgroup, or under induction (I shall not say what it means at the moment) to a larger group, what their matrix elements, characters, and infinitesimal characters are, and so on.

There arise as many problems as there are groups. Since groups occur in all areas of mathematics and its applications, these problems also arise everywhere.

They are solved by different methods in different areas, and the solutions are sometimes quite unlike. Thus, there is one set of theorems for compact groups, a second set of theorems for semisimple groups, a third set of theorems for nilpotent groups, and so on.

The orbit method, which I shall talk about, has the advantage that it considers all groups at once and gives a universal recipe for describing the set of irreducible representations and answering most of the related questions. I shall not describe this method in detail.[3] I shall only mention the related notions and explain what the method does. The most popular among all my fascinating enterprises is the so-called User's Guide. When you buy an electric iron or something else, an owner's manual is included. Such a manual exists for the orbit method too. It says what you must know and do to obtain an answer – what you must evaluate, multiply, add, and so on.

I shall list the basic ingredients and, then, proceed to the main topic of my lecture, which is how to apply the orbit method to finite groups. For beginning mathematicians, it is always more pleasant to deal with something tangible and finite. When I was a first-year student, I believed that I could solve any problem concerned with a finite set of objects. This turned out to be not very true; there are quite finite problems which people cannot solve so far. I shall try to formulate some of such problems later on. Right now, I shall finish the description of the orbit method and say how it can be applied to finite groups.

Our next object is a Lie algebra $\mathfrak{g} = \mathrm{Lie}(G)$. The letter $\mathfrak{g}$, lowercase Gothic g, has become a conventional notation for the Lie algebra associated with a Lie group G. The theory of Lie groups and representation theory have been reformed many times, and each epoch had its own notation. At present, it is more or less conventional to denote Lie groups by capital Latin letters and Lie algebras by lowercase Gothic letters. Although, there are retrograde scientists, who use obsolete notation, and progressive scientists, who invent their own new notation. But I shall adopt the most common notation.

A Lie algebra is what remains of a group when only infinitesimal neighborhoods of the identity are considered. Namely, we take the tangent space T_eG to a Lie group G at its identity e and interpret the vectors of this tangent space as points infinitely close to the identity. What does remain of the group law when only the points very close to the identity are considered? If we introduce local coordinates with origin at the identity, each element of the group will be represented by a vector. Take two such vectors $\vec{x} = (x_1, \ldots, x_n)$ and $\vec{y} = (y_1, \ldots, y_n)$ and consider their product $\vec{x} \cdot \vec{y} = \vec{z}$ in the Lie group. The coordinates of the

---

[3] See A. A. Kirillov. *Elements of Representation Theory* (Moscow: Nauka, 1972) [in Russian]. (*Editor's note*)

obtained vector $\vec{z}$ are functions of $\vec{x}$ and $\vec{y}$:

$$z_k = \phi_k(\vec{x}, \vec{y}).$$

It turns out that the associativity of multiplication and the condition that the multiplication of any element by the identity yields the same element imposes strong constraints on the function ϕ_k, and under a suitable choice of local coordinates, it has the following remarkable form:

$$\phi_k(\vec{x}, \vec{y}) = \vec{x} + \vec{y} + [\vec{x}, \vec{y}] + \dots$$

(the ellipsis denotes terms of the third and higher orders). This formula means that, on any Lie group, the group law is commutative in the first approximation and determined by a bilinear skew-symmetric expression $[\cdot, \cdot]$ satisfying the Jacobi identity in the second approximation. This bilinear operation transforms the tangent space to the group at the identity into the so-called Lie algebra. To every Lie group, a Lie algebra is associated.

Suppose that a Lie algebra has basis $X_1, \dots, X_n$; then, to specify the structure of this Lie algebra, it is sufficient to specify the products of basis vectors: $[X_i, X_j] = c_{ij}^k X_k$. I use standard Einstein's rule that the presence in some expression of the same index as a superscript and as a subscript mean summation over this index. Thus, a Lie algebra is simply a set of structural constants c_{ij}^k. The great discovery of Sophus Lie is that this set of structural constants contains all the information about the Lie group. In principle, you can extract everything you want to know about this group from the set of structural constants. A little later, I shall say how this affects the problems which we shall deal with.

Geometrically, what we need is not the space $\mathfrak{g}$ itself but its dual space, which is denoted by $\mathfrak{g}^*$. This is the space of linear functionals on $\mathfrak{g}$. If the initial space has basis $X_1, \dots, X_n$, then we can introduce dual functionals $F^1, \dots, F^n$ in such a way that each functional takes the value 1 at the vector with the same number, vanishes at the remaining basis vectors, and is extended to the other vectors by linearity.

The space $\mathfrak{g}^*$, as opposed to the Lie algebra $\mathfrak{g}$, carries no multiplication – functionals cannot be multiplied. Although, we can introduce multiplication by force, thus obtaining a very interesting construction known as a bialgebra. It is related to quantum groups, Drinfeld algebras, and other things. This is a separate science, and I shall not touch it here. We are interested in $\mathfrak{g}^*$ as a linear space.

The last ingredient from the general theory of groups and Lie algebras which I need today is the action of a Lie group G on the algebra Lie $\mathfrak{g}$ and the dual action on the space $\mathfrak{g}^*$. This action can be defined in many equivalent ways.

The simplest way to show that such an action exists is as follows. For each $x \in G$, consider the transformation $A(x)\colon g \mapsto xgx^{-1}$. This transformation is an automorphism of the group. In addition, this is a smooth transformation of the manifold G leaving the point e fixed. Hence there is a derivative mapping, which acts in the tangent space and is denoted by $\mathrm{Ad}\,x$. It takes an element X of the tangent space to another element of the tangent space, which I conventionally denote by xXx^{-1}. Such a notation is justified, because the overwhelming majority of Lie groups can be realized as subgroups of matrix groups (if infinite matrices are allowed, then all Lie groups can be realized so; if only matrices of finite order are considered, then there are exceptions), and for a matrix group, this formula can be understood literally. This is what most physicists do, because for physicists, any group consists of matrices.

We have described the action on the Lie algebra itself, while we need an action on the dual space. I denote the action of an element x on the dual space $\mathfrak{g}^*$ by $K(x)$ (K is an abbreviation for the Russian term *коприсоединенное*, which means *coadjoint*: the representation $\mathrm{Ad}\,x$ is said to be adjoint, and the representation $K(x)$, coadjoint). The action of $K(x)$ on a functional F is defined by

$$\langle K(x)F, X \rangle = \langle F, \mathrm{Ad}\,x^{-1}X \rangle.$$

The matrix of the coadjoint representation differs from the matrix of adjoint representation in that it is replaced by the inverse matrix and transposed. Seemingly, the difference is minor. But it turns out that the adjoint representation and the coadjoint representation look quite different. Roughly speaking, the coadjoint representation has much more properties than the adjoint one.

One of the evident geometric properties is that all coadjoint orbits are symplectic manifolds, and the action of the group preserves the symplectic structure. "Coadjoint orbits" is a jargon term used in place of "orbits of the coadjoint representation." An orbit is obtained by applying $K(x)$ to a functional F for all $x \in G$.

Now, I shall make yet another brief digression, this time on symplectic manifolds. This notion came from mechanics. In a certain sense, symplectic manifolds are an odd analogue of Riemannian manifolds. Maybe, Riemannian manifolds are even closer to beginning mathematicians. These are manifolds on which lengths can be measured; i.e., for each of them, a positive definite quadratic form on the tangent space is defined. For a symplectic manifold, instead of a quadratic form on the tangent space, a skew-symmetric bilinear form, i.e., a skew-symmetric inner product, is defined. Roughly speaking, this is an odd analogue of a metric. In general, at present, people are inclined to believe that each notion must have an odd analogue, and we can gain a proper understanding of various phenomena only by considering objects and their odd analogues simultaneously.

A more accurate analogy between Riemannian manifolds and symplectic manifolds is obtained when taking into account that the very basic attribute of Riemannian manifolds is curvature. Spaces may be flat or nonflat, and this nonflatness manifests itself in that the metric has curvature. When the metric is replaced by a skew-symmetric quadratic form, the notion of curvature persists. So, symplectic manifolds are an odd analogue of flat Riemannian manifolds, i.e., they have curvature zero. This can be expressed in the coordinate from as follows. If we write locally the skew-symmetric form in the form $\omega = \omega_{ij} dx_i \wedge dx_j$, then the flatness condition takes the form

$$d\omega = \frac{\partial \omega_{ij}}{\partial x_k} \, dx_i \wedge dx_j \wedge dx_k = 0.$$

As is known, any flat metric can be made constant in a suitable coordinate system. A symplectic structure also becomes a constant symplectic structure in a suitable coordinate system; i.e., we can choose a local coordinate system in such a way that the coefficients ω_{ij} are constant numbers. Symplectic geometry has no local invariants. But it does have global invariants, and this is a very interesting science – symplectic topology. Everybody can become acquainted with it in the recently published collection of works of Arnold.[4] *A propos*, it is interesting in many other respects too.

Each orbit of the coadjoint representation has the canonical structure of a symplectic manifold, and the symplectic form ω on the orbit is invariant with respect to the action of the group. This property, whose formulation seems to be very complicated, has a clear easy-to-formulate consequence; namely, all coadjoint orbits necessarily have even dimension.

The simplest example is as follows. If the initial group is the rotation group of three-dimensional space, then the corresponding Lie algebra is the usual three-dimensional space, the adjoint representation is the so-called tautological representation (each rotation of three-dimensional space acts precisely as a rotation of three-dimensional space), and the orbits are two-dimensional spheres or the origin. As you see, only dimensions 0 and 2 occur.

The basic principle of the orbit method is that the following two sets, although not precisely coinciding, are in a certain sense close to and responsible for each other. These are the set $\widehat{G}$ (the equivalence classes of unitary irreducible representations, briefly unirreps) and the set $\mathfrak{g}/G$ of coadjoint orbits:

$$\widehat{G} \cong \mathfrak{g}/G.$$

---

[4] See the reference on p. 3. (*Editor's note*)

This correspondence looks like something very universal; it must be valid for any Lie group. Therefore, it is natural to regard this relation as some general principle, which is valid whenever its both sides make sense. For example, a Lie group can be understood in a somewhat generalized sense. We can assume that it is not a classical smooth manifold with real coordinates but, say, a complex manifold, or a manifold over another field (an algebraic variety). We can assume also that it is an infinite-dimensional manifold. Finally, we can assume that this is a quantum group, which is not a group at all but still does have an associated Lie algebra and a coadjoint representation. Of most interest to me today is the case where the group is an algebraic variety over a finite field. Then it itself is finite as a group. All theorems and conjectures which I am going to talk about refer to the very simple and comprehensible object, finite groups. Remarkably, some of the results obtained by applying the orbit method turn out to be well-known correct theorems, some are new (although still correct) theorems, some turn out to be false, and, finally, some remain open problems. The problems are the first thing I want to talk about.

The general principle of the orbit method is that this method is a linear or, differently, quasiclassical approximation to the true representation theory. Namely, instead of a nonlinear manifold (Lie group), we take a linear manifold (its Lie algebra) and consider the multiplication law only up to quadratic terms, which is the first step of approximation. Thus, the closer the Lie group to its Lie algebra the better the method works.

Each Lie group is related to its Lie algebra by a natural mapping from the Lie algebra to the Lie group. Given a tangent vector, i.e., a direction of motion from the identity, you can move along this direction in the most natural way, so that the trajectory of motion is a one-parameter subgroup. The parameter on this curve can be chosen in such a way that the following two conditions hold:

(1) $g_t g_s = g_{t+s}$ (this can be achieved by virtue of the general theorem that any one-dimensional group is locally the additive group of real numbers);

(2) $\dot{g}_0 = X$, where X is a given vector.

This system of ordinary differential equations with such an initial condition has a unique solution, and g_1 is denoted by $\exp X$. This notation is chosen because, for the matrix Lie group, we have $\exp X = \sum_{k \geqslant 0} X^k / k!$, i.e., the true exponential is obtained.

The mapping exp is defined on the entire Lie algebra. But the image of this mapping covers the Lie group not entirely. Therefore, the inverse mapping (it would be natural to call it logarithm) is defined not anywhere, is one-to-one not anywhere, and, in general, has much worse properties than the direct mapping exp.

The orbit method works most simply when the mapping exp is one-to-one. The most remarkable example for which this mapping is one-to-one is my main object of consideration today.

Let G_n denote the algebraic group of upper triangular matrices. I shall make a brief digression on algebraic varieties. This is a very important notion; to be more precise, of most importance is the related ideology rather than the notion itself. At the dawn of this science, the term *algebraic variety* was used for the set of consistent solutions to an algebraic system of equations. In due course, people understood that this point of view was incorrect. For example, if you consider the real numbers, then the equations $x^2 + y^2 = -1$ and $x^2 + y^2 = -3$ are equivalent, because each of them has no solutions. At the same time, clearly, these are slightly different equations, and they must describe different sets. Indeed, if we consider the complex numbers, then these two equations have different solutions. Yet, consider the equation $x + y = x + y + 2$. It also has no solutions and, therefore, must be considered equivalent to the first two equations. But, clearly, it is impossible to think up a transformation of these equations into each in other. These are different objects. An whereas the first two equations can be rescued by considering complex solutions, the third one has no complex solutions. But when we consider solutions in, e.g., the field of residues modulo 2, this equation determines the entire plane, i.e., any pair (x, y) is its solution.

Systems of equations have different solution sets when we consider them over different fields or, more general, over different algebras. Little by little, people understood that a system of equations is a functor from the category of algebras to the category of sets. Namely, given a system of equations with coefficient from some field k and an algebra A over this field, the solution set over this algebra is some set X_A. Thus, we have a correspondence $A \mapsto X_A$, and this correspondence is a functor. I shall not explain what a functor is; ask your clever neighbor about it.

This understanding of an algebraic variety is already correct. Another question is what algebras should be considered. Sometimes, we must consider the complex numbers, and sometimes narrower structures are sufficient. In some situations, it is useful to consider noncommutative algebras A. This refers to the science called noncommutative algebraic geometry.

An algebraic group is simultaneously a group and an algebraic variety. But we must remember that an algebraic variety is not a set. It becomes a set only after some algebra A is chosen. Thus, an algebraic group is not a group in the usual sense, merely because it is not a set, whereas group theory teaches us that groups are sets.

An algebraic group is something that becomes a group after you choose an algebra A and consider the solution set of the corresponding system of equations over the algebra A. As you understand, this is a whole family of groups depending on the choice of the algebra.

After this prelude, I shall write an example of an algebraic group. This group G_n is an algebraic subvariety in the set of all matrices: $G_n \subset \mathrm{Mat}_n$. And it is determined by the system of equations $x_{ij} = 0$ for $i > j$ and $x_{ij} = 1$ for $i = j$. We can consider solutions to this system over any field and over any algebra. I claim that G_n is an algebraic group. I shall not carefully define what it means; I shall only verify that, as soon as we consider solutions over some algebra, we obtain a group. For simplicity, let $n = 3$. Then the solution set of this system over a given algebra A looks like

$$\begin{pmatrix} 1 & a_{12} & a_{13} \\ 0 & 1 & a_{23} \\ 0 & 0 & 1 \end{pmatrix},$$

where $a_{ij} \in A$. Such matrices are said to be strictly upper triangular (strictly because of the ones on the diagonal). For any algebra A, the set of strictly upper triangular matrices is a group under the usual matrix multiplication.

Now, I can already specify the subject matter of today's lecture. Namely, I can describe what the orbit method can say about the representations of this group in the case where $A = \mathbb{F}_q$ is a finite field with q elements. Here $q = 2, 3, 4, 5, 7, 8, 9, 11, 13, \ldots$. There exist no fields comprising 6, 10, or 12 elements. The most interesting test example is the field with 4 elements: there exists a field with 4 elements, but it is not the field of residues modulo 4.

Interesting problems arise when the so-called asymptotic representation theory is engaged. This means that we consider a whole infinite series of finite groups, rather than one finite group, and examine its asymptotic properties. For example, we might ask whether these asymptotic properties can be interpreted as some results of the theory of representations of some ideal object at infinity.

The group $G_n(\mathbb{F}_q) \subset \mathrm{Mat}_n(\mathbb{F}_q)$ has two parameters, n and q. They play quite different roles, although there is a curious argument implying that there must be a relation between these two parameters. We can proceed in two ways. First, we can fix n and vary q. This means that we consider the same algebraic group over different fields. It turns out that the orbit method is perfectly adjusted to such a variation and gives a universal description for the representations of all these groups with certain additional corrections depending on n. But when we fix q and let n tend to infinity, the problem becomes much more interesting, and even the simplest questions still have no answers. It is this problem that I want to discuss: What does the representation theory of the group of upper triangular matrices of very large order with elements from a given finite field look like? As to the field, it often affects hardly anything, and the results are approximately the same for all fields. For this reason, we only consider the simplest field, namely, the field $\mathbb{F}_2 = \{0,1\}$ with two elements.

The field $\mathbb{F}_2 = \{0,1\}$ is the simplest one by virtue of the axiom $0 \neq 1$. This and only this axiom prohibits a field consisting of only one element. In some cases, it makes sense to consider a field consisting of one element. The axiom is violated, but there are no other obstructions. Interestingly, some formulas make sense even when q is equal to 1. More than that, there exists a metamathematical philosophy which says that the case of $q = 1$ corresponds in a certain sense to an infinite field, namely, to the complex field $\mathbb{C}$. Furthermore, $q = -1$ corresponds to the field $\mathbb{R}$. But this is another topic, which I do not want to discuss because of time limitations. Those who wish to learn more about it might want to look up Arnold's *Selected Works*.[5]

In what follows, I shall often restrict myself to the simplest field with two elements, although there are interesting questions for general fields too.

We start with describing coadjoint orbits. As we believe that the orbit method works, i.e., that irreducible representations correspond to coadjoint orbits, it would be good to learn to somehow describe the coadjoint orbits. This gives rise to the following question. In algebraic geometry, there is a notion of generic elements. Generic orbits are usually easy to describe; at least, they are easy to describe for the triangular group. Next are degenerate orbits, and the higher the degree to which a class of orbits is degenerate the more difficult the problem of classifying them. This problem is not completely solved as yet. Moreover, it is not solved for any field. For the real field, the problem arose very long (about thirty years) ago. It was known even then. And even then, it was unclear how to classify the representations of the triangular group with real elements. After the orbit method had been invented, it became clear that classifying representations is the same thing as classifying coadjoint orbits. Seemingly, the problem becomes substantially simpler, because it now deals with quite a finite object – the finite-dimensional triangular matrices and a finite-dimensional group acting on them in a certain way – instead of infinite-dimensional representations in a Hilbert space. Nevertheless, the orbits still resist classification. Moreover, there are considerations which suggest that the solution must be nontrivial. On the other hand, everyone who thought of this classification became convinced that, even if such a classification could be obtained, it should not strongly depend on the field. Whatever the field (complex, real, or finite), the classification is approximately the same. It is only unknown what it is. Thus, the case of the simplest field is of greatest

---

[5] V. I. Arnold. *Selected Works–60* (Moscow: Fazis, 1997). (*Editor's note*)

interest.

Now, I shall write several simple formulas which show what the coadjoint representation for the triangular group looks like. The group G_n itself consists of elements

$$g = \begin{pmatrix} 1 & * & * & * & * \\ 0 & 1 & * & * & * \\ 0 & 0 & 1 & * & * \\ 0 & 0 & 0 & 1 & * \\ 0 & 0 & 0 & 0 & 1 \end{pmatrix}.$$

(We take $n = 5$ by way of example.) The entries on the diagonal are ones, under the diagonal zeros, and above the diagonal arbitrary elements.

The Lie algebra is the tangent space. The set G_n is no longer a manifold, and it has no formal tangent space, but we can define a tangent space for an arbitrary algebraic variety. In the case under consideration, the tangent space consists of the same elements with ones replaced by zeros on the main diagonal:

$$\mathfrak{g}_n \ni X = \begin{pmatrix} 0 & * & * & * & * \\ 0 & 0 & * & * & * \\ 0 & 0 & 0 & * & * \\ 0 & 0 & 0 & 0 & * \\ 0 & 0 & 0 & 0 & 0 \end{pmatrix}.$$

It is convenient to represent the elements of the dual space in the form of lower triangular matrices:

$$\mathfrak{g}_n^* \ni F = \begin{pmatrix} 0 & . & . & . & . \\ * & 0 & . & . & . \\ * & * & 0 & . & . \\ * & * & * & 0 & . \\ * & * & * & * & 0 \end{pmatrix}.$$

There are zeros on the diagonal and some elements under the diagonal; I do not want to write zeros above the diagonal, so I put dots. Let me explain why. If some space is a subspace in the space of matrices, then the dual space is a quotient space of the dual space of matrices. The passage to the dual space is a contravariant functor, which interchanges subspaces and quotient spaces. The space of matrices is self-dual. The trace $\mathrm{tr}(F \cdot X)$ is a bilinear function depending on matrices F and X. For a fixed F, it is a linear functional of X, and for a fixed X, it is a linear functional of F. And each linear function of X can be represented in this form for a suitable matrix F. Thus, the space $\mathrm{Mat}_n(\mathbb{F}_q)$ is dual to itself. Moreover, most importantly, this duality is invariant with respect to conjugation: the replacement of X and F by xXx^{-1} and xFx^{-1}

does not affect the result. Therefore, the duality survives the adjoint action on X and, simultaneously, the coadjoint action on F. But if X ranges only over some (upper triangular) matrices rather than over all of them, then as F we must take the quotient space of all matrices modulo the matrices that kill the upper triangular matrices. The dots above the main diagonal represent arbitrary numbers, which we ignore.

Now, the coadjoint action can be written as

$$K(x)F = [xFx^{-1}]_{\text{lower part}}.$$

The matrix F can be thought of as the lower part of some matrix also, because the lower part of a transformed matrix depends only on the lower part of the initial matrix.

After this preliminary, I can write out explicit formulas for the orbits of the action of the coadjoint representation. We multiply the matrix F by an upper triangular matrix g on the left and by the upper triangular matrix g^{-1} on the right:

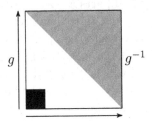

As is known, multiplication by a triangular matrix g of the specified form on the left leaves the bottom row intact, adds the bottom row with some coefficient to the row next to bottom, adds a linear combination of the two bottom rows to the row which is next to the next to the bottom row, and so on. The left arrow shows that the procedure goes on from bottom to top. Similarly, multiplication by the matrix g^{-1} on the right leaves the first column intact, add the first column multiplied by a coefficient to the second column, and so on. The bottom arrow indicates the direction of the process. We see that the changes propagate from the bottom left corner to the right and upward. In particular, the bottom left element remains unchanged. This is an example of what is called invariants. Similarly, the minor composed of the last two rows and the first two columns also remains unchanged. To its second column, we add the first with some coefficient, but this does not affect the determinant. The same refers to the rows. The bottom left minors, which I denote by $\Delta_1, \Delta_2, \ldots, \Delta_{[n/2]}$, are invariants; that is, they do not change under the coadjoint action. It turns out that the system of equations

$$\Delta_1 = c_1, \quad \Delta_2 = c_2, \quad \ldots, \quad \Delta_{[n/2]} = c_{[n/2]},$$

obtained by equating these invariants to constants determines a set which is invariant with respect to the coadjoint action and consists of precisely one orbit for generic (typical) values of the constants. What does "typical" mean? For example, it is sufficient that all these constants, except the last one when n is odd, be nonzero. If some constants vanish, then this system of equations still determines an invariant set, but it may decompose into smaller orbits. The pattern of the decomposition of a large invariant set into small orbits becomes more complicated as the number of vanishing invariants increases, i.e., as the dimension of the corresponding coadjoint orbit decreases. The structure resembles a branching tree. There is a generic invariant set; as a rule, this is simply one orbit. Some exceptional sets turn out to be not orbits; in these sets, additional invariants arise. They are not invariants in general, but they are invariants on the particular set under consideration. This gives rise to a further partitioning, where typical parts are again orbits and exceptional parts provide additional invariants of the third level, etc. This branching process continues while orbits of dimension zero (i.e., points) are obtained. An adequate apparatus for describing the whole process has not yet been invented.

Now, I enter a new domain, which is related to two things. The first is experimental mathematics. Somebody sits and calculates something and obtains some result. Then he calculates something else and obtains another result. When several results are obtained, he figures whether they can be fitted into some theory. Usually, the success depends on the amount of preliminary calculations. In the nineteenth century, people were coverings kilograms of paper with writing for years. Now people work on computers, and for months rather than for years, because they must write papers, compete for positions, etc. The life accelerates, and we have less and less time for thinking. But nevertheless, experimental mathematics flourishes. Maybe, this is partly because there are people who do not worry about their positions and can spend unlimited time thinking. There are at least two eminent mathematicians, Conway and Coxeter, who sit in their positions so firmly that they do not have to think about the bare necessities, and they can think about the purport of life and of mathematics. Recently, they both invented several perfectly unexpected things, which could be thought up only by someone having unlimited free time. The last invention of Coxeter is very suitable for a mathematical contest.

Experimental mathematics is only one part. The second is a science which, maybe, does not even exist as a science but only as a direction. This is the theory of partly completely integrable systems. Integrable systems are mechanical systems for which solutions, roughly speaking, can be written explicitly. When quantum, rather than classical, systems are considered, matrices whose eigenvalues and eigenvectors can be written explicitly are meant. Such matrices are few, and they are usually related to some symmetry. The presence of symmetry facilitates finding eigenvalues and eigenvectors. But about ten years ago, it was discovered that there are matrices for which only the first several eigenvalues and eigenvectors can be written and the others resist all efforts – not because of human stupidity but because of the nature of things: the first eigenvalues are good, after which a chaos sets in and nothing can be said.

A similar phenomenon is observed in the science which I am talking about. It manifests itself as follows. Let us do a piece of experimental mathematics, namely, take the simplest field with two elements and perform routine calculations. Taking matrices of orders 1, 2, 3, etc. and calculating the numbers of adjoint orbits for them, we obtain some sequence. I shall write out its first several terms. To be more convincing, I shall perform the calculations until they become too cumbersome.

Let us start with matrices of order 1. The Lie group consists of one matrix with element 1, and the Lie algebra consists of one matrix with element 0. This matrix is the unique orbit.

For matrices of order 2, the Lie group consists of the matrices

$$\begin{pmatrix} 1 & a \\ 0 & 1 \end{pmatrix},$$

and the dual space to the Lie algebra consists of the matrices

$$\begin{pmatrix} 0 & \cdot \\ x & 0 \end{pmatrix}.$$

Over the field with two elements, we obtain two orbits, because the group is then commutative and its action is trivial, and the element x itself takes two values.

The case of order 3 involves some computing. The Lie group consists of the matrices

$$\begin{pmatrix} 1 & a & b \\ 0 & 1 & c \\ 0 & 0 & 1 \end{pmatrix}.$$

The dual space consists of the matrices

$$\begin{pmatrix} 0 & \cdot & \cdot \\ x & 0 & \cdot \\ z & y & 0 \end{pmatrix}.$$

The transformation is $(x, y, z) \mapsto (x + az, y - bz, z)$.

I shall draw the set orbits. We are interested in the field with two elements, but the picture is the same for all fields, so I shall draw it over the field of real numbers. The orbits look as follows. Since z is an invariant (this is the invariant Δ_1 that we have considered in the general case), each plane $z = \text{const}$ is an invariant set. If $z \neq 0$, then we can make x and y arbitrary by selecting a and b, i.e., the entire plane consists of one orbit. This agrees with the general geometric fact that all orbits must be even-dimensional. If $z = 0$, then all additions are also zero, and all points are fixed. This means that the coordinate plane $z = 0$ is partitioned into one-point orbits. The dimension of a point is also even.

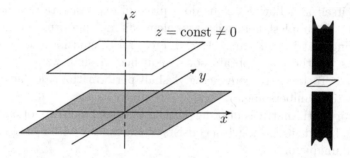

The set of orbits can be represented visually as follows. We draw a thick axis z consisting of thick points; they correspond to orbits. The point $z = 0$ must be removed and replaced by a whole plane consisting of meager points. The result is the orbit space together with its natural topology: approaching zero along the thick z-axis, we obtain all meager points as a limit. Such a topology is typical of orbit spaces. *A propos*, it is used in representation theory too. The set of irreducible representations also has its own natural topology, and it turns out to coincide with the topology of the orbit space.

Now, we return to the finite field. The number of thick orbits equals $q - 1$, and the number of meager orbits equals q^2. In all, we have $q - 1 + q^2$ orbits. Note that the number of orbits is a polynomial in q. This is a general fact: for any n, the number of orbits is a polynomial in q.

Continuing the calculations, we obtain the table

n	-1	0	1	2	3	4	5
$O(n)$	1	1	1	2	5	16	61

I have supplemented the table by the two initial terms, the numbers of orbits for $n = 0$ and for $n = -1$. The point is that such a sequence arises also from quite different considerations, and the first three terms of this other sequence are ones.

Now, I shall make a digression on partly exactly solvable problems. It turns out that the fragment of the sequence written above coincides with the respective fragment of another sequence, which is very remarkable and occurs in many other situations. The trouble is that the next position in this well-known sequence is occupied by 272, while the number of orbits for $n = 6$ is 275. The formula for the sequence K_n mentioned above is

$$\sum_{n \geqslant 0} K_n \frac{x^n}{n!} = \tan x + \sec x.$$

The numbers of orbits start well, but then they unexpectedly deviate from this good sequence.

This *status quo* remained unchanged for a comparatively long time; I gave my first talk on this subject about two years ago. At that time, these two sequences and their divergence in the sixth term had already been known. It turns out that, when the field with q (rather than two) elements is considered, there arises a q-analogue of the sequence K_n. The terms of this sequence are polynomials in q. As well as the numbers K_n, they are determined from the so-called Euler–Bernoulli triangle. It is described in Arnold's *Selected Works*.[6] The Euler–Bernoulli triangle has a remarkable q-analogue, whose elements are polynomials in q. Actually, it involves yet another auxiliary parameter t, and polynomials in two variables are obtained; they become polynomials in q at $t = 1$. So, it turned out that the difference, when considered over the field with an arbitrary number q of elements, is divisible by $(t - q)(t - q^2)$. In particular, if the parameter t takes one of the exceptional values q and q^2, the discrepancy vanishes.

Now, I shall explain this in detail. We partition the set $\mathfrak{g}_n^*(\mathbb{F}_q)$ into orbits (here n is the order of the triangular matrices). This set is a linear space over the field $\mathbb{F}_q$ with dimension equal to the number of nonzero elements in the triangular matrices. Therefore, $|\mathfrak{g}_n^*(\mathbb{F}_q)| = q^{n(n-1)/2}$. This finite set decomposes into orbits. Each orbit Ω is even-dimensional (over a finite field, this assertion has a quite definite meaning, namely, that the orbit Ω consists of q^{2r} points). We can classify orbits according to their dimensions. This is a fairly contrived classification, but it makes some sense. Let $a_n^r(q)$ be the number of $2r$-dimensional orbits in $\mathfrak{g}_n^*(\mathbb{F}_q)$. For the sequence $a_n^r(q)$ depending on the parameters n and r, it is natural to introduce the generating function

$$P_n(t, q) = \sum a_n^r(q) t^r.$$

Now, we have a sequence of polynomials which depend on two variables, t and q. These polynomials can be included in the generalized Euler–Bernoulli

[6] V. I. Arnold. *Selected Works–60* (Moscow: Fazis, 1997), p. 688. (*Editor's note*)

triangle up to $n = 5$. For $n = 6$, there is a difference, but it is divisible by $(t - q)(t - q^2)$. At $t = q$ or q^2, this difference vanishes. This is not surprising for $t = q^2$. Roughly speaking, t classifies the orbits according to dimension, and setting $t = q^2$, we replace each orbit of dimension $2r$ by a q^{2r}-point set. This is merely the number of points in the orbit, and the total number of points is easier to count than the number of orbits; it is known. Therefore, divisibility by $t - q^2$ is proved very simply. But when $t = q$, we obtain the square root of the number of points in an orbit. According to the orbit method, this number can be interpreted as the dimension of the irreducible representation. Therefore, we obtain the sum of the dimensions of irreducible representations rather that the number of orbits. This sum of dimensions behaves better than the number of orbits. For example, it better agrees with the predictions of the theory. It turned out that, for $t = q$, the coincidence of the two sequences extends further; it had been verified up to $n = 11$. This required a large amount of computation; personally, I am uncapable of doing such a work, but I have found a coauthor, Anna Mel'nikova from the Weizman Institute, who was able to accomplish the task. She performed all the calculations and found that the formulas coincide up to $n = 11$. We were about to rejoice when the news that the formula cannot be valid for $n = 13$ arrived. The point is that this formula has a corollary, which I shall state below, and this corollary was disproved by a certain example.

This is not quite a counterexample; it contradicts the much more general assertion, I would even say the bold assertion, that the orbit method applies *literally* to the sequence of groups under consideration. If this were so, then the orbit method would give, in particular, an explicit formula for the characters of irreducible representations. And if this formula were valid, then all the characters of irreducible representations would be real (for the field with two elements). As every expert in group theory knows, this implies that each element of the group is conjugate to its inverse: $g \sim g^{-1}$. The conjugacy classes differ in characters, and the value of a character at an element g is the conjugate value of this character at the element g^{-1}. For the real numbers, they coincide.

This gives rise to a problem which can be formulated in a way easy to understand for a first-year student. Given a triangular matrix with elements from $\mathbb{F}_2$, determine whether or not it is conjugate to its inverse matrix inside the triangular group. If we considered the group of all matrices, then the answer would be positive. The point is that a triangular matrix reduces to a set of Jordan blocks, and each matrix under consideration has the same Jordan blocks as its inverse. But inside the triangular group, the answer is not that obvious. For matrices of order 13, it has been shown with the aid of a computer that there is precisely one counterexample. Therefore, only the formula for characters that is obtained by literally applying the orbit method is incorrect. It is a pity, because if such an application were valid, we could prove the coincidence of all the polynomials.

In fact, I made a little hasty conclusion. The existence of a counterexample does not imply that the polynomials are different for $t = q$. It only implies that the proof which I have in mind does not work. The polynomials themselves are so beautiful that, maybe, they do work.

2

Although this lecture is a continuation of the first one, it can be listened to independently. I hope to tell about more specific things. I want to tell about problems related to representations of finite groups which arise from the orbit method. Although the orbit method deals with infinite (and even infinite-dimensional) groups, it turns out that, when being interpreted correctly, it applies to finite groups too.

I shall start with a conjecture; understanding it requires nothing at all. This is in the spirit of the young mathematician's studies here, at Independent University, and follows the traditions of school study groups on mathematics. I shall tell about one remarkable sequence of polynomials. I believe that it is remarkable, and one of the reasons for this belief is that this sequence has at least five different definitions, which have not been proved to be equivalent.

We shall consider sequences of polynomials denoted by $A_n(q)$, $B_n(q)$, $C_n(q)$, $D_n(q)$, and $Y_n(q)$. These polynomials bear no relation to simple Lie algebras of series A, B, C, and D. I remind those who have attended the past lecture that we considered representation theory. One of the representation-theoretic characteristics of a finite group G is the set of dimensions of the irreducible complex representations of this group. For the set of irreducible complex representations of a group G considered up to equivalence I use the notation $\widehat{G}$. I denote an element of the set $\widehat{G}$ by λ and a representation from the equivalence class λ by π_λ. Thus, π_λ is a homomorphism $G \to \mathrm{GL}(d(\lambda), \mathbb{C})$. Since the group is finite, each of its finite-dimensional representation is equivalent to a unitary representation, and we can assume that $\pi_\lambda \colon G \to U(d(\lambda)) \subset \mathrm{GL}(d(\lambda), \mathbb{C})$.

The set of positive integers $d(\lambda)$ has several remarkable properties. For example, all these numbers divide the order of the group. In addition, there is the celebrated Burnside identity

$$\sum d^2(\lambda) = \#G.$$

(Here $\#G$ denotes the number of elements in the group G.) And if we take the sum of zeroth powers, we obtain

$$\sum d^0(\lambda) = \#\widehat{G}.$$

It would be good to interpolate these two identities and insert the sum of $d(\lambda)$ between them: $\sum d(\lambda) = ?$ In reality, it is not known whether this sum has some good properties. But there is a special case in which it does have good

54

properties. This is the case where all the representations are real (*a priori*, they are complex).

Now, I shall make a digression to remind those who know it and inform those who do not that each complex irreducible representation π_λ of a group has one of the following three types:

real type: the representation π_λ is equivalent to a real representation, i.e., in a suitable basis, all operators of the representation are represented by matrices with real coefficients;

complex type: the complex conjugate representation $\bar{\pi}_\lambda$ is not equivalent to the representation π_λ (for each representation, we can construct its complex conjugate representation by choosing a basis and replacing each element in each matrix of the representation by its complex conjugate: complex conjugation is an automorphism of the complex field, so all identities are preserved and the representation remains a representation);

quaternion type: the representation π_λ is not equivalent to a real representation but $\bar{\pi}_\lambda \sim \pi_\lambda$.

I would say that the quaternion type is the most interesting type of representations. It arises when the complex dimension of the representation is even and this even-dimensional complex space is obtained from the quaternion space of half the dimension by restricting the scalar field. The operators of the representation can then be written with the use of quaternion matrices.

The best-known example is the group $SU(2)$, which is itself the group of quaternions with modulus 1. This group is infinite, but it has many finite subgroups. The tautological representation of the group $SU(2)$ by one-dimensional quaternion matrices is an example of a quaternion representation.

As is known, each quaternion can be represented by a complex matrix of order 2; therefore, each quaternion representation of dimension n can be considered as a complex representation of dimension $2n$, if desired. This is a representation of quaternion type.

We define the index of a representation λ as

$$\operatorname{ind} \lambda = \begin{cases} 1 & \text{if the representation } \lambda \text{ is of real type,} \\ 0 & \text{if the representation } \lambda \text{ is of complex type,} \\ -1 & \text{if the representation } \lambda \text{ is of quaternion type.} \end{cases}$$

There exists a remarkable formula due to Hermann Weil which computes this index, namely,

$$\operatorname{ind} \lambda = \frac{1}{\#G} \sum_{g \in G} \operatorname{tr} \pi_\lambda(g^2).$$

I suggest that you prove this equality as an exercise. It is not easy. Even some of the experts in representation theory whom I asked could not prove this equality at once (unless they already knew the proof). Thus, I should give a hint. I do not know whether there exist many different proofs. I like the proof that I invented myself very much, and I want to advertise it.

Consider the space $\mathbb{C}[G]$ of all complex-valued functions on our group. For any finite group, there exists a Fourier transform which maps this space to the space of matrix functions on the dual object, i.e., on $\widehat{G}$. Let us denote this space by $\mathbb{C}\mathrm{Mat}[\widehat{G}]$. Being the Fourier transform, this correspondence is one-to-one. The precise formula is as follows: a function $f \in \mathbb{C}[G]$ is transformed into the function

$$\tilde{f}(\lambda) = \sum_{g \in G} f(g)\pi_\lambda(g).$$

For every class λ, the representation π_λ assigns a matrix $\pi_\lambda(g)$ of order $d(\lambda)$ to each element g. It can be verified that, if G is the circle or the line rather than a finite group, then this procedure yields the Fourier series or integral, respectively.

The injectivity of the Fourier transform implies, in particular, the coincidence of the dimensions of the spaces $\mathbb{C}[G]$ and $\mathbb{C}\mathrm{Mat}[\widehat{G}]$. *A propos*, this is precisely the Burnside identity. The Fourier transform is not only an isomorphism of spaces but also an isomorphism of algebras: it takes a convolution of functions to a product of matrices.

In the space $\mathbb{C}[G]$, consider the linear operator V that maps each function f to the function $\check{f}$ defined by $\check{f}(g) = f(g^{-1})$. Let us calculate the trace of this operator in two different ways. First, we calculate the trace of the operator in the initial space $\mathbb{C}[G]$ by choosing the natural basis $\{\chi_g\}$, where each function χ_g takes the value 1 at the element g and 0 at all other elements. The trace is the sum of the diagonal elements; therefore, it is equal to the number of the elements in the group that coincide with their inverses, because only these basis elements are mapped to themselves; all the other basis elements are mapped to different basis elements and correspond to nondiagonal elements of the matrix. The equality $g = g^{-1}$ can be written in the form $g^2 = 1$. The transformations whose squares are the identity transformation are called involutions. Thus, if $\mathrm{Inv}(G)$ is the set of all involutions in the group G, then $\mathrm{tr}\, V = \#\,\mathrm{Inv}(G)$.

The trace of the operator V can also be calculated by applying the Fourier transform. When this method is used, the elements of the space are sets of matrices rather than functions on the group. The number of these matrices is equal to the number of irreducible representations, and their orders are equal to the dimensions of the irreducible representations. A function f corresponds to a set of matrices $\tilde{f}(\lambda_1), \ldots, \tilde{f}(\lambda_n)$ of orders $d_1 = d(\lambda_1), \ldots, d_k$. Let us determine how the operator V acts on these matrices.

If the representation is of real type, then the operator V takes the matrix to its transpose: $A \mapsto A^t = {}^T\!A$. The contribution of each real representation to the trace is easy to evaluate. The nondiagonal elements are mapped to other nondiagonal elements, and they make no contribution, while each diagonal element contributes 1. Therefore, each real representation makes a contribution equal to its dimension.

The representations of complex type are even easier to handle: they make no contribution at all. The representations π_λ and $\pi_{\overline{\lambda}}$ are not equivalent. The transformation V simply interchanges values at different points: the representations λ and $\overline{\lambda}$ are not equivalent, these are two different points in the space $\widehat{G}$. So the transformation V interchanges different elements of the matrix, which gives no contribution to the trace.

Evaluating the contribution to the trace for representations of quaternion type is more difficult. It involves computing, and we must know precisely how quaternions transform into matrices of the second order under complexification and what happens to them. I shall say the result without proof. Each complex representation of quaternion type makes a contribution equal to the negative of its dimension.

Comparing these results with the definition of the index, we obtain the following final result:

$$\# \operatorname{Inv}(G) = \operatorname{tr}(V) = \sum_{\lambda \in \widehat{G}} \operatorname{ind}(\lambda) d(\lambda).$$

The number of involutions in a group is expressed as the alternating sum of the dimensions of its irreducible representations. This formula takes an especially simple form in the case where all the representations are real. Then the subscript is identically equal to 1, and we obtain simply the sum of the dimensions of the irreducible representations. Thus, if all the irreducible representations are real, then

$$\# \operatorname{Inv}(G) = \sum_{\lambda \in \widehat{G}} d(\lambda).$$

Everything said above is a prelude to the definition of the first sequence of polynomials. In today's lecture, I shall define five sequences of polynomials and state the conjecture that these five sequences coincide. I am almost ready to define the first sequence of polynomials, but still only almost. Some more information about finite fields is needed. In the past lecture, I introduced the notation $\mathbb{F}_q$ for the finite field with q elements, where $q = p^k$ and p is a prime. In what follows, we shall consider polynomials in q. It would be rather strange if an independent variable could take only several values. There have been various attempts to interpret the values of these polynomials for $q \neq p^k$. But the problem is not only unsolved, it is not even stated. For this reason, I shall not discuss it. Although, as Arnold mentions in his book, the most interesting problems are those which are not even stated. This is an example of such a problem.

Consider the group $G_n(\mathbb{F}_q)$ consisting of the upper triangular matrices g of order n which have ones on the diagonals and arbitrary elements of the field $\mathbb{F}_q$ above the diagonal. The order of this group equals $q^{n(n-1)/2}$, because each of the $n(n-1)/2$ elements above the diagonal can take precisely q different values.

Generally, this group admits representations of all the three types – real, complex, and quaternion. But if $q = 2^l$ and n is small, then all representations of this group are real; at $n = 13$, there is an example of a nonreal representation. We ignore this example and assume that all the representations are real (in the case of $q = 2^l$) as a first approximation. Then the equality

$$\# \operatorname{Inv}(G) = \sum_{\lambda \in \widehat{G}} d(\lambda)$$

gives an expression for the number of involutions in the group in terms of the dimensions of its irreducible representations. For fields with even numbers of elements, the involutions are very easy to calculate. Namely, let us write the matrix g in the form $g = 1_n + X$, where 1_n is the identity matrix of order n. Then $g^2 = 1_n + 2X + X^2 = 1_n + X^2$, because $2X = 0$ in a field with even number of elements. Therefore, the equation $g^2 = 1_n$ is equivalent to the equation $X^2 = 0$.

Consider the problem: How many upper triangular matrices X with elements from the field $\mathbb{F}_q$ such that $X^2 = 0$ exist?

Now, we can give the definition of the first sequence of polynomials. We define $A_n(q)$ as the number of solutions to the equation $X^2 = 0$ among the upper triangular matrices of order n with elements from the field $\mathbb{F}_q$. Here q is an arbitrary number for which a field with q elements exists.

Problem. Prove that $A_n(q)$ is a polynomial in q.

This is a good combinatorial problem. Given a finite object, it is required to calculate the number of points in this object. This is the fundamental problem of combinatorics. In this respect, combinatorics in not quite a science, because for every problem a separate theory is invented. But many other sciences are not sciences in this sense either. Yurii Ivanovich Manin once explained to me that algebraic geometry is not a science, because it consists of a set of solved problems, and each problem requires its own method. Certainly, there are some general notions and tricks, but, as a rule, they work in two or, at most, three cases; methods that apply to a larger number of problems occur very rarely. The most interesting results in algebraic geometry involve individual tricks. The same applies to combinatorics. As well as algebraic geometry, it is only able to solve several special problems. Several may be several thousand, but this means only that there are several thousand particular problems which combinatorics is able to solve.

How can our particular combinatorial problem be solved? It is natural to try to obtain a recursive expression of $A_{n+1}(q)$ in terms of $A_n(q)$. But I cannot do this, and, as far as I known, nobody can. We have no choice but apply one of the few standard tricks, namely, to complicate the problem so as to make it easier to solve. Let us divide the set of all solutions into types according to the ranks of the matrices:

$$A_n^r(q) = \{X \in A_n(q) \mid \operatorname{rk} X = r\}.$$

Seemingly, the problem becomes more complicated, because we must calculate many numbers at once instead of only one number. On the other hand, as in many other cases, such a complication results in simplification, because for the numbers $A_n^r(q)$ a recursive relation exists.

This recursive relation is obtained as follows. Consider the matrix $\mathcal{X} = \begin{pmatrix} X & x \\ 0 & 0 \end{pmatrix}$. Clearly, $\mathcal{X}^2 = \begin{pmatrix} X^2 & Xx \\ 0 & 0 \end{pmatrix}$. Therefore, $\mathcal{X}^2 = 0$ if and only if $X^2 = 0$ and $Xx = 0$. By assumption, we already know the number of solutions to the equation $X^2 = 0$: it has been obtained at the preceding step. And the number of solutions to the equation $Xx = 0$ depends on the rank of the matrix X. Therefore, in the general case, the number of solutions to this equation is not known. But if we know that $\operatorname{rk} X = r$, then we know the number of solutions too. Moreover, knowing something from linear algebra, we can easily figure out the rank of the matrix $\mathcal{X}$. It either does not change or increases by 1. This depends on the relation between X and x, which is easy to control. At this point, I conclude my explanation and write out the recursive relation:

$$A_{n+1}^{r+1}(q) = q^{r+1} A_n^{r+1}(q) + (q^{n-r} - q^r) A_n^r(q).$$

Now, we can forget that q takes only very special values and use the recursive relation. It defines polynomials A_n^r, which can be evaluated in turn provided that the very first polynomial A_0^0 is known. And this first polynomial is, naturally, equal to 1. We obtain a table of polynomials of very general form; all of their coefficients are nonzero, and we can say nothing good about them. But considering $\sum_r A_n^r(q) = A_n(q)$, we see that a magical reduction occurs – almost all terms cancel each other. By way of proof, I shall draw the table of polynomials A_n:

$$A_0 = 1,$$
$$A_1 = 1,$$
$$A_2 = q,$$
$$A_3 = 2q^2 - q,$$
$$A_4 = 2q^4 - q^2,$$
$$A_5 = 5q^6 - 4q^5,$$

$$A_6 = 5q^9 - 5q^7 + q^5,$$
$$A_7 = 14q^{12} - 14q^{11} + q^7,$$
$$A_8 = 14q^{16} - 20q^{14} + 7q^{12},$$
$$A_9 = 42q^{20} - 48q^{19} + 8q^{15} - q^{12},$$
$$A_{10} = 42q^{25} - 75q^{23} + 35q^{21} - q^{15},$$
$$A_{11} = 132q^{30} - 165q^{29} + 44q^{25} - 10q^{22}.$$

I recall that $A_2(q)$ is the number of solutions to the equation $X^2 = 0$ among the second-order upper triangular matrices. The equality $X^2 = 0$ holds for all second-order upper triangular matrices (with zeros on the diagonal). This means that $A_2(q) = q$.

In the table, first are three polynomials with one monomial, then three polynomials with two monomials, then three polynomials with three monomials, and then three polynomial with four monomials. I was looking at this table of polynomials for two or even three days (it was written out up to dimension 20) and found many properties which could be used to extend it unlimitedly. And only after that, I guessed what answer was correct.

The leading coefficients of the polynomials are Catalan numbers. Everybody knows how the Pascal triangle is constructed. We construct a similar triangle but place a mirror and prohibit going beyond this mirror:

```
    1
      1
    1     1
      2     1
    2     3     1
      5     4     1
    5     9     5     1
```

The elements of the first column are precisely the Catalan numbers. But the mirror hinders. It would be good to remove it while retaining the law. As is well known from physics, a mirror can be replaced by a reflection. In the mirror world, we must in addition change the signs. Then we obtain a triangle constructed by precisely the same rule as the Pascal triangle. The very mirror is filled with zeros:

$$
\begin{array}{ccccccccccccccc}
 & & & & & & -1 & & 1 & & & & & & \\
 & & & & & -1 & & 0 & & 1 & & & & & \\
 & & & & -1 & & -1 & & 1 & & 1 & & & & \\
 & & & -1 & & -2 & & 0 & & 2 & & 1 & & & \\
 & & -1 & & -3 & & -2 & & 2 & & 3 & & 1 & & \\
 & -1 & & -4 & & -5 & & 0 & & 5 & & 4 & & 1 & \\
-1 & & -5 & & -9 & & -5 & & 5 & & 9 & & 5 & & 1
\end{array}
$$

It is easy to figure out that this is merely the difference of two Pascal triangles, one growing from 1 and the other from -1.

Now, I suggest to extend the Catalan triangle and compare it with the table of polynomials. This is an excellent example of how a formula can be revealed in experimental mathematics. But this does not yet mean that it can be proved. It was proved comparatively recently, and this required efforts of two prominent experts in combinatorics. They proved the formula by using their recent theory. At present, there is a book entitled "$A = B$." The contents of this book are a construction of an algorithm; given a formula $A = B$, this algorithm outputs a proof of this formula provided that the formula is true. One of the authors of this book is my colleague at the university in Philadelphia. When I informed him that I has a formula which I could not prove, he said that they had a book in which this formula was proved. I suggested that they processed my formula with their book. They did. The algorithm did not work at once, but they exerted themselves and carried through the proof.

Assuming that the first row of the Catalan triangle contains the numbers $c_{1,-1} = -1$ and $c_{-1,1} = 1$, we can write the recursive formula

$$c_{k,l} = c_{k-1,l-1} + c_{k-1,l+1}$$

for the numbers $c_{k,l}$. To draw a triangular table, we heed two indices, and it would be good if these indices had the same parity.

Now, let us introduce a new sequence of polynomials; I denote it by C_n in honor of Catalan:

$$C_n(q) = \sum_{s \equiv n+1 (2) \equiv (-1)^n (3)} c_{n+1,s}\, q^{\frac{n^2}{4} + \frac{1-s^2}{12}}.$$

This is an experimental rule, which says what numbers from the Catalan triangle give the coefficients of the polynomials A_n.

Theorem 1. $A_n(q) = C_n(q)$.

This theorem is proved with the help of the book "$A = B$."

We have already accomplished a significant part of the program, namely, defined two sequences of polynomials. We know an explicit formula for the polynomials C_n, but we know no other interpretation of these polynomials. We know an interpretation of the polynomials A_n (if q is a power of two, then $A_n(q)$ is the number of solutions to the equation $X^2 = 0$ among the upper triangular matrices over the field $\mathbb{F}_q$), but we know no explicit formula for these polynomials. The theorem gives both an explicit formula and an interpretation. *A propos*, I am not quite satisfied with the proof of this theorem, at least because I cannot understand a single word in it. Thus, if somebody will think up a more intelligible proof, this shall be quite a step in science, and we shall publish it with pleasure.

In addition to the Pascal triangle, there is a much smarter triangle about which I personally learned from a lecture delivered by Arnold four years ago. And from the new book of Arnold, I learned that this triangle has been known for more than 100 years. But the book did not say to whom and where. Implicitly, it was not Arnold who invented the triangle, because he is less than 100 years old yet. This is the so-called Euler–Bernoulli triangle. It is constructed similarly to the Pascal triangle, but with the use of a shuttle motion. The rule is as follows: we move from left to right along the odd rows and from right to left along the even rows, every time starting with zero. Otherwise, the rules of the Pascal triangle apply.

The zeroth row consists of one element 1. The first row comprises two elements, 0 and 1. The second row begins from the right. We put 0 and move to the left; at each step, we take the sum of the preceding number in the same row and the preceding number in the preceding row, i.e., of the nearest right and upper-right numbers. The next row starts from the left. The first element is always 0, and then we write the sums of pairs of numbers already obtained:

$$
\begin{array}{ccccccccccccc}
 & & & & & & 1 & & & & & & \\
 & & & & & 0 & & 1 & & & & & \\
 & & & & 1 & & 1 & & 0 & & & & \\
 & & & 0 & & 1 & & 2 & & 2 & & & \\
 & & 5 & & 5 & & 4 & & 2 & & 0 & & \\
 & 0 & & 5 & & 10 & & 14 & & 16 & & 16 & \\
61 & & 61 & & 56 & & 46 & & 32 & & 16 & & 0 \\
\end{array}
$$

On the left-hand side of the triangle, every other element is 0 by definition. The remaining numbers $(1, 1, 5, 61, \dots)$ are well known. They are called the Euler numbers. Any handbook on special functions contains a table of Euler numbers. They are denoted by E_n and defined by the relation

$$\sum_{n \geqslant 0} E_n \frac{x^n}{n!} = \frac{1}{\cos x}.$$

I believe, this is the original definition of Euler.

Taking the right-hand diagonal, we obtain different numbers, namely, the normalized Bernoulli numbers. As is known, the Bernoulli numbers proper are rational. But we obtain the so-called normalized Bernoulli numbers, which are integer. The normalized Bernoulli number $\widetilde{B}_n$ is obtained from the usual Bernoulli number B_n by the formula

$$\widetilde{B}_n = B_n \frac{2^{2n}(2^{2n} - 1)}{2n}.$$

The normalized Bernoulli numbers are defined by the relation

$$\sum_{n > 0} \widetilde{B}_n \frac{x^n}{n!} = \tan x.$$

So (this fact is usually concealed from the students of the mechanics and mathematics faculty) not only do sine and cosine have simple Taylor expansions; secant and tangent also have Taylor expansions which can be described explicitly (in terms of the Euler and Bernoulli numbers).

It is hard to compute the Euler and Bernoulli numbers separately, while the whole triangle is computed very easily.

A generating function for the entire triangle can be written. This is also a good exercise.

My contribution to this science is that the correct understanding of the Euler–Bernoulli triangle requires replacing numbers by polynomials. Moreover, the polynomials must be in two variables, q and t. Replacing numbers by polynomials is a well-known principle of mathematics. It is sometimes called the principle of q-analogue. Many mathematical values, which seem to be integers, must be interpreted as polynomials. One of the explanations is as follows. Most frequently, integers arise in mathematics as dimensions of some spaces. On the other hand, spaces are often graded, i.e., represented as sums of subspaces enumerated by integers. To take grading into account, we must consider the dimension of each homogeneous component separately rather than only the dimension of the entire space. Such dimensions can be coded by polynomials: the free term is equal to the dimension of the zeroth-grade space, the coefficient of q is the dimension of the first-grade space, etc. Thus, whenever we have a graded space, instead of its dimension, we can consider the graded dimension, i.e., a polynomial.

A well-known example is $n!$. The correct interpretation of the number $n!$ is the polynomial $\prod_{k=1}^{n} \frac{1-q^k}{1-q}$. The degree of this polynomial equals $n(n-1)/2$. At $q=1$, this polynomial takes the value $n!$. (Considering the fractions $\frac{1-q^k}{1-q}$, we must calculate the limit as q tends to 1.) As opposed to $n!$, the expression $\prod_{k=1}^{n} \frac{1-q^k}{1-q}$ has a deep well-known geometric meaning: this is the dimension of a homology group of the flag space. The homology groups are graded by dimension; therefore, instead of numbers, the polynomials defined by the above formula arise. The flag space is one of the most remarkable existing manifolds. Very much is known about it. To a certain degree, this explains why $n!$ occurs in various formulas. For example, why does x^n like being divided by $(n!)$? Because the flag space exists. I shall not explain this now. There exist numerous books where this is explained.

Let me give yet another example where numbers are replaced by polynomials. Consider the number of combinations

$$C_n^k = \binom{n}{k} = \frac{n!}{k!(n-k)!}.$$

This expression has a q-analogue; for example, we can replace each factorial by the polynomial specified above. In fact, the q-analogue of the binomial coefficient was invented earlier than the q-analogue of the number $n!$. It was invented by Gauss, who argued as follows. What is the number of combinations of k objects out of n? This is the number of k-point subsets in an n-point set. But this is a bad object, because finite sets have no structure. This object is too weak for being considered by mathematicians. Let us take a more interesting object, e.g., a linear space. But if we take a real or complex space, then the object is infinite, and it is impossible to count its points. The way out is to take a space over a finite field. It has finitely many points. What is the right analogue of q-point subsets in an n-point set? This is a q-dimensional subspace in the n-dimensional space over the field $\mathbb{F}_q$. The number of q-dimensional subspaces in the n-space over $\mathbb{F}_q$ is a polynomial in q. At $q=1$, this polynomial equals the binomial coefficient. To interpret it as a limit, we can define the n-dimensional linear space over the field with one element as a mere n-point set. This is one of the possible interpretations of the limit as $q \to 1$, but it is far from being unique.

There are at least two or three other ways of replacing numbers by polynomials. By tradition, the polynomials are in q, but the letter q denotes completely different things under different approaches. In the last example, q was the number of elements in a finite field. In the example about the flag space, q was the dimension of a homology group. Other situations also occur. Sometimes, q is a certain root of unity. Sometimes, it is a small parameter. Perhaps, most remarkable fact is that different approaches lead to the same expressions. This refers to experimental mathematics. My explanation is that the reasonable expressions are few, while the nature is immense. Therefore, the nature is bound to use the same expressions many times, simply because there are not enough reasonable expressions to serve all natural phenomena.

After this digression, I want to tell how to turn the Euler–Bernoulli triangle into a triangle whose elements are polynomials in q. A precise definition involves polynomials in two variables, q and t. But the considerations which I have briefly mentioned in my preceding lecture allow us to put $t = q$ and deal with polynomials in one variable. Now, I shall write the rule according to which the q-analogue of the Euler–Bernoulli triangle is constructed. A minor modification of the rule for constructing the Euler–Bernoulli triangle yields the rule

$$b_{k,l} = q^{-1}b_{k-1,l+1} + (q^{l+1} - q^l)b_{l,k-1}$$

for $k > 0$. In addition, $b_{0,l} = q^l b_{l-1,0}$ and $b_{0,0} = 1$.

The set of these rules makes it possible to reconstruct the entire triangle. Note that, in the recursive formula, the indices k and l of the last term are interchanged. This is a consequence of shuttling: every number depends on numbers either on the left or on the right. For this reason, k and l interchange.

At this point, I want to state two unsolved problem.

1. We have a recursive relation, i.e., can write the triangle, but we do not know a formula for it.

2. Even the asymptotic growth of the elements of the triangle is unknown.

The second problem is, perhaps, more interesting: How rapidly do the elements of the triangle grow at a fixed q?

The third sequence of polynomials is defined as $B_n(q) = b_{n-1,0}(q)$.

Thus, we have defined three out of five sequences of polynomials. There remain the most interesting polynomials D_n and Y_n. To intrigue the listeners, I shall write the formula for D_n at once:

$$D_n(q) = \zeta_{G_n(\mathbb{F}_q)}(-1), \quad \text{where } q = 2^l.$$

Here $G_n(\mathbb{F}_q)$ is the group of strictly upper triangular matrices of order n with elements from $\mathbb{F}_q$.

Let me remind you of the definition of the classical Riemann zeta-function: $\zeta(s) = \sum_{n \geqslant 1} n^{-s}$. This function is very famous. The last unsolved great problem of mathematics is to find the zeros of the zeta-function. The well-known conjecture of Riemann is that they lie on the line $\operatorname{Re} s = 1/2$. To share in Riemann's fame, many people tried to generalize the zeta-function. For an expert in representation theory, generalizing the zeta-function is a piece of cake, because the positive integers are the dimensions of irreducible representations of the simplest noncommutative compact group, namely, of $SU(2)$. This group has precisely one irreducible representation in each dimension (up to equivalence). Therefore, given an arbitrary group G with finite-dimensional irreducible representations (e.g., a finite or compact group), we can define a zeta-function related to this group as

$$\zeta_G(s) = \sum_{\lambda \in \widehat{G}} d(\lambda)^{-s}.$$

Strange as it seems, this function has not attracted much attention so far. Maybe, this is because all related meaningful examples about which something making sense can be said reduce to the usual zeta-function. Although, certainly, some known identities can be rewritten as identities for the values of the zeta-function. For example, the Burnside identity, which I have written out above, means that, for a finite group G, we have

$$\zeta_G(-2) = \#G.$$

At the beginning of this lecture, I mentioned that, for finite groups, the expression

$$\zeta_G(0) = \#\widehat{G}$$

makes sense. Sometimes, the expression

$$\zeta_G(-1) = \sum d(\lambda)$$

also makes sense. Namely, in some cases, the equality $\zeta_G(-1) = \#\operatorname{Inv} G$ holds. The definition of the polynomials D_n involves numbers of precisely this form, $\zeta_G(-1)$.

Under our definition, it is quite unclear why D_n are polynomials.

The last thing which I want to talk about is the definition of the fifth sequence of polynomials Y_n. These polynomials are related to the description of coadjoint orbits for the group of upper triangular matrices over a finite field. I explained in the first lecture what a coadjoint orbit is. This is some trajectory described by a point under the action of a group. If the field were real, these trajectories would be fairly simple algebraic varieties. And there is a remarkable theorem, which is not proved in full generality, that if we have a good algebraic variety, i.e., a set determined by algebraic equations, and if these equations make sense over a finite field (e.g., if their coefficients are integer), then the number of solutions to these equations over the field $\mathbb{F}_q$ is a polynomial in q.

For example, the equation $x^2 + y^2 = 1$ over the field $\mathbb{F}_q$ has a finite number of solutions, and this number of solutions is a polynomial in q. This is a very beautiful problem. I highly recommend to those who wish to better acquaint themselves with finite fields that they try to solve this problem.

There is yet another conjecture; for varieties of dimension 1, it was stated by Andre Weil, one of the most outstanding mathematicians of our century, and proved comparatively recently by another outstanding mathematician, Pierre Deligne. The conjecture is that the number of solutions is a polynomial whose coefficients are expressed in terms of the structure of the solution set over the complex field.

The same equation $x^2 + y^2 = 1$ over the complex field determines a set homeomorphic to $\mathbb{C} \setminus \{0\}$. A natural compactification of this set is the two-dimensional sphere $S^2 \simeq \mathbb{C}P^1$.

Unfortunately, this theorem is not proved in full generality. It is not known what algebraic varieties have the remarkable property that, when considered over the field $\mathbb{F}_q$, they have cardinalities polynomial in q. The meaning of the coefficients of these polynomials is not known either. But there is a mnemonic rule. These coefficients must correspond to the dimensions of the homology groups of the variety over the real or complex field. To be more precise, $q = 1$ corresponds to the complex field and $q = -1$ corresponds to the real field. An approximate informal explanation is as follows. A variety is in a certain sense made up of cells which have the structure of usual affine space. Over the real field, a cell of dimension n makes a contribution of $(-1)^n$ to the homology in dimension n. Over the complex field, a cell always makes a contribution of 1 to the homology in dimension $2n$. And over a finite field, a cell of dimension n makes a contribution equal to q^n. Therefore, if a variety admits a cell decomposition consistent with its structure of an algebraic manifold (for instance, the circle admits a decomposition into a point and a straight line), then we obtain a proof. This "proof" is very simple and convincing, but it is incorrect, because the varieties admitting such cell decompositions are too few. But the assertion itself is valid in much more general situations, when there are no cell decompositions.

I do not know whether or not this theorem is true in the case of interest to me. The experts whom I asked about this could not say anything at once. So, the last polynomial $Y_n(q)$ describes the number of coadjoint orbits for our group. This number can be interpreted as the number of points in some algebraic variety over a finite field. If certain known theorems from this area apply to the variety, then Y_n is a polynomial in q.

The polynomial Y_n can be interpreted in the spirit of modern quantum statistical physics. This theory is rather poor in terminology. One must know the words state, Gibbs distribution, energy, and statistical sum. Nothing else is required. A system of states is a lattice, a system of nodes, or something with cells, edges, and vertices; this is a two- or, in the last resort, three-dimensional formation. States are usually obtained by assigning some particular value, most frequently 0 or 1, to each element of the system.

In the case under consideration, as a system we can treat a strictly upper triangular matrix; a state of the system is obtained when each cell above the diagonal is occupied by an element of the finite field. It is also interesting to consider a composite system made up of upper triangular matrices X and Y and a lower triangular matrix F. The number of states for this system is equal to $q^{3n(n-1)/2}$. This number is fairly large. The first interesting case is that of $q = 2$ and $n = 13$. In this case, there are 2^{228} states.

Given a system and states, it is required to think up some function being evidently better than all other functions. Usually, when someone devises such a function, he (or she) calls this function energy, denotes it by H, and takes the sum

$$\sum_s e^{-\beta H(s)}$$

over all states s. Here β is a parameter usually interpreted as the reciprocal temperature, i.e., $\beta = 1/kT$, where T is the absolute temperature on the Kelvin scale and k is the Boltzmann constant. Calculating such a sum, we obtain a function of β. While the whole process is completely uncontrollable, it is sometimes possible to say something about this function of β. For example, sometimes, it can be proved to be smooth. In this case, the system is of no physical interest, for it admits no phase transitions. Sometimes, this function is not smooth. Then the system has phase transitions. In very rare cases, we can calculate this function explicitly and see the phase transitions with the naked eye: the answer is then specified by one formula in one interval and by another formula in another interval.

Let us return to our example. What function is most interesting among those determined by three matrices (two upper triangular and one lower triangular)? I claim that the most interesting function is $\text{tr}(F \cdot [X, Y])$. The values of this function belong to a finite field. But we must attach a meaning to the exponential of this function rather than to the function itself. It is well known what the exponential is. The exponential is a function satisfying the functional condition that addition must become multiplication. For a finite field, this is the so-called additive character χ. We obtain the sum

$$\sum_{X,Y,F} \chi(\text{tr}(F \cdot [X, Y])).$$

There is a conjecture that this sum equals $q^{n(n-1)/2}Y_n(q)$. This means that the function very strongly oscillates, and almost all terms cancel each other. For this reason, it is extremely difficult to evaluate the sum. Usually, such sums are computed by the Monte Carlo method, but it does not apply to strongly oscillating functions.

Now, I can state the general conjecture: All the five sequences of polynomials coincide.

What is known in this connection? The results are as follows.

1. $A_n = C_n$ for all n. I repeat that the proof of this assertion exists and is published.[7] The fashionable modern journal in which it is published has no physical body; it is not printed on paper and exists only in electronic files. But all subscribers can have it on their computers. I printed out the proof but did not understand anything, I must confess. Thus, if somebody would explain the proof to me without referring to that paper, I would be pleased greatly.

2. The assertion $A_n = B_n$ has been checked experimentally for $n \leqslant 26$. From my point of view, this leaves no doubts in its validity. But there are examples of similar coincidence. For instance, the polynomial $n^2 - n + 41$ gives primes for the first 40 values of n.

3. The equality $A_n = B_n = C_n = D_n = Y_n$ has been verified experimentally for $n \leqslant 11$.

4. Apparently, $A_n \neq D_n$ for $n = 13$. This was explained in the preceding lecture. Recently, it has turned out that there exists a triangular matrix of order 13 over the field with two elements which is not conjugate to its inverse. It has Jordan blocks of sizes 4, 5, and 4.

At this point, I want to conclude the lecture. In case somebody becomes interested and ready to discuss what I was talking about, I am always available by electronic mail at

`kirillov@math.upenn.edu`

---

[7] S. B. Ekhad and D. Zeilberger. The number of solutions of $X^2 = 0$ in triangular matrices over GF(q). *Electronic J. Combin.*, **3** (1996), #R2.

The interested reader can find details in:

A. A. Kirillov. On the number of solutions to the equation $X^2 = 0$ in triangular matrices over a finite field. *Funktsional. Anal. i Prilozhen.*, **29** (1) (1995), 82–87.

A. A. Kirillov. Merits and demerits of the orbit method, *Bull. Amer. Math. Soc.*, **36** (4) (1999), 433–488. (*Editor's note*)

D. V. Anosov

On the development of the theory of dynamical systems during the past quarter century

Extended notes of three lectures delivered in the spring of 1998

Preface

This essay on the achievements of the theory of dynamical systems (DS) and related areas of the theory of ordinary differential equations and other mathematical disciplines over the past approximately 25 years[1] is very brief and incomplete, especially in comparison with the surveys of the same subject covering earlier periods that were published in the well-known VINITI series "Progress in Science and Technology" and "Current Problems in Mathematics. Fundamental Directions" (the most recent of them partially cover the period considered in this essay). In this connection, I refer to the recent survey of Yoccoz [1], which considers the subject from a different angle and substantially supplements the list of topics considered in this paper.[2] An extensive material, including quite fresh results, is contained in the voluminous book [2], which has recently been translated into Russian.

Not only Yoccoz but also a number of other speakers who delivered plenary and large sectional talks at international mathematical congresses told about dynamical systems. All these reports can be recommended as authoritative surveys of various aspects of the subject, which give both prospects and the most current state of the art. I distinguish Yoccoz's report because of its broad scope. Later, a fairly extensive report was made by J. Moser [3].

The features according to which the material for the first two sections was selected are evident from the titles. The selection was based on clear formal criteria; I believe, it is free of subjectivity in this respect. These criteria are reliable in the sense that the corresponding results deserve some discussion anyway; certainly, it was necessary to say about the emergence or renaissance of whole large branches of the DS theory; and when I mention a name of a prominent

---

[1] In the course of our exposition, we sometimes touch upon earlier works; yet, some topics fit more naturally in longer time intervals, and it would be inexpedient to artificially "cut off" their beginnings.

[2] First, I did not write or wrote more briefly about the topics covered in Yoccoz's survey. Besides, clearly, I know works of Russian mathematicians better than Yoccoz knows them, while Yoccoz knows works of western mathematicians better than I do.

scientist together with a problem, this is most frequently not merely the statement that the problem was posed by this scientist but also an acknowledgement of its importance.[3] (Celebrated names are rarely mentioned in relation to not very significant problems, although there is hardly a celebrity who did not state such problems for students. Certainly, any problem has its author (in the last resort, the author of a problem is the author of its solution), but the authors of the problems from Section 2 are very prominent mathematicians, and these problems are customarily mentioned along with their names.)

As to more or less broad new (or renewed) directions, I believe, all of them are mentioned in Section 1; but significant achievements not always consist in solving some named problem and are far from being exhausted by what is covered in Section 2.[4] In Section 3, I briefly mention some of these achievements. I did not have such plain criteria for selecting material as those used in the preceding sections, so the choice was unavoidably incomplete (I did not even write about what I much dealt with myself in recent years, that is, about flows on surfaces and related geometric questions) and, probably, subjective to some extent. Besides, with the best will the world, I could not write in Section 3 about everything as thoroughly as in the first sections (if the exposition of Sections 1 and 2 can be called thorough) remaining within reasonable space and time limits. Yet, I hope that, even when I, in essence, only briefly mention results and names and give a slightly, so to speak, annotated (and very incomplete) list of references, this may still be useful as an initial information about what is going on in the world. If the reader becomes interested in something, even the incomplete bibliography which I could provide will help to at least start more seriously studying the subject.

I tried to take into account works (known to me) concerning not only smooth but also topological or purely metric (in the sense of measure) systems. But the topics considered in most detail largely refer to smooth systems. This is not because such systems are the center of my interests; I hope that I am familiar with at least the most striking results about other systems (because striking results are often mentioned in conversations with colleagues). If I were writing about earlier works, the balance would be different, but in the past quarter century, neither new directions (with one exception) nor named

---

[3] When talking about "named" problems, I also mention some related questions. Sometimes, I even allot more space for these questions, so the named problem is used as a cause for reviewing a whole domain.

[4] And they must not be less important than the results given in Section 1, because the intrinsic content of events of lower "taxonomy" level may be as significant as that of high-level events (it suffices to recall the creation of KAM (Kolmogorov–Arnold–Moser) theory and the "hyperbolic revolution" in the 1960s, which did not go beyond the framework of smooth DSs).

problems related to topological or purely metric systems arose. Something about them appears only at the end of Section 3, not counting the subsection about group actions in Section 1.

During my work on this paper, I asked various questions of my acquaintances among mathematicians (and, sometimes, of mathematicians with whom I was not acquainted in person). Of course, they were largely participants of my seminars at the Steklov Mathematical Institute and Moscow State University and mathematicians from Nizhnii Novgorod,[5] but I also obtained some information from other, including foreign, colleagues. I thank all of them and hope that, using this information as I thought best and in accordance with my plans, I did not garble it. (If I did, this is my fault.)

Now, let me comment the bibliography. I tried to refer to most recent publications, to surveys, or to papers containing more or less detailed (not necessarily original) presentations of certain circles of questions (that is, to what is called expository papers). This allowed me to save space, and I ask the reader to regard with understanding the fact that I sometimes economize on references to pioneering works.

Reading this paper requires, first, the general mathematical grounding that is usually provided for university students of mathematics during the first three years of study (after that, the study usually becomes specialized) and, secondly, some knowledge of DSs, mainly smooth ones.[6] I had presented the contents of this paper in lectures for students who had special grounding in smooth DSs. For this reason, talking about ergodic theory, I made more explanations than in other places. The paper inherited this special feature of the lectures.

A fairly comprehensive, although not new, textbook on ergodic theory is [4].

In the time elapsed since the publication of the Russian version of this article, two fundamental reference books dealing with all the various branches of DSs, of a total length of over 2200 pages, have appeared:

Handbook of Dynamical Systems, vol. 1a, eds. B. Hasselblatt and A. Katok (Amsterdam: North-Holland, 2002).

Handbook of Dynamical Systems, vol. 2, ed. B. Fiedler (Amsterdam: North-Holland, 2002).

One more volume (1b) is expected.

---

[5] It is not by accident that S. Smale "enrolled" me in the Nizhnii Novgorod school, although he knew that I am a Muscovite.

[6] In addition to textbooks, I recommend reading the surveys from the VINITI series already mentioned above (some, but not all, of them are included in the bibliography) and articles in *Mathematical Encyclopaedia*, which contain not only definitions and brief summaries but also fairly extensive references.

I shall also mention here S. Smale's article "Mathematical problems for the next century" in *Mathematics. Frontiers and Perspectives*, eds. V. Arnold, M. Atiyah, P. Lax, and B. Mazur (Providence, RI: Amer. Math. Soc., 2000), pp. 221–244. A considerable part of this paper is devoted to DSs. It does not only summarize results of development but rather poses problems for the future research; of course, discussing them, Smale talks about the contemporary state of the art.

Unfortunately, I did not have the time to give an account of new works, even in the succinct, "first approximation," style in which my article was written. I had therefore to limit myself to adding a few disparate remarks, but I hope, they will not be superfluous.

1 New or "renewed" directions

1.1 Symplectic geometry

By the beginning of the past quarter century, DS theory (including related questions) had steadily branched into four main parts, each characterized by the presence (and use) of a certain structure on phase spaces that in a sense determines the DSs under consideration; these are differential dynamics (the theory of smooth DSs),[7] topological dynamics (the theory of topological DSs), ergodic theory (in which the phase space is assumed to be a measurable space or, even more frequently, a space with measure), and the analytic theory (in which the phase space and independent variable, "time," are assumed to be complex). Certainly, other structures emerged too, but they were taken into account, so to speak, within the framework of one of these four theories. Thus, Hamiltonian systems have their own specific features related to the symplectic

---

[7] Essentially the same branch of DS theory is known as the "qualitative theory of ordinary differential equations"; the choice of a name depends on the traditions of scientific schools. In what follows, the term "local theory" is used many times. This is the conventional name of the part of smooth DS theory that deals with the behavior of the trajectories of a flow near an equilibrium point or with the behavior of a periodic solution or trajectories of an iterated mapping near a fixed or periodic point. The "global theory" considers the behavior of trajectories on the entire phase space or, at least, in some "more or less extensive" domain.

As the difference between the local and global theories was mentioned, it should be said that this difference is partly conventional: if the coordinates of the phase velocity $\mathbf{v}(x)$ of a (local) flow on $\mathbb{R}^n$ are homogeneous polynomials of degree k in the coordinates of x, then studying the behavior of the trajectories near the origin (being an equilibrium point) is largely equivalent to a global study of a certain flow on $(n-1)$-dimensional projective space $\mathbb{RP}^{n-1}$ onto which the trajectories are "projected" along radii. (Strictly speaking, when the trajectories are projected onto $\mathbb{RP}^{n-1}$, a field of directions (tangent lines), rather than a vector field, is obtained. However, this does not affect our further considerations, although requires a certain attention.) In this way, we can obtain (for a suitable k) flows on $\mathbb{RP}^{n-1}$ of a form general enough to realize (for $n \geqslant 4$) various patterns of the complex behavior of trajectories known in the global theory. We can also make the corresponding trajectories on $\mathbb{R}^n$ to lie entirely in a neighborhood of 0 and behave as complexly as those on $\mathbb{RP}^{n-1}$. Thus, the n-dimensional local theory, so to speak, includes the $(n-1)$-dimensional global theory (or, at least, a very significant part of it), so it cannot be simpler that the global theory.

But, in practice, this fundamental observation is not that important. The point is that, for $k > 1$, the equilibrium point 0 is degenerate, and the codimension of degeneracy (the "degree" characterizing it) grows rapidly with k. The local theory hardly deals with cases so degenerate. What is said above means only that (at least, in the multidimensional case) the local theory cannot pretend to completely investigate all the possibilities, because they are immense in the global situation. But such pretensions have never been advanced.

structure, but they were always studied as a part of the theory of smooth DSs.

During the past 20 years, a new discipline, symplectic geometry (or symplectic topology), has arisen; in the hierarchy of mathematical disciplines, it is at at least the same level as the four traditional branches of DS theory and even partially goes beyond it. New disciplines of such a high level emerge rarely, and this event attracts special attention.

A symplectic manifold is a smooth manifold M together with a closed exterior differential 2-form ω defined on M and nondegenerate at all of its points. As it is usual in the theory of exterior differential 2-forms, the closedness of ω means that $d\omega = 0$, where d is the exterior differential. The notion of nondegeneracy can be introduced for arbitrary (not only skew-symmetric) bilinear forms on a vector space (on the tangent space T_xM in the case under consideration). Such a form is said to be nondegenerate if its coefficient matrix is nondegenerate. This definition presumes the use of coordinates; in linear algebra, equivalent coordinate-free formulations are given. I mention at once that a skew-symmetric form can be nondegenerate only in the even-dimensional case $(\dim T_xM = 2n)$. Finally, we assume that, in terms of local coordinates, the coefficients of ω are smooth functions of their arguments (the degree of smoothness depends on a particular problem, but very often, or even in most cases, C^∞-smoothness can be assumed).

Thus, a symplectic manifold is not merely a manifold M but a pair (M, ω). Although, an explicit mention of ω is often missing from the notation of a symplectic manifold, and it is denoted simply by M.

The definition given above is similar to the definition of a Riemannian manifold. The latter involves a symmetric bilinear differential form g instead of skew-symmetric form ω; g is usually not only nondegenerate but also positive definite[8] (both conditions can hold in odd dimensions as well); no differential conditions are imposed on it (while ω must be closed).

However, the similarity between Riemannian and symplectic geometry does not go beyond the initial definitions. In the Riemannian case, different manifolds may have different local structures; there is an extensive system of local invariants – the curvature tensor and its covariant derivatives. In the symplectic case, according to a theorem of G. Darboux,[9] near any point $x \in M$, there exist so-called symplectic, or canonical, coordinates (Darboux coordi-

---

[8] This means that the corresponding quadratic form is positive definite. If the form is not positive definite, then the corresponding manifold is often said to be pseudo-Riemannian and the inner product in the tangent spaces, pseudo-Euclidean.

[9] It was also proved by G. Frobenius, but, for some reason, he is not usually mentioned in the standard references to the theorem.

nates) $p_1, \ldots, p_n, q_1, \ldots, q_n$ in terms of which ω is locally expressed in the form

$$\omega = \sum_{i=1}^{n} dp_i \wedge dq_i. \tag{1}$$

Therefore, any two symplectic manifolds (M_1, ω_1) and (M_2, ω_2) of the same dimension have the same local structure; namely, any two points $x_1 \in M_1$ and $x_2 \in M_2$ have neighborhoods U_1 and U_2 between which there is a diffeomorphism $f \colon U_1 \to U_2$ such that f transforms the forms into each other (that is, $f^* \omega_2 = \omega_1$).[10] *A propos*, such a diffeomorphism is called a symplectomorphism. In the case under consideration, it is local in the obvious sense, but there exist global symplectomorphisms too; these are diffeomorphisms $f \colon M_1 \to M_2$ for which $f^* \omega_2 = \omega_1$.

Although symplectic manifolds themselves have no local invariants (except dimension), submanifolds of a symplectic manifold may be different even when their dimensions coincide. Thus, an important role is played by the Lagrangian submanifolds, that is, submanifolds N of dimension equal to half the dimension of the ambient manifold and such that the restrictions $\omega \restriction N$ of the form ω to these submanifolds vanish identically;[11] at the same time, there exist submanifolds of the same dimension with nonzero restrictions $\omega \restriction N$. Still, even for submanifolds, the number of invariants is much less than that in Riemann or Euclidean geometry. For example, if N_1 and N_2 are submanifolds of a symplectic manifold (M, ω) and $U_i \subset N_i$ are neighborhoods of points $a_i \in N_i$, and if there exists a diffeomorphism $f \colon U_1 \to U_2$ for which $f(a_1) = a_2$ and $f^*(\omega \restriction U_2) = \omega \restriction U_1$, then there exists a symplectomorphism $g \colon V_1 \to V_2$ between some neighborhoods V_i of the points a_i in the entire M which locally extends f in the sense that it coincides with f on $V_1 \cap U_1$. In the case under consideration, locally, the intrinsic geometry of a submanifold completely determines its extrinsic geometry (compare this with curves in $\mathbb{R}^n$!).

I believe that this scarcity of local invariants delayed the emergence of symplectic geometry, because it was not clear in advance what to study.

---

[10] In this respect, the situation is similar to that for flat Riemannian manifolds. Thus, there is some analogy between the closedness of the form ω and vanishing of the curvature tensor. However, it does not seem to be deep, if only because closedness is a differential condition of the first order, while zero curvature is that of the second order. This distinction may appear too formal, but there is also a quite meaningful difference: the group of isometries of a connected Riemannian manifold is finite-dimensional, while the group of symplectomorphisms of a symplectic manifold is infinite-dimensional.

[11] Lagrangian submanifolds, as well as some related objects, are present implicitly in the mathematical apparatus of classical analytical mechanics, but the explicit definition is due to V. P. Maslov and V. I. Arnold.

Symplectic manifolds arise in mathematics largely in relation to Hamiltonian systems of classical mechanics and to Kähler manifolds. A usual Hamiltonian system is a system of differential equations with $2n$ unknowns $p_1, \ldots, p_n$ (momenta) and $q_1, \ldots, q_n$ (coordinates) of the form

$$\frac{dp_i}{dt} = -\frac{\partial H}{\partial q_i}, \qquad \frac{dq_i}{dt} = \frac{\partial H}{\partial p_i}, \tag{2}$$

where H is a function of the unknowns and, possibly, time (it is called the Hamiltonian of the system). The vector field V of the phase velocity of this system is related to the differential dH and symplectic form (1) as follows: for any vector U,

$$\omega(V, U) = -dH(U) = -U \cdot H,$$

where $U \cdot H$ means that the vector U acts on H in a definite way as a first-order linear differential operator (when this operator is to be distinguished from the vector, it is denoted by D_U or $\mathcal{L}_U$). In other words, with the help of ω, we pass from the vector V to the covector (i.e., a linear functional, or a linear form on vectors) defined by $U \mapsto \omega(V, U)$; this covector coincides with dH up to sign. In the general case, with a function H on a symplectic manifold (M, ω) one can associate a vector field V on M by precisely the same rule; V is called the globally Hamiltonian vector field (with Hamiltonian H). In the Darboux coordinates, the coordinates (components) of this field are precisely the right-hand sides of system (2). In classical mechanics, the initial form ω often has form (1), but even then further transformations (especially those reducing dimension with the use of symmetries) may change ω. A (local) flow with Hamiltonian vector field of phase velocity is also called a Hamiltonian. Such a flow $\{\phi_t\}$ preserves the form ω, i.e., the mappings ϕ_t are symplectomorphisms ($\phi_t^* \omega = \omega$). Conversely, if a flow $\{\phi_t\}$ with phase velocity field preserves ω, then both the flow and the field are locally Hamiltonian: each point of the manifold M lies in some domain W in which the field V is obtained by the method described above from some function (local Hamiltonian) H_W defined in W. But generally, it is impossible to pass from a system of local Hamiltonians $\{H_W\}$ defined in domains W covering M to one "global" Hamiltonian (defined on the entire M and determining the field V as specified above).

There is a different source of symplectic manifolds. I assume that the reader is familiar with the notion of a complex analytic manifold. On such manifolds, it is natural to consider so-called Hermitian metrics instead of usual Riemann metrics, because the tangent space to a complex manifold M is naturally a complex vector space, and instead of a Euclidean inner product on this space, it is natural to consider a Hermitian inner product (a sesquilinear form in the terminology of Bourbaki). If such a form is defined in each tangent space and has sufficiently smooth coefficients in local coordinates, then we say that M has a Hermitian structure, or that it is a Hermitian manifold. The real part g of the Hermitian form is a Euclidean inner product; thus, any Hermitian manifold is automatically Riemannian. (The prefix "pseudo" can be inserted everywhere.) But the imaginary part of a Hermitian form is a nondegenerate skew-symmetric 2-form ω. If it is closed, then the Hermitian manifold is called Kähler. On the one hand, this definition is short; on the other hand, it is this definition that is largely used in dealing with Kähler manifolds. But at first sight, it seems somewhat formal and unmotivated. For this reason, I shall give a different definition, which is natural enough geometrically. A Riemannian metric g determines a Levi-Cività connection on M in a definite way. This allows us to perform a parallel transport of vectors from the tangent space along any smooth curve $\gamma(t)$ on M. We obtain linear mappings $T_{\gamma(t_1)}M \to T_{\gamma(t_2)}M$. They preserve the Euclidean inner product; but we deal with the complex situation, and it is desirable that they preserve the entire Hermitian inner product and be linear not only over $\mathbb{R}$ but also over $\mathbb{C}$ (i.e., that they preserve multiplication of vectors by complex numbers in the obvious sense). It turns out that this desire is precisely equivalent to the property of being Kähler.

There are fairly many Kähler manifolds. Complex projective space has a natural Kähler metric (the Fubini–Study metric); it induces Kähler metrics on algebraic subvarieties of this space. (On the other hand, there exist Kähler manifolds not being algebraic varieties.)

That "symplectic" considerations and results constitute a relatively independent and unified complex of notions and methods was repeatedly (starting with the mid-1960s) emphasized by V. I. Arnold, who often dealt with this complex and (as opposed to other researchers) in relation to various topics (such as Hamiltonian DSs, Lagrangian surgery (or perestroikas), and asymptotic methods in the theory of partial differential equations).[12] A kind of "forerunner," which has later become an organic component, of the new discipline was the theory of Lagrangian and Legendrian perestroikas and singularities together with a number of related questions developed in the preceding two decades. An especially significant progress in this area was made by V. I. Arnold and his school and by A. Weinstein. The theory is exposed in [5]. The fundamental distinguishing feature of this stage in comparison with the current state of the art was the lack of the discovery of global invariants of symplectic manifolds themselves (not only of some objects related to them).

A stimulating role in the development of symplectic geometry was played by

---

[12] Naturally, the ultimate sources of the symplectic manifolds arising in his work were in the same spirit as the two sources specified above, but immediate reasons for considering such manifolds were different.

a paper of C. Conley and E. Zehnder in which the following Arnold's conjecture was proved for the 2n-torus $\mathbb{T}^{2n}$ considered with the natural symplectic structure (which comes from the standard symplectic structure in $\mathbb{R}^{2n}$ under the standard factorization): for a symplectic diffeomorphism g_1 of a compact symplectic manifold M homotopic to the identity diffeomorphism g_0 in the class of symplectic diffeomorphisms by means of a deformation $\{g_t\}$ whose rate $\dfrac{dg_t}{dt}$ has single-valued Hamiltonian for all t,[13] there exist at least as many fixed points as the number of critical points for a smooth function on M (for $M = \mathbb{T}^{2n}$, there are $2n + 1$ of such points, and if they are nondegenerate, then their number is equal to 2^{2n}). The proof combined two earlier ideas (which had already been used to solve other problems), namely, P. Rabinowitz's idea of a new variational approach to periodic solutions to Hamiltonian systems and C. Conley's idea of a topological characterization of the behavior of a flow near certain (so-called isolated or locally maximal) invariant sets by means of a generalization of the classical Morse index of equilibrium points of the gradient flow (Morse considered critical points of functions, which is equivalent).

At present, Arnold's conjecture is proved in a number of other cases, and several authors announced its proof in the general form.

The last step, after which symplectic topology had become an undoubtedly autonomous discipline, was made by M. Gromov [6]. Before proceeding to his approach, I shall mention some of his results. Gromov introduced several symplectic invariants (i.e., invariants with respect to symplectomorphisms). The only invariant known earlier was the volume. The usefulness of the new invariants is seen from the following statement due to Gromov,[14] which seems surprising and has an evident formulation; it is known under the expressive name "theorem about the impassability of a symplectic camel through the eye of a needle": For an $n > 1$, there exist no continuous family of symplectomorphisms $\{\psi_t;\ 0 \leqslant t \leqslant 1\}$ in $\mathbb{R}^{2n}$ with coordinates p_i, q_i $(i = 1, \ldots, n)$ that preserve (1) (see p. 76) and transfer a ball B of radius R from the half-space $p_1 < 0$ into the half-space $p_1 > 0$ is such a way that, in the process of deformation, the intersection

$$\psi_t B \cap \{(p, q);\ p_1 = 0\}$$

is strictly inside some $(2n - 1)$-dimensional ball of radius $r \leqslant R$. (Certainly, an octopus, which is subject only to the condition of volume preservation, can

---

[13] This is one of the equivalent conditions on g_1 imposed in this conjecture. More frequently, it is given in a different formulation involving so-called asymptotic cycles, which are a homological analogue of the Poincaré rotation number (it refers to homeomorphisms of the circle).

[14] The first outline of the proof was given by Gromov together with Y. M. Eliashberg.

push its way through a small hole; this is confirmed by the experience of keeping these animals in confinement.) I also mention, not going into details, that this approach allowed Gromov to obtain an unexpected information about the global properties of Lagrangian submanifolds and even of submanifolds in the usual Euclidean space (with coordinates p_i, q_i and form (1)), although it was unclear *a priori* why they would have some special properties.

As mentioned, a Kähler manifold M has a symplectic structure "nicely" related to the corresponding complex and Riemann structures. In the terminology of Gromov, the corresponding pseudo-complex structure[15] on M is "tamed" by the symplectic structure (Gromov describes precisely what relations he means by taming). Such pseudo-complex structures are diverse, and everybody knows that the true complex Kähler structure is much better than all the others. It turns out that a symplectic manifold admits many pseudo-complex structures tamed by the symplectic one; not trying to distinguish one "good" structure among them, Gromov suggested to consider all these structures at once (Riemannian metrics then arise naturally). In a certain sense, all the "bad" structures together proved a suitable substitute for one "good" structure! For each of them, pseudo-holomorphic mappings of the disk to the manifold under consideration can be considered (for the sake of brevity, they are referred to as pseudo-holomorphic disks); they are defined by directly generalizing the usual holomorphy. The pseudo-holomorphy condition is written in the form of a quasilinear system of partial differential equations generalizing the classical Cauchy–Riemann system. Gromov studied this system and proved, roughly speaking, that it has many solutions, as in the classical case. Gromov's symplectic invariants are defined with the use of the pseudo-holomorphic disks corresponding to all possible pseudo-complex structures tamed by the initial symplectic structure.

As we see, Gromov's approach involves partial differential equations instead of DSs. But some of the results (as well as ideas and methods) proved useful for DS theory. Other authors gave a different interpretation of this approach (or, more precisely, of something similar to it), which is closer to DS theory. In this way, the best results on periodic solutions to Hamiltonian systems were obtained (they are due to C. Viterbo, H. Hofer, and E. Zehnder; see also the theorem of Hofer cited at the end of Section 2.5). This aspect of the matter is well presented in [7]. The collection of papers [8] considers a wider circle of questions of symplectic topology, including some relations to DSs not mentioned in [7] (especially as far as invariant Lagrangian manifolds, tori largely, are con-

---

[15] That is, the structure of complex space on each tangent space $T_x M$ smoothly depending on x in the obvious sense; in the case under consideration, it is determined by the initial complex structure of M, but in the general case, it may be obtained otherwise.

cerned; for them, a number of qualitative problems was not even stated until the new symplectic ideas penetrated DS theory (not counting, of course, the special case of invariant curves with two-dimensional diffeomorphisms, which were mentioned even by G. Birkhoff). As far as I know, there are no surveys on this topic. In addition to the referenced cited above, there are [9, 10].) A textbook on symplectic topology proper is [11]. In recent years, reports on symplectic geometry (both in relation to DSs and independently of them) were delivered at international and European mathematical congresses, and in their proceedings many information on the state of the art in this area can be found.

I cannot discuss here Floer's theory of cohomologies, which appeared in connection to symplectic geometry and part of which (but, of course, only part) is to some extent sketchy at present. As far as I know, there is no detailed textbook exposition of this material.

Let us mention two more somewhat related cycles of papers (they, especially the second one, are indeed related to the topic only to some extent).

Continuing and developing the study initiated earlier, V. I. Arnold and his collaborators tackled a series of questions known as nonlinear Morse theory [12]. It considers the variability and intersection properties of certain curves or more general manifolds, which can be regarded as generalizations of the well-known theorems of Sturm and Morse; the latter are then related to the special cases where these curves are graphs of certain functions or even of solutions to linear differential equations.

The second cycle consists of two proofs of the theorem that, on a two-dimensional sphere with an arbitrary Riemann metric, there exist infinitely many closed geodesics (the earlier result, whose proof was essentially contained in works of L. A. Lusternik and L. G. Schnirelmann of the 1920s but finished off only half a century later [13, 14], was concerned with three closed geodesics; though, those three geodesics had no self-intersections). One proof starts with V. Bangert's reduction of the problem to two cases, which were proved by Bangert himself and J. Franks [15, 16]. A gap in [16] was filled in J. Franks' paper "Area preserving homeomorphisms of open surfaces on genus zero." *New York J. Math.*, **2** (1996), 1–19. This proof of the theorem is distinguished by a remarkable combination of novelty and tradition (and of a kind of "elementarization"). The case treated by Franks was noticed even by G. Birkhoff. This is the case in which there is a "simple" (having no self-intersections) closed geodesic L such that any geodesic intersecting L necessarily intersects L again. (This case includes, in particular, the quite classical case of a metric of positive curvature.) Birkhoff observed that, in this case, the problem reduces to studying a certain self-mapping of an annulus-like domain. Franks proved that this mapping has infinitely many periodic points, which is equivalent to the existence of infinitely many closed geodesics. (As is known, Birkhoff himself obtained a result about self-mappings of an annulus, namely, a proof of the conjecture stated by A. Poincaré and known as "Poincaré's last theorem," although Poincaré published it as a conjecture. *A propos*, this theorem stimulated Arnold to state the conjecture mentioned above, while Franks started his work in this domain with obtaining a different proof of the Birkhoff theorem and related results.) Franks used some results of M. Handel (related to Arnold's conjecture mentioned above; since Handel did not published the proofs, Franks gave them under additional assumptions sufficient for his purposes. Then Franks returned to this circle of questions again; this time, in addition to the Handel theorem, he proved Arnold's conjecture for diffeomorphisms of surfaces no referring to general symplectic theory[16]) [17]. (Later, S. Matsumoto, continuing this line of study, suggested a complete proof in the general case of homeomorphisms both for Arnold's conjecture – for surfaces[17] – and for the Handel theorem announced in [18].) Soon after the appearance of the papers of Franks and Bangert, N. Hingston suggested a second proof of the theorem about an infinite number of closed geodesics, which was based on variational considerations [19]. It is closer in spirit to the traditions of differential geometry (and the further from the topic of this section).

I must say that, in Russia, the cycle of studies of A. T. Fomenko and his students on the topology of integrable Hamiltonian systems (see their recent book [20]) is also referred to symplectic geometry. Certainly, this is a geometry, and it is related to the corresponding symplectic structure (without which Hamiltonian systems cannot be considered), but the contents of this cycle of studies are fairly far from the topic of this section; I would rather classify them with the theory of integrable systems (some other aspects of this theory are considered in Section 3.5).

S. Kobayashi writes almost at the very beginning of his book *Transformation Groups in Differential Geometry*: "Not all geometric structures are created equal: some of them are creations of the nature, while the others are products of the human mind. Among the former, Riemannian and complex structures are distinguished by their beauty and wealth." This book was published in 1972. It seems that, at that time, symplectic manifolds were rather products of the human mind, but now the efforts of the human mind turn them gradually

---

[16] Here I mean this conjecture for area-preserving homeomorphisms of orientable surfaces of genus > 1. A. Floer had already proved it by "general symplectic" methods at that time.

[17] In the two-dimensional case, the symplecticity of a mapping reduces to preservation of area, and the latter property can be considered not only for diffeomorphisms but also for homeomorphisms.

into creatures of nature "refreshingly unlike"[18] its other creations.

1.2 Conformal dynamics

Another direction, being extensively developed in these years but not going beyond the limits of the theory of dynamical systems and constituting a part of differential dynamics, is conformal dynamics, which studies iterations of analytic functions in the complex domain. The phase space is some domain D in the complex plane (even the cases in which D is the entire plane $\mathbb{C}$ or the plane extended to the Riemann sphere are very interesting) and the dynamical system under consideration is a system defined in D and having discrete time $\{f^n\}$, where f is an analytic function (it may be polynomial, and even polynomial of the second degree!).

This direction is not new; it goes back to the classical works of P. Fatou and G. Julia in the beginning of this century (not counting the earlier works on local questions[19]). Even then, the complex behavior of trajectories resembling that in hyperbolic theory was noticed. This theory did not exist yet, but the complex behavior of geodesics on surfaces of negative curvature had already been discovered (by J. Hadamard for the first time); however, at that time, these two cases of complex behavior were not compared. In the course of time, the work in conformal dynamics ceased, and this direction fell in a long "hibernation" (although, there appeared some occasional papers; especially worthwhile is the 1942 brilliant local result of C. Siegel[20]); it awoke about 1970, apparently because of the recognition of relations or analogies with hyperbolic theory, which had already emerged at that time. Of great importance were also numerical experiments; they helped to discover a number of new phenomena and state a number of conjectures, working on which meant very much for the whole theory. The initiative in the experiments was taken by B. Mandelbrot; later, due to the widespread use of sufficiently powerful personal computers, performing such experiments became commonly available (although requiring certain care sometimes).

In conformal dynamics (as well as in studying real one-dimensional mappings, which were also very popular during the period in question), we have

---

[18] These are words of V. I. Arnold said on a humbler occasion, namely, when comparing linear algebra in the symplectic space $(\mathbb{R}^n, \omega)$ where ω is a constant (has constant coefficients) with the usual Euclidean geometry.

[19] The basic local question is: Let $f(a) = a$; under what conditions is the conformal mapping f conjugate to its linear approximation, i.e., to the linear mapping $x \mapsto a + f'(a)(x - a)$ near a? and, if there is no conjugation, what is the obstruction? If $|f'(a)| \neq 1$, the conjugacy is comparatively easy to prove; otherwise, the situation is much more complicated (see [1]).

[20] That was the first positive result for $|f'(a)| = 1$.

an almost unique opportunity to fairly completely study the complex behavior of DSs, both in the sense of qualitative patterns in the phase space and in the sense of dependences on parameters. It is possible to investigate DSs for which opposite types of behavior of trajectories are realized in different parts of the phase space: in one part, it may be hyperbolic (conformal dynamics often deals with weakened versions of hyperbolicity, which are being successfully studied in the case under consideration), while in another part, it may be quasiperiodic or something like that.

Some notion of this direction is given by [21]. This book is unlike usual mathematical books, where figures illustrate the text; [21] most resembles an album with beautiful computer pictures, which are explained in the text. As a rule, precise statements of theorems are given, but the proofs are sometimes omitted. The book contains an extensive bibliography and detailed historical comments; in this respect, [21] does meet the traditionally high requirements to a textbook-survey. (In addition, it contains remarks on computational difficulties arising in some cases.) We also mention surveys [22, 23] and book [24]; finally, the newest information is contained in the proceedings of the last international mathematical congresses, at which conformal dynamics was invariably paid a considerable attention.

A few pictures from [21] can give a much better idea of sets arising in conformal dynamics than several lectures without pictures. I only mention the following, by way of example. In some sense, all the "complexity" of the dynamics $\{f^n\}$ is concentrated on the so-called Julia set $J = J_f$, which may look fairly complicated and strikingly beautiful. Putting aside beauty, other sections of DS theory also deal with objects of complicated structure, which play a similar role. Now, let $f = f_c$ depend on a parameter c (a quite meaningful and not yet completely investigated example is $f_c(z) = z^2 + c$). Then there arises the question about bifurcations, i.e., changes of the qualitative pattern under variation of c. The corresponding values of c are mostly elements of the so-called Mandelbrot set M. At this point, all the other sections of DS theory lag far behind conformal dynamics, for they have nothing like the detailed images of M from [21].

In some sense, the set M bears a striking resemblance to the Julia sets, although this similarity is somewhat imperceptible; and it cannot have a simple description, because there are infinitely many Julia sets and only one M for the family $\{f_c\}$ specified above. Globally, M and J are quite different, and the similarity manifests itself in their local structures. In a paper of Lei Tan (she is a student of one of the well-known figures in this area, A. Douady), it is explained that the local structure of M near certain points $c \in M$ and the local structure of J_{f_c} near certain points $z \in J$ are asymptotically the same: examining these sets through a microscope, we would see that the higher magnification the closer the similarity in the field of vision of the microscope. This is an example of a theorem suggested by computer experiments. Lei Tan accomplished her work almost 15 years ago; this is a long period for the domain under consideration, and her theorem may be considered easy according to current standards. But the following statement is an example of a very difficult "computer-stimulated" conjecture: For the given (very simple outwardly!) family $\{f_c\}$, the Mandelbrot set is locally connected (this statement is known as the *MLC conjecture*). It has attracted a very close attention of experts (the statement is interesting in its own right, and it turns out that, in addition, it has interesting consequences, but I shall not dwell on this), and still, the existence of arbitrarily small connected neighborhoods is proved not for all points of M. The existing proof for part of the points of M again turns to account for some local similarity between M and J (the first significant progress was made by J.-C. Yoccoz, and the most recent results known to me are due to M. Lyubich [25]).

B. Mandelbrot rejoiced very much when he obtained the first (poor from the modern point of view) images of Julia sets and of the set M (which he introduced himself and which now bears his name), because they provided new and important examples of what he called fractal sets, or, briefly, fractals; this had strengthened his confidence in their importance and gave a valuable material for propaganda. The word "fractal" comes from "fraction" and is used because fractals usually have fractional Hausdorff dimensions. But, whereas Hausdorff dimension is a precise mathematical notion, fractals are not such. Mandelbrot said that all the figures which he had studied and called fractals had the property of being "irregular but self-similar" to his mind. According to Mandelbrot, the word "similar" not always has the classical meaning of being linearly dilated or contracted, but it always harmonizes with the convenient and broad meaning of the word "alike." ("Self-similarity" is understood in approximately the same sense in which "the same local structure" was understood above.) This is, certainly, an informal description rather than a definition. Mandelbrot explains that a set of fractional Hausdorff dimension is, of course, "irregular," but it is not necessarily self-similar, and a set of integer Hausdorff dimension may be "irregular" (say, "indented") and have or have not the property of local self-similarity; thus, "fractality" must not be understood literally. So fractality is a property defined not quite precisely; rather, is it described in more or less detail by employing numerous examples both from mathematics and from natural sciences. Fractals are a rocky coast, a chain of mountains, a fluttering flame, a cloud, a mold colony... Mandelbrot even specifies quantitative characteristics of self-similarity for various natural-science examples. Certainly, in these examples, as opposed to mathematical ones, self-similarity occurs only within a certain scale range. According to Mandelbrot, in inanimate nature, the extreme scales of this range can differ approximately by a factor of 10^4, and in biological examples, by a factor of 10^2.

These ideas and observations of Mandelbrot became rather popular, and, in his wake, many other authors began to discover fractals in nature. Apparently, indeed, together with objects whose shapes are fairly accurately described by familiar "Euclidean" figures, there are natural objects that have "fractal" shapes. Certainly, this is a mere phenomenology; the question arises, why a given object has fractal shape, and why this shape has given quantitative characteristics. As far as I know, in the "mathematical natural sciences" (understood in a very broad sense), things with this question are not going too well. But it is worthwhile to recall that understanding the nature of the familiar "Euclidean" shapes has progressed fairly slow and is not completed so far;[21] thus, it is not surprising that the corresponding questions as applied to the new forms still remain unanswered. At the same time, fractals have became "fashionable," with all the usual consequences[22] (cf. catastrophe theory mentioned in Section 1.4).

Let us return to mathematics. Stressing the "self-similarity" of fractals, Mandelbrot drew the attention of mathematicians to the essential property of a number of irregular sets with complex structure occurring in mathematics. Certainly, self-similarity may be an artifact, i.e., something made by human beings; this is the case for classical examples such as nowhere differentiable functions of K. Weierstrass and B. van der Waerden, the curves of G. Peano, H. Koch, and D. Hilbert, and the carpet of W. Sierpinski. In all these examples, some objects were constructed in infinitely many steps and, certainly,

---

[21] For instance, the regular shape of crystals is known since ancient times; even in the past century, their quantitative characteristics and the fact that a given substance always crystallizes into one crystal form or into one several crystal forms with characteristics inherent in the substance were known; it was conjectured long ago that the reason for that was the location of atoms or molecules at the nodes of some crystal lattice; a quantitative substantiation of this conjecture was given by E. S. Fedorov and Shönflies over one hundred years ago. But only about 90 years ago, this conjecture was proved by means of X-ray diffraction on crystals; as to the crystal phase transitions of most substances at low temperatures, the first results in this direction were obtained in statistical thermodynamics only comparatively recently (*a propos*, they bear some relation to DS theory; see the mention of phase transitions in Section 1.3). Finally, the experimental study of the crystal growth mechanism, in particular, the discovery of the dislocation growth mechanism dates almost exclusively from the after-war period (the important idea of dislocation emerged earlier, but on a different account). I believe, obtaining a growth mechanism "from first principles" has not been considered at all so far.

[22] Perhaps, fractals had reached the summit of their fame in this walk of life when the following words were written: "It can be said that Jesus Christ is at the center of a multidimensional fractal that propagates according to certain generation rules, which can be described in binary terms" (M. Eliadis and I. Kuliano. *Dictionary of Religions, Fancies, and Beliefs* (Moscow: Rudomino; St.-Petersburg: Universitetskaya Kniga, 1997)).

the authors strove to make these steps similar to each other, in order that the construction be easier to describe. But in other cases, such as those of J and M, self-similarity is not implied directly by the initial definition or construction and is an important feature.

At the beginning of this century, when conformal dynamics originated, the main tool was the methods of the "pure" theory of functions of a complex variable. In studying local questions, expansions in series and majorants were used, and in global problems, an important role was played by considerations related to the compactness of various sets of analytic functions (P. Montel shaped them into the theory of normal families of analytic functions, which has various applications, at approximately the same time). Starting with K. Siegel, many authors employed various tricks to fight small denominators, such as more involved constructions of majorants used by Siegel himself and, later, the KAM (A. N. Kolmogorov–V. I. Arnold–J. Moser) method. Instead of expansions in series, this method uses a more complicated process with infinitely many changes of variables, which converges more rapidly; another aspect of the method is somewhat analogous to Newton's method for solving nonlinear equations (Kolmogorov said that he was partly guided by this analogy). These two tricks are of "general mathematical" value; later, for the same purpose, more special methods (related to what is known as renormalization and to the "geometry" of the pertinent number-theoretic problem of small denominator) were developed. Being specialized, they made it possible to obtain better results, but (so far?) only for one-dimensional problems (real or complex). The extensive development of new methods in the 1970s was related partly to the progress in the theory of functions of a complex variable (e.g., the Teichmüller spaces, about which much had been known at that time, proved useful). Employment of quasiconformal mappings turned out to be a very happy idea. At first glance, the trick looks like a foolish escapade. Wishing to construct a conformal mapping with certain properties, we take a mapping f which is good in all respects except it is not conformal; however, it is conformal with respect to some nonstandard complex structure deserving offensive epithets among which the mildest would be "absurd" and "useless." If only we managed to correct f so as to make it conformal in the usual sense... And here, a *deux ex machina* comes on; this is the measurable Riemann mapping theorem[23] asserting that f is quasiconfor-

---

[23] This theorem has this appellation in conformal dynamics; it is generally referred to as the existence theorem for quasiconformal homeomorphisms (the appellation does not specify precisely what homeomorphisms are meant, neither shall I). The theorem has been generalized gradually; what conformal dynamics needs is the latest version, which is due to L. Ahlfors, L. Bers, I. N. Vekua, and C. Morrey. The idea to use it in conformal dynamics is due to D. Sullivan. Apparently, the possibility of applying it to the theory of Klein groups (published somewhat earlier) was also noticed by Sullivan. In the meantime, starting with 1981, the

mally conjugate to some truly conformal mapping. Finally, there is the already mentioned renormalization trick, which is not related in a specific way to the one-dimensional complex situation.[24] To a certain extent, renormalization is related to the general idea of local self-similarity, but it is applied not only to sets but also (and largely) to mappings. Though, I do not think that those who invented this trick have been influenced by the general idea of Mandelbrot. It seems that the idea must have had some effect, because this method was used to study the above-mentioned questions concerning J and M, but the point is that the method appeared much earlier, first in relation to mathematical questions of statistical physics and then in works on one-dimensional dynamics in a real domain.

By the way, I said nothing about one-dimensional dynamics in a real domain. It has also been extensively developed, starting with the 1970s. But at present, I believe, conformal dynamics is a direction more significant in all respects, and it exercises more influence; thus, although I want by no means to diminish the importance of one-dimensional real dynamics, I shall confine myself to merely mentioning it and refer the reader to book [26] (see also the end of Section 2.1). Comparing the two theories, we cannot help but perceive that both of them essentially use the special features of objects under examination, but in conformal dynamics these specifics constitute the subject of the rich and deep theory of functions of a complex variable, while in real dynamics they are no more than properties of the straight line. It is amazing that a theory as rich in content as real dynamics has arisen and continues being successfully developed on such a humble foundation. It is worth mentioning that some recent advances in this theory were made by carrying over considerations to the complex domain for a while[25] (see, e.g., the end of Section 2.1). Even earlier, the passage to the complex domain was used in relation to Feigenbaum universality[26] [26]. At large, such a trick is, to put it mildly, not new (it has

---

measurable Riemann mapping theorem had found independent applications to local questions (S. M. Voronin, B. Malgrange, and J. Martinet with J. P. Ramis). The idea of Sullivan was used by A. Douady and J. Hubbard in other problems of conformal dynamics, after which it, so to speak, "took root" in minds of researchers in this domain.

[24] Seemingly, this contradicts what was said above; but above we meant a particular trick using renormalization together with the "geometry" of a certain number-theoretic problem, rather than the general idea of renormalization; it is understanding of this "geometry" that the multidimensional situation lacks.

[25] Sometimes, it is possible to apply this trick even to nonanalytic mappings, which can be extended over the complex domain to mappings that are not analytic but, roughly speaking, differ little from analytic mappings.

[26] This is a certain law (which is also a kind of self-similarity) of the infinite sequence of period-doubling bifurcations of a one-dimensional mapping. (When such a bifurcation occurs, the periodic point loses stability and gives birth to a stable period-doubled point.) It is

been known for almost 450 years in algebra), but in DS theory, complexification worked well only in local and related questions.[27] Apparently, here, on the bounded "spot" of one-dimensional dynamics, we are witnessing the beginning of the penetration of this old but eternally youthful idea into the much younger theory of dynamical systems.

1.3 Nonclassical transformation groups

In most of this paper, by a DS we usually mean an action of the group or semigroup $\mathbb{Z}$, $\mathbb{R}$, $\mathbb{Z}_+$, or $\mathbb{R}_+$[28] on a phase space X (where X has a certain structure and the action is compatible with this structure in a certain sense). But we might consider actions of other groups as well. In this case, we say that we consider "transformation groups"; but if we wish to emphasize that we are interested in questions similar in many respects to those about the usual DSs, then we say that we deal with a "DS with nonclassical time" (ranging over the group or semigroup G), while the usual DSs are "DSs with classical time." Such a terminology is justified, the more so that, for historical reasons,

---

proved that this law is a "generic property"; M. Feigenbaum discovered it in the course of numerical experiments. (As Feigenbaum said somewhat humorously, it helped him a lot that he performed his computations on a programmable pocket calculator, for this made him to organize computations more carefully than when using a more powerful computer). Although it involves directly only one-dimensional mappings, the corresponding sequences of bifurcations occur for multidimensional DSs (including flows) too, because it is quite likely to happen that the "main events" are still "one-dimensional." See the story about this discovery and its (partly hypothetical) role told by Feigenbaum himself in [27].

[27] By questions related to the local ones I mean here questions concerning bifurcations which are more or less global in the phase space but local with respect to a parameter; see the end of Section 2.3.

[28] A left action of a (semi)group G on X is a mapping

$$G \times X \to X \qquad (g, x) \mapsto \phi_g(x)$$

such that

$$\phi_e(x) = x \quad \text{for all } x$$

(e is the identity element) and

$$\phi_g \circ \phi_h = \phi_{gh} \quad \text{for all } g \text{ and } h.$$

A right action is defined similarly. This is the same thing as a left action for a commutative G, but in the general case, left and right actions are different. It should be mentioned that, following Bourbaki, we use the notations $\mathbb{Z}_+$ and $\mathbb{R}_+$ for the sets of all nonnegative elements of $\mathbb{Z}$ and $\mathbb{R}$, respectively (thus, the sign $+$ is not to be understood too literally). We also use the notation $\mathbb{N} = \mathbb{Z}_+ \setminus \{0\}$.

By analogy with the case of classical time, the set $\{\phi_g(x); \ g \in G\}$ is called the trajectory (or orbit) of the point x (under the action of G); this will be needed later on.

the term "transformation groups" is mainly used when talking either about a continuous action of a compact group or about an algebraic action of an algebraic group satisfying a strong regularity condition; in both cases, the questions under examination are of a different character.

The direction which we shall consider is the ergodic theory of DSs with nonclassical time. But we shall start with a digression.

In what follows, amenable groups are mentioned. Although their definition was given by J. von Neumann as early as 1929, it is not generally known so far. For this reason, I shall give two later equivalent definitions[29] only for the case of a discrete group G; this group is even assumed to be finitely generated in the second definition. (The general case is that of a locally compact topological group or semigroup.)

The definition of E. Fölner postulates the existence of a certain system $\{F_n\}$ of subsets of G, which is now known as a Fölner system. They play approximately the same role as the intervals $[-n, n]$ play for $\mathbb{Z}$: as n increases, they cover the entire G; we can average over them; and under a shift by a fixed element of the group, the averages over sufficiently large sets from the Fölner system almost do not change.

Another definition (going back to H. Kesten and improved by R. I. Grigorchuk) is suitable for a finitely generated group G. If G has n generators, then it can be represented as the quotient group $G = F_n/H$ of the free group with n generators by a normal subgroup H. Let $f(k)$ denote the number of all irreducible words from F_n having length $\leqslant k$, and let $h(k)$ be the number of irreducible words that belong to H. It is almost obvious that $f(k) = 2n(2n-1)^{k-1}$. The group G is amenable if $\lim\limits_{k\to\infty} \sqrt[k]{h(k)} = 2n - 1$, i.e., if the number of words of length $\leqslant k$ in H increases (with k) at approximately the same rate as that in F_n. Thus, the factorization of F_n modulo H is very substantial: very many words in the generators are equal to 1 in G! Nevertheless, there does remain something: all commutative, nilpotent, and even solvable groups are amenable.

Amenable groups are considered in [28] (although, the second definition is more recent than the book).

Now, we can proceed to the topic. Soon after the first ergodic theorems (the "statistical" theorem of J. von Neumann and the "individual" theorem of G. Birkhoff) were proved, efforts to transfer them to other transformation group and semigroup with invariant measure were made. Analogues of ergodic theorems for $\mathbb{Z}^n$, $\mathbb{R}^n$, $\mathbb{Z}^n_+$, and $\mathbb{R}^n_+$ were obtained fairly quickly; next was the turn of different or more general groups. Apparently, a natural class of groups (semi-

---

[29] They are more classical in spirit than the original definition of Neumann involving the possibility of a certain transfinite construction (which yields the so-called left-invariant Banach mean) in the space of bounded functions on G.

groups) for which there is a hope to obtain ergodic theorems more or less parallel to classical time theorems is the class of locally compact amenable groups (semigroups). The statistical ergodic theorem is proved in precisely this generality. As to the individual ergodic theorem, the situation is more complicated. We shall consider only discrete groups.[30] Let (X, μ) be a space with finite[31] measure on which a discrete amenable group G acts by measure-preserving transformations $\{\phi_g\}$. By analogy with the classical case, the individual ergodic theorem must involve the means $f_n(x) := \dfrac{1}{\#F_n} \sum_{g \in F_n} f(\phi_g(x))$, where F_n is a set from the Fölner system $\{F_n\}$ and $\#F_n$ is the number of its elements. It turns out that we cannot always assert that the sequence $f_n(x)$ converges almost everywhere. A sufficient condition is that the system $\{F_n\}$ increases (i.e., $F_n \subset F_{n+1}$) and, which is most important,

$$\sup_{n \in \mathbb{N}} \frac{\#(F_n^{-1} F_n)}{\#F_n} < \infty$$

(the latter condition was specified by A. Calderon). Fölner systems satisfying the Calderon condition exist for almost nilpotent groups (i.e., for groups having nilpotent subgroups of finite index).[32] They also exist for countable locally finite groups. It is not known whether there exists another significant class of groups admitting such Fölner systems and whether the Calderon condition can be replaced by something more general. Thus, an individual ergodic theorem whose statement is fairly close to that of the classical theorem is proved for almost nilpotent and for locally finite groups G. For other groups, there are only some announcements, which remain not supported by detailed publications suspiciously long.

---

[30] In the literature, actions of locally compact groups and semigroups were considered as well. The discreteness assumption is made for the sake of simplicity (at this point, there is no noticeable difference between the discrete and locally compact cases, although such a difference arises at some places further on).

[31] It is well known that, for classical time, there are ergodic theorems in spaces with infinite measure too, and their formulations and proofs for such spaces are almost the same; however, for the further development of the theory, the cases of finite and infinite measure turn out to be fundamentally different, and now this development deals mainly with spaces of finite measure. In the nonclassical case, the situation with individual ergodic theorem is still unclear; thus, it is natural to consider first spaces of finite measure, as is usually done.

[32] In [29], it is said that the countable groups admitting Fölner systems satisfying the conditions specified above, of which the Calderon condition is most important, are essentially groups of polynomial growth, i.e., finitely generated groups for which the function $f(k)$ defined above has polynomial growth, and M. Gromov proved that such groups are almost nilpotent. It is not specified in [29] what "essentially" means; since polynomial growth can be considered only for finitely generated groups, possibly, it is these groups that are meant. But, as far as I know, that the Calderon condition implies a polynomial growth of $f(k)$ is not proved even for them.

Ergodic theorems for amenable groups are presented in [29, 30].

Recently, E. Lindenstrauss announced the following general result. In any countable amenable group, there is a Følner system $\{F_n\}$ such that, for any action with finite invariant measure, the individual ergodic theorem holds for the averages over $\{F_n\}$. (Thus, "some Følner systems are more Følner than others.") There is no detailed publication so far, but the announcement is recent, so "suspiciously long" does not apply.

It turns out that ergodic theorems may have place beyond the class of amenable groups. Thus, R. I. Grigorchuk proved an individual ergodic theorem for the finitely generated free groups (semigroups) and some groups close to them, whereas it is believed (and partly seen from the second definition) that the free groups are most unlike amenable groups.[33] The point is that Grigorchuk changed the method for averaging "in time," that is, over the group. I would compare this change to the passage from the usual convergence to Cesàro means in the theory of series. As is known, there are many different summation methods. Something similar may take place in ergodic theory; or, maybe, there is a universal method, at least for finitely generated groups.

But I mentioned nonclassical time not only in order to discuss ergodic theorems (if it were so, the title of this subsection would hardly suit). Ergodic theorems constitute only a part of the ergodic theory of systems with classical time. Historically, this part was developed first, and it gave the name for the entire theory; but in the past half century, it became only a part, not even half, of the theory. The "abstract" – "purely metric" – sections of ergodic theory consider various properties of ergodic systems, including their partial classification. Usually, instead of arbitrary spaces with measure (even if normalized), Lebesgue spaces are considered;[34] although, this hardly restricts the generality

---

[33] Initially, Grigorchuk published a communication about his results only in the proceedings of a provincial conference [31] (and considered only free groups; though, he did not require the finiteness of the invariant measure). It received no attention, and, later, other authors published similar papers [32]. Analogous theorems for action of some semisimple Lie groups were also considered [33, 34].

I shall not discuss the earlier papers [35, 36]. That they are valuable as pioneering works is undoubted, but they consider special situations and have specific features; it is desirable to understand them from general positions, but I (and, I believe, not only I) am not ready for this. (For example, in [36], averaging over a solvable group is performed in such a way as if the group were free.) I shall only mention the statistical ergodic theorem for free groups proved in [37].

[34] A Lebesgue space is a space with measure isomorphic (in the sense natural for spaces with measure) to the standard object, interval $[0, 1]$ with normalized Lebesgue–Stieltjes measure. Equivalently, a Lebesgue space is isomorphic to the interval $[0, a]$ with a Lebesgue measure to which no more than a countable number of "atoms" with measures p_n are added; in addition, the normalization condition $a + \sum p_n = 1$ must hold. This definition takes into account some

of the results from the point of view of their application to particular examples. Other sections, which can be conventionally called "applied," study the properties of special systems or of some classes of such systems. (These systems are frequently smooth or topological; thus, in principle, studying them could be referred to topological or smooth dynamics, but when we consider properties related to "abstract" ergodic theory, the studies of the corresponding (classes of) particular systems are usually also included in ergodic theory.)

For systems with nonclassical time, a few studies of this kind have been done before 1970, but an extensive development started about 1970 or a little later. In some cases, a similarity to the classical case was revealed, while other cases turned out to be different from it; moreover, the interface between common and unusual situations depends on the question under consideration. Frequently, the case of amenable groups is similar to the classical case, as above, but sometimes, even the case of $\mathbb{Z}^2$ fundamentally differs from that of $\mathbb{Z}$.

As above, it is impossible to tell about a whole science in a few words. I shall only give several examples (since I am trying to characterize the area as a whole, some of these examples date from a period earlier than the past quarter century). I shall consider only actions of discrete groups (although many (but not all) considerations apply to separable locally compact semigroups[35]); moreover, in the first examples ((a)–(c)), I shall assume that the group G acts on a Lebesgue space (X, μ) and that all transformations under consideration preserve the measure μ.

(a) One of the sections of ergodic theory, spectral theory, starts with assigning operators U_g in $L^2(X, \mu)$ to elements $g \in G$:

$$(U_g f)(x) = f(\phi_g^{-1}(x)). \tag{3}$$

The operators U_g are unitary, and $U_{gh} = U_g U_h$; thus, we have a unitary representation[36] of the group G. This gives rise to the problem of investigating the

---

useful special features of the majority of particular examples encountered in ergodic theory, which allow a much further progress in analyzing purely metric questions than for general spaces with measure. That the class of Lebesgue spaces is sufficiently large is witnessed by the fact that any metrizable compact set with normalized measure (defined on its Borel subsets) is a Lebesgue space.

The theory of Lebesgue spaces was developed by J. von Neumann and P. Halmos (they used a different term) and, especially, by V. A. Rokhlin.

[35] Even for noncommutative topological groups, difficulties may arise because of the difference between left- and right-invariant Haar measures. This is something new in comparison with the case of discrete groups.

[36] If we took ϕ_g instead of ϕ_g^{-1} on the right-hand side of (3), we would obtain an antirepresentation, for which $U_{gh} = U_h U_g$; generally, it is no worse than a usual representation, but it is less familiar. Dealing with semigroups, we would have to consider antirepresentations, and (which is more essential) the operators U_g would be isometric rather than unitary.

properties of this representation and determining to what extent they reflect the properties of the initial DS. When we change a DS for a metrically isomorphic one, the corresponding representation is replaced by a unitary equivalent representation. In functional analysis, there is a system of notions that refer to the properties of unitary representations invariant with respect to unitary equivalence. In the simplest case of the powers U^n of a unitary operator U (i.e., when we deal with a representation of $\mathbb{Z}$), everything reduces to the spectrum of this operator; for this reason, the corresponding notions, invariants, and properties are called spectral, and they retain this name in more general cases. The constants are always invariant with respect to (3), and they form a one-dimensional subspace; for this reason, consideration of this representation is tacitly restricted to the subspace $H \subset L^2(X, \mu)$ orthogonal to the constants; the corresponding spectral properties are called also properties of the DS $\{\phi_g\}$. I must say that, in the case under consideration, even for one operator U, the spectrum is understood in a more delicate sense than in the elementary courses of functional analysis, where the spectrum of a linear operator A is defined as the set of those $\lambda \in \mathbb{C}$ for which the operator $A - \lambda I$ has no bounded inverse defined on the entire space. Under this definition, the spectrum of the DS $\{\phi^n\}$ coincides with the unit circle in all practically interesting cases; thus, the invariant which it provides is almost always the same and, therefore, useless. But for various special (and yet important) classes of operators considered in functional analysis, there are more delicate versions of the notion of spectrum; most important for us is such a refined notion of spectrum for a unitary operator U and a self-adjoint operator A (the latter appears in studying flows as the generating operator of the corresponding one-parameter group of unitary operators). Because of space limitations, I refer the reader to advanced textbooks on functional analysis for this notion (it is also given, at least partially, in some books on ergodic theory).[37] I shall only mention that the spectrum is said to be continuous when U or A has no eigenfunctions (on H), discrete when the eigenfunctions form a complete system (in H and, hence, in the entire $L^2(X, \mu)$), and mixed otherwise.

Note that the unitary equivalence of the representations corresponding to two DSs does not generally imply that the DSs are metrically isomorphic. Therefore, generally, the properties of a DS are not completely determined by its spectral properties. But sometimes they are. In the classical situation of a metric automorphism ϕ or a flow with invariant measure ϕ_t, this happens when the DS is ergodic[38] and the spectrum of the corresponding U or A is

---

[37] See also Section 3.9.

[38] In the "abstract" ergodic theory of DSs with classical time, it is proved that each system uniquely decomposes into ergodic components in a certain natural sense; for this reason, it

discrete. In this case, the spectrum completely determines the DS up to metric isomorphism. In the nonclassical situation, G. Mackey[39] distinguished a case similar in many respects to that specified above; because of this similarity, it is given the same name. This is the case where the representation under consideration decomposes into a discrete direct sum of irreducible finite-dimensional representations. The description of this decomposition is conventionally called a spectrum. If G is commutative, then the situation is quite similar to the classical one, but for general groups, this is not so; in particular, a discrete spectrum may not determine a DS up to metric isomorphism.

What we known about the case of nondiscrete spectrum (even for classical time) can hardly be called a general theory; this is rather a set of examples and classes of examples. For nonclassical time, even less is known. I shall mention only one fact, which contrasts with the classical situation. S. Banach once noticed that, in all classical examples with Lebesgue spectrum known to him, the spectra are countably multiple, and asked whether this is always the case. At present, more examples are known, and all of them confirm the observation of Banach, but the answer to his question is still unknown.[40] At the same time, for the multiplicative group of nonzero rational numbers, there is an example of an action with Lebesgue spectrum of multiplicity 1 (M. E. Novodvorskii).

Let me mention, at the same time, an example which shows the contrast between the classical and nonclassical amenable cases, although this example is not "spectral" (at least, it is not directly related to spectrality).[41] V. A. Rokhlin conjectured that, if a DS with classical time has the mixing property, then it has the property of mixing of any multiplicity. This problem also remains open,

---

is accepted that the properties of "general" DSs, as it were, reduce to the properties of their ergodic components, and it is ergodic DSs that should mainly be studied (whereas in the "applied" theory, we must try to determine whether each particular DS is ergodic or not). For a DS with nonclassical time, this is the case if G is a locally compact group with countable base (and the measure in the phase space may be only quasi-invariant rather than invariant). For more general groups G, different definitions of ergodicity and of decomposition into ergodic component become nonequivalent (see the article "Metric transitivity" in *Mathematical Encyclopaedia*); still, the aforesaid approach does apply to a certain extent.

[39] It should be mentioned that, during the first two thirds of the 1960s, there was hardly anyone but Mackey who propagated a study of DSs with nonclassical time beyond the framework of ergodic theorems.

[40] If the spectrum has not only a Lebesgue component but also a singular one, then the multiplicity of the Lebesgue component may be finite; see [38], where references to earlier examples of this kind are also given.

[41] The mixing involved in this example is a "spectral" property, that is, it is determined by some property of the spectrum of the corresponding $\{U\}$ or $\{U_t\}$, while multiple mixing is not characterized in terms of spectra (although, the validity of the Rokhlin conjecture stated below would eventually imply the existence of such a characteristic).

whereas for $G = \mathbb{Z}^2$, F. Ledrappier constructed an example of a DS with mixing but without multiple mixing.

(b) Another large section of ergodic theory is entropy theory, which not only defines a metric invariant called entropy and studies its properties but also considers a number of related questions. It can be carried over to amenable groups to a large extent; the first steps in this direction were made by A. M. Stepin, who defined an entropy h_μ for an action of an amenable group G with invariant normalized measure μ in the late 1960s. Later, together with A. T. Tagi-zade, he defined a topological entropy for a continuous action of G on a metric compact set and proved that it coincides with $\sup_\mu h_\mu$, where the least upper bound is taken over all invariant normalized measures of this action. (Thereby, it is implied that such measures exist; see (d).) This generalizes the theorem of E. I. Dinaburg, T. Goodman, and L. Goodwin, which refers to classical time; see also (f).

(c) Yet another example where the class of amenable groups is a natural class to which a well-known result obtained initially for classical time carries over is as follows. Any two measure-preserving ergodic actions of two countable discrete amenable groups on a Lebesgue space X are trajectory equivalent, i.e., there exists a metric isomorphism $X \to X$ which takes the trajectories of one system to trajectories of the other (A. Connes, J. Feldman, and B. Weiss, 1981). Thus, from the purely metric point of view, an ergodic action admits only one partitioning into trajectories; as a "standard pattern," we can take, say, the partitioning of the circle into the trajectories of rotation through an "irrational" angle.[42]

A nonamenable group has trajectory nonequivalent ergodic actions. The following 1980 result of R. Zimmer contrasts with the Connes–Feldman–Weiss theorem especially sharply. Let G and H be connected semisimple Lie group of rank larger than 1 without center and finite quotient groups. Suppose that $\{\phi_g\}$ and $\{\psi_h\}$ are their actions on a Lebesgue space (X, μ) such that they preserve the measure μ, are ergodic, and remain ergodic when being restricted to an arbitrary nonidentity normal subgroup; suppose also that any nonidentity element of each of the groups moves all or almost all[43] points of X. If these actions are trajectory equivalent, then G and H are isomorphic, and the appropriate identification of their elements makes the actions metrically isomorphic.

---

[42] That is, through an angle incommensurable with the "full" angle of 360°. When the circle is represented as $\mathbb{R}/\mathbb{Z}$, this is a shift by an irrational (in the usual sense) number. Note that the partitioning of the circle into the trajectories of this shift is involved in constructing the best-known example of a nonmeasurable set.

[43] In the first case, the action is said to be free. In the second case, Mackey and Zimmer call the action "essentially free."

In other words, there exist an isomorphism of Lie groups $f: G \to H$ and a metric automorphism $h: X \to X$ such that $h\phi_g = \psi_{f(g)}$ for all $g \in G$.

When something admits only one realization (up to obvious modifications such as "twistings" by f and h described above), the term "rigidity"[44] is often used. When the paper of Zimmer appeared, a number of results (mainly due to G. Mostow and G. A. Margulis) on the rigidity of discrete subgroups of Lie groups had already been obtained; Zimmer himself mentioned that they had an influence on his work (possibly, in respect of conceptual associations more than in the sense of direct logical dependence)[45]). Later, papers (mainly by A. Katok and his collaborators) on some rigidity phenomena in a smoother situation bordering on hyperbolic theory were published.

(d) There is a well-known theorem of N. M. Krylov and N. N. Bogolyubov on the borderline between topological dynamics and ergodic theory. It asserts that a classical-time topological dynamical system with compact phase space has at least one normalized invariant measure. Soon after it was published, in 1939, Bogolyubov noticed that precisely the same theorem is valid for continuous actions of amenable groups[46] on compact spaces (see [39] for more details). It can be proved that, conversely, if any continuous action of a locally compact group on a compact space admits a finite invariant measure, then the group is amenable.

(e) This theorem of Bogolyubov, as well as the "classical" Krylov–Bogolyubov theorem, says nothing about the properties of invariant measures. These properties may be different in different examples; it may even happen that one system has many normalized invariant measures, including ergodic ones, with essentially different properties. One measure may be concentrated on one point (fixed for all ϕ_g), while another may be positive for all open sets. The former measure is ergodic,[47] but, certainly, no meaningful assertions about a system with such a measure (except that it has a fixed point) can be made. In the latter case, the measure may or may not be ergodic; if it is ergodic, then it may have or have not stronger properties of "quasi-random" character (such

---

[44] Apparently, this term came from the geometry of surfaces; however, some of the most respectable experts in this area prefer the term "single-valuedness" and use "rigidity" only for a certain infinitesimal version of single-valuedness.

[45] However, Zimmer has proved a theorem (unfortunately, its formulation is too cumbersome) which implies both his rigidity result and some results of Mostow and Margulis; thus, a formal link also takes place.

[46] At that time, Bogolyubov called such groups Banach, because they are characterized by the existence of a Banach mean.

[47] When different invariant measures on one topological DS are considered simultaneously, the ergodicity of the DS with respect to a measure μ is often referred to as the ergodicity of a measure μ.

as mixing, positive entropy, etc. in the classical situation). Sometimes, the existence of an invariant measure which definitely deserves special attention is known in advance (such is the case we run into in the classical situation when considering Hamiltonian systems); sometimes, there is no *a priori* "privileged" invariant measure, and the problem of the existence of not merely invariant measures but measures with certain interesting properties is to be solved separately. Naturally, it is considered in much more special situations than the general situations of the Krylov–Bogolyubov and Bogolyubov theorems.

An interesting and important class of examples for which this question is studied arises from statistical physics, or is at least suggested by it. Let us imagine that, in each point of the lattice $\mathbb{Z}^m$, a particle in one of k possible states is placed (this is a "lattice system"). We denote the phase space of states of one particle by A (for example, we can take $A = \{1, \ldots, k\}$, if there is no particular reason for using some other notation for particle states). Then the state of the entire infinite system of particles is described by the function $\xi \colon \mathbb{Z}^m \to A$, where $\xi(g)$ is the state of the particle in the point $g \in \mathbb{Z}^m$. I expressly denote a point of $\mathbb{Z}^m$ by g, in order to pass at once from $\mathbb{Z}^m$ to an arbitrary group G (in fact, in the larger part of this section, a semigroup is sufficient, but we consider groups to obviate the necessity for making stipulations in what follows). Certainly, in the case of $G \neq \mathbb{Z}^m$, we can no longer imagine a crystal placed in $\mathbb{R}^m$, but still, we can consider functions $\xi \colon G \to A$. The set $\Omega := A^G$ of all such functions is the phase space of our infinite system. It is endowed with a natural topology (the Tychonoff product topology of the direct product of an infinite number of copies of the space A)[48] and with a somewhat less natural metric. The metric is somewhat less natural because there exist many metrics generating this topology, and *a priori*, there are no reasons for preferring one of them. The experience shows that, in the case of $G = \mathbb{Z}^m$, the following metric is well suited for our purposes. We take some (no matter which one) metric d on A (for example, we can assume that the distance between different points of A equals 1 or set $d(i,j) = |i - j|$) and an arbitrary number $a \in (0,1)$ and put

$$|g| = \sum_r |g_r| \quad \text{for } g = (g_1, \ldots, g_k) \in \mathbb{Z}^m$$

and

$$\rho(\xi, \eta) = \sum_{g \in \mathbb{Z}^m} a^{|g|} d(\xi(g), \eta(g)).$$

---

[48] A reader not familiar with the Tychonoff product topology may assume that this is the topology generated by the metric ρ introduced below; this can be regarded as its definition in the case under consideration.

It is easy to verify that the topology on Ω generated by this metric is precisely the Tychonoff product topology. For other groups, we can take something similar; instead of $a^{|g|}$, we must take a suitable function of g which, so to speak, sufficiently rapidly decreases as g moves away from the identity element of the group; I shall not dwell on this. Thus, Ω is a metric compact space.

Finally, there is a natural action of G on Ω, namely,

$$(g, \xi) \mapsto \phi_g(\xi), \quad \text{where} \quad (\phi_g(\xi))(h) = \xi(gh) \quad \text{for all } h \in G. \tag{4}$$

Here we write the group operation as multiplication $((g, h) \mapsto gh)$ rather than as addition $((g, h) \mapsto g + h)$, because G may be noncommutative. (Of course, when $\mathbb{Z}^m$ is considered, the sign $+$ is used, because the group operation is the usual addition). The mappings ϕ_g are homeomorphisms of Ω.

The object obtained, that is, the metrizable compact space Ω together with the group G acting on it as specified, is called a Bernoulli topological DS, or the topological Bernoulli action of the group G. It is called so because, for $G = \mathbb{Z}$, an element $\xi \in \Omega$ can be interpreted as the record of the results of an infinite sequence of trials with the same possible outcomes that form the set A; $\xi(n)$ is the outcome of the trial performed at time n (time is discrete and ranges over the entire $\mathbb{Z}$). The $\xi(n)$ with $n \leqslant 0$ are the outcomes of the already performed trials (in particular, $\xi(0)$ is the outcome of the trial performed "now"), and the $\xi(n)$ with $n > 0$ are the outcomes of future trials; the former are known to us, while the latter are not. More frequently, the two-sided infinite sequence

$$\ldots, \xi(-n), \xi(-n + 1), \ldots, \xi(-1), \xi(0); \xi(1), \xi(2), \ldots, \xi(n), \ldots, \tag{5}$$

is considered; its elements are written precisely as above, from left to right in the order of increasing "time." To distinguish the position of the element number zero, a semicolon is used. The action of $\mathbb{Z}$ on Ω reduces to iterations of the "Bernoulli topological shift"

$$\sigma : \Omega \to \Omega, \quad \xi \mapsto \sigma\xi, \quad (\sigma\xi)(n) := \xi(n + 1)$$

and its inverse mapping σ^{-1}. Under the shift, sequence (5) shifts to the left with respect to the semicolon; the previous $\xi(1)$ becomes the element number zero, so we can imagine that we have repeated the trial and know this element now. Let us denote the projection of the infinite product

$$\Omega = A^{\mathbb{Z}} = \ldots A \times A \times \ldots \times A \times \ldots$$

onto its "zero" factor A by π_0, so that $\pi_0\xi = \xi(0)$. Then we can say that, in successive trials, the outcome $\pi_0(\sigma^n\xi)$ is observed at time n.

Thus, we see that, for $G = \mathbb{Z}$, the same object (Ω, σ) can be interpreted two-fold. From the point of view of statistical physics, with which we started, the object under consideration is the states of an infinite chain. No dynamics in the usual sense (in the sense of the evolution of a system with changing time) is involved, because true time is not mentioned at all. In (5), n is the number of an element of the chain, and the action σ corresponds to the spatial shift of the chain by one "link." From the probabilistic point of view, to which we proceed, n can be regarded as "true," "physical" time: in time n, a point ξ of the phase space is transformed into $\sigma^n \xi$. This corresponds to the representation of the trials included in the given sequence of trials as being performed one after another; in addition, in each trial, only $\pi_0 \xi$ (rather than ξ) is observed, so the quantities observed with changing time are $\pi_0 \sigma^n \xi$, which coincide with $\xi(n)$ in the case under consideration. (In more general cases, the phase space Ω and mapping $\sigma \colon \Omega \to \Omega$ may be different, and the observed quantity may be some function $f \colon \Omega \to \mathbb{R}$, so that, in the course of time, observations give the sequence $f(\sigma^n \xi)$.) The statistical physics approach is also suitable for $G = \mathbb{Z}^m$ with $m > 1$; again, "dynamics" means the action of spatial shifts (translations) on the system under consideration rather than a change of state with time (which is absent). But a sequence of experiments can no longer be considered, although we could imagine trials numbered for some reason by elements of $\mathbb{Z}^m$ (perhaps, this even has a (quasi)realistic interpretation) or by elements of a general group G. For $G \neq \mathbb{Z}^m$, the statistical physics interpretation also becomes a conventionality suggested by the analogy with the physically real cases of $G = \mathbb{Z}$, $\mathbb{Z}^2$, and $\mathbb{Z}^3$.

Although we have partly used the language of probability theory and even mentioned Bernoulli, we have used no probabilities so far. As is known, the name of Jacob Bernoulli is connected with the study of a sequence of identical independent trials. In this case, each possible outcome $a \in A$ has some probability $p(a)$, and the independence of the trials manifests itself in that the outcome at time n does not depend on the outcomes at different moments of time. Therefore, the probability that the outcomes at times $n_1, \ldots, n_l$ are $a_1, \ldots, a_l$ is equal to $p(a_1) \cdot \ldots \cdot p(a_l)$. In fact, we have described a measure μ on Ω with respect to which the sequence of random variables $\pi_0(\sigma^n \xi)$ describes a sequence of independent identical trials. Let us repeat the description once more and, simultaneously, transfer it to the general case of $\Omega = A^G$, where the probabilistic interpretation (in terms of a sequence of trials) loses its original meaning, although, conventionally, we can continue using the probabilistic language. (In such cases, the term "random field" is used in probability theory; this agrees with the "intuitive meaning" of these words when $G = \mathbb{Z}^m$.)

The elements $a \in A$ must be assigned probabilities or measures $p(a) \geqslant 0$

in advance; the sum of the probabilities or measures must equal 1. Thereby, a subset $B \subset A$ is assigned the measure $p(B) = \sum\limits_{a \in B} p(a)$, and we obtain a measure on the finite set A, which is, of course, trivial. The measure on Ω constructed below depends on these $p(a)$, i.e., on this measure p on A.

A cylindrical subset of the space Ω is a set of the form

$$C = \{\xi;\ \xi(g_1) \in B_1,\ \ldots,\ \xi(g_l) \in B_l\}, \tag{6}$$

where $g_1, \ldots, g_l \in G$ and $B_1, \ldots, B_l \subset A$. It is assigned the measure

$$\mu(C) = p(B_1) \cdot \ldots \cdot p(B_l). \tag{7}$$

In measure theory, it is proved that, on the σ-algebra of Borel subsets of a compact space Ω, there exists a unique measure μ which takes value (7) for each cylindrical set (6). From the point of view of measure theory, it is the direct product

$$p^G = \ldots \times p \times p \times \ldots \times p \times \ldots$$

of the measures p on the factors A of the direct product A^G. It is almost obvious that this measure is invariant with respect to the action (4) of the group G and that the action is ergodic with respect to this measure.

In the classical case, the shift σ considered with respect to the constructed measure is called the Bernoulli metric automorphism. It is natural to call the measure itself also Bernoulli. The properties of the DS $\{\sigma^n\}$ with this measure differ drastically from the properties of ergodic DSs with discrete spectra. The system $(\{\sigma^n\}, \mu)$ has mixing of all degrees, Lebesgue spectrum of countable multiplicity, and positive but finite (metric) entropy. In general, the Bernoulli automorphism is, as it were, a specimen of a DS with "quasi-random" properties (as it should be because of its origination); moreover, it is, so to speak, an extreme such specimen, not counting DSs with infinite entropy. In the case of nonclassical time, the Bernoulli DSs play a similar role.

The definition of a Bernoulli DS and a Bernoulli measure as such does not depend on whether G is amenable. But the further considerations in this and the next subsections require amenability.

A significant achievement of DS theory is the theorem of D. Ornstein, according to which Bernoulli automorphisms with equal entropies are metrically isomorphic. D. Ornstein and B. Weiss showed that this theorem can be carried over to Bernoulli actions of amenable groups [40, 41].

(f) The measure on A (i.e., the system of numbers $p(a) \geqslant 0$ satisfying the condition $\sum p(a) = 1$) can be chosen arbitrarily; thus, for a Bernoulli topological DS, there are continuum many invariant ergodic measures. But it turns out that, in addition to the measures constructed above, this DS has very many other invariant normalized measures, including ergodic ones. Some of them undoubtedly deserve special attention.

For $G = \mathbb{Z}$, some measures, as Bernoulli ones, are distinguished (from the very large set of invariant measures) or introduced (independently of what we know about other measures) on the basis of probabilistic considerations. Probability theory considers some sequences of identical experiments which are not independent. Especially important are the cases where $\{\pi_0(\sigma^n \xi); \; n \in \mathbb{Z}\}$ is a Markov process. They correspond to new invariant normalized measures μ on A, which may be ergodic or not. But, under the passage from $\mathbb{Z}$ to G, the definition of the Markov property loses sense. A reasonable modification of this definition exists for $G = \mathbb{Z}^m$, but at present, as far as I know, the corresponding objects play a noticeable role in the theory of random fields rather than in ergodic theory. We leave them aside and turn from probability theory to statistical physics, which we started with but abandoned.

From the physical point of view, the Bernoulli measures correspond to the situation where the states of the particles at the vertices of the lattice $\mathbb{Z}^m$ do not depend on each other. Certainly, this occurs when the particles do not interact, but of interest is the case where an interaction does take place. In this case, considerations borrowed from statistical physics lead to new normalized invariant measures. As we shall see, they can be defined in such a way that the definition make sense not only for a topological Bernoulli DS but also for an action of a group G (amenable, as previously) on a compact metric space. Although, it may happen that there exist no measures satisfying this condition. But if such measures exist, then they deserve attention. I must warn the reader that, since our subject-matter is DS theory rather than statistical physics after all, the constructions presented below may involve some inconsistencies with statistical physics, even as applied to lattice systems. It can be proved that these inconsistencies do not affect the final results for such systems, but we shall simply ignore them, because what is borrowed from statistical physics plays a rather heuristic role in our considerations.

In statistical physics, for a system having a finite number of states $\alpha_1, \ldots, \alpha_N$ with energies $E(\alpha_j)$, the so-called statistical sum

$$Z = \sum_{j=1}^{N} e^{-\beta E(\alpha_j)} \tag{8}$$

is introduced. Here β is the reciprocal temperature expressed in suitable units; since temperature is measured in degrees from time immemorial, to pass to "suitable" units, we must multiply it by the Boltzmann constant k; thus, $\beta =$

$\frac{1}{kT}$. For a macroscopic system, its free energy F considered in phenomenological thermodynamics is expressed in terms of Z as

$$F = -\frac{\ln Z}{\beta}. \tag{9}$$

In thermodynamical equilibrium (corresponding to a given temperature T), the state α_j has probability

$$p(\alpha_j) = e^{-\beta E(\alpha_j)}/Z. \tag{10}$$

One of the possible characterizations of this probability distribution is as follows: given β and $E(\alpha_j)$, the value

$$\sum p_i E(\alpha_i) + \frac{1}{\beta} \sum p_i \ln p_i \tag{11}$$

attains its minimum over all distributions of the probabilities $(p_1, \ldots, p_N)$ at $p_i = p(\alpha_i)$. Substituting these p_i into (11), we obtain precisely F.

Naturally, we assume that, in (11), $p_i \ln p_i = 0$ at $p_i = 0$, because $\lim_{p \to 0} p \ln p = 0$. Thus, we can assume that (11) involves only positive p_i's. When we remove the terms with those j for which $p_j = 0$ from (8), the statistical sum Z can only decrease, and if we prove that

$$\sum p_i E(\alpha_i) + \frac{1}{\beta} \sum p_i \ln p_i \geqslant -\frac{1}{\beta} \ln Z \quad \text{for } p_i > 0 \text{ such that } \sum p_i = 1 \tag{12}$$

for this decreased Z, we will thereby prove it for the initial Z. The function $\ln x$ is concave, i.e., its graph is convex upward; hence, we have

$$\sum p_i \ln x_i \leqslant \ln \sum p_i x_i$$

for any $x_i > 0$. In particular,

$$\sum p_i \ln(e^{-\beta E(\alpha_i)}/p_i) \leqslant \ln \sum p_i e^{-\beta E(\alpha_i)}/p_i = \ln \sum e^{-\beta E(\alpha_i)},$$

which is equivalent to (12).

Note *a propos* that, if we substitute $p_i = p(\alpha_i)$ in (11), then the first term will become equal to the energy of our system in the given equilibrium state and the second will be $-TS$, where $S := -k \sum p_i \ln p_i$ is the entropy (measured in the macroscopic units used in phenomenological thermodynamics). Thus, $F = E - TS$. This relation is the definition of F in phenomenological thermodynamics.

After this brief excursus in statistical physics, we return to our lattice system or, more generally, to a Bernoulli DS (with an amenable group G). First, let me make one terminological comment. It happens that, to the same term, quite different meanings are attached in different sciences. If the sciences are far from each other, this does not lead to misunderstandings; for example, it is hard to imagine how topological cells could be confused with cells in biology. But sometimes the same word means different things in fairly close sciences, which are adjacent and even partially overlap. Then the terminology should be made more precise. Such is the case of the term "state." In DS theory, a state is a point in the phase space. In statistical physics, a state is a probability distribution (measure) on the phase space. (At the end of the preceding paragraph, the word "state" was used in precisely this sense one time.) If we adhere to the statistical physics terminology, then we must not regard points of the phase space as states. In the theory of lattice systems, they are often called configurations.[49] But this paper is written from different positions, so – why not call a measure a measure?

The total energy of the entire infinite lattice system is generally infinite and cannot be dealt with. But it is reasonable to assume that we can, so to speak, separate out a part corresponding to one particle, and that for a particle located at a point g, this contribution is equal to the value of some function $E(g,\xi)$ depending on g and on the state ξ of the entire lattice system. We assume that the interaction is invariant with respect to the left group shifts. Therefore, if the particle at a point g is in a state $\xi(g)$ and the particles at the points gh (with all possible h) are in states $\xi(gh)$ (as is the case for the state ξ of the entire system), then the energy contribution corresponding to the "gth" particle is equal to the energy contribution corresponding to the "eth" particle (e is the identity element of the group) provided that it is in the state $\xi(g)$ and the "hth" particles are in the states $\xi(gh)$. This takes place when the state of the entire system is $\phi_g(\xi)$. Thus, $E(g,\xi) = E(e, \phi_g(\xi))$, and it is sufficient to consider the function $f(\xi) := -E(e,\xi)$; we assume it to be continuous. (The minus sign has no physical motivation, but it is commonly used in the relevant mathematical

---

[49] There are certain grounds. In classical mechanics, the configuration of a system is the arrangement of its parts without taking into account velocities. Now, we deal with equilibrium statistical physics, where no motion is involved. (If desired, we can imagine that, even if there are some motions, they are "concealed," "hidden inside states," and that they are effectively taken into account only in the energy characteristic f mentioned below. Though, physical problems involving lattice particles with finite numbers of states refer to spin, and spin is not at all described in the classical terms, which distinguish between coordinates and momenta and where energy is divided into potential and kinetic. As we have touched upon this topic, it should be mentioned that, actually, the quantum mechanical description of a spin is much more complicated than the primitive situation considered here, where it is said only that a particle has finitely many states (with corresponding energy characteristics). Nevertheless, sometimes (and, apparently, more frequently than it might be expected) the primitive model described above gives a sufficiently good approximation).

literature, because it somewhat simplifies some formulas. The theory being discussed, although suggested by statistical physics, is mathematical after all, and, in the general case, the values it involves by no means can have a physical meaning such as in the example under consideration.) We then have

$$E(g, \xi) = -f(\phi_g(\xi)). \tag{13}$$

(Under a more realistic approach, in the case of $G = \mathbb{Z}^m$, the consideration starts with reducing the interaction to pair, triple, ..., r-fold, ... interactions; however, under the assumption that all interactions rapidly decrease with increasing the number of particles and the distance, the final result is the same.)

Our plan is as follows. First, we consider a finite "piece" S_n of the lattice system that is formed by the particles located at those points of the lattice $\mathbb{Z}^m$ that lie in a cube or, more generally, by the imaginary particles enumerated by the elements of the Fölner set F_n. Now, we want to pass to the limit of $n \to \infty$. Formula (9) suggests that it is reasonable to take

$$-\frac{\ln Z_{F_n}}{\beta \, \# F_n}, \tag{14}$$

where $\# F_n$ is, as above, the number in elements of F_n. In the limit of $n \to \infty$ (of course, its existence is to be proved), we obtain the free energy per one particle. (The "total" free energy of the entire infinite lattice system, which should be understood as the limit of $-\ln Z_{F_n}/\beta$, is, most likely, infinite).

In the accomplishment of this plan, the implementation of some details is quite different from what might seem more natural.

First, in (14), the minus sign is omitted, and the corresponding limit is called "pressure" rather than "free energy" (per one particle); accordingly, it is denoted by P. To leave aside the "everyday" meaning of the word "pressure," in thermodynamics, the pressure of a macroscopic continuous physical system is understood as follows: we must express F in terms of the volume V and temperature T of the system (possibly, this expression includes also some parameters); then $P = -\partial F(V, T)/\partial V$. Apparently, limit (14) with the negative sign is called pressure because, for a lattice system, it is natural to replace the derivative of F with respect to volume by the increment of F obtained by adding one particle; this gives (14) (without the minus sign, if the increment, as well as the derivative in the case of a continuous system, is taken with the minus sign). However, I believe that pressure for lattice systems is not defined in statistical physics.

Secondly, $\beta = 1$ is usually taken. This corresponds to multiplication of the function $f(\xi)$ by a constant factor, which is not essential. (This becomes essential when, for some reasons, we need to consider a family βf with parameter β

rather than a single function f. In physics, such a need has solid grounds: we must bear in mind that the same system is to be considered at various temperatures. In DS theory, there are no reasons for paying special attention to the family βf; if such a reason arises (this happens sometimes), nothing prevents us from replacing f by βf.)

Thirdly (this is more serious), Z_{F_n} is defined as follows. The states S_n are elements of A^{F_n} (i.e., functions $F_n \to A$). We somehow extend every such function α to a mapping $\xi_\alpha \colon F \to A$ and set (cf. (13))

$$Z_{F_n} := \sum_{\alpha \in A^{F_n}} e^{-\sum\limits_{g \in F_n} E(g, \xi_\alpha)} = \sum_{\alpha \in A^{F_n}} e^{\sum\limits_{g \in F_n} f(\phi_g(\xi_\alpha))}. \tag{15}$$

In the exponent, the interaction between the "gth" particle (where $g \in F_n$) and all remaining particles of the "lattice," not necessarily corresponding to elements of F_n, are taken into account. From the point of view of statistical physics, we could take into account only the interactions between particles from S_n, considering the distinguished "piece" S_n as an isolated system. (Then it would be unnecessary to extend $\alpha \colon F_n \to A$ to $\xi_\alpha \colon F \to A$.) I repeat that, actually (under natural assumptions about interactions), this does not affect the final result and is not that important for us. The quantity Z_{F_n} depends on the particular choice of the extensions of α to ξ_α. But it turns out that, under any choice, the limit

$$P(f) = \lim_{n \to \infty} \frac{1}{\#F_n} \ln Z_{F_n} \tag{16}$$

exists and depends neither on the particular choice of ξ_α nor on the Følner system. The number $P(f)$ is called the topological pressure corresponding to the continuous function $f \colon \Omega \to \mathbb{R}$. (Certainly, it depends on G, on A, and, in a more general situation which we shall consider soon, on the DS under consideration; but usually, G, A, and the DS are assumed to be fixed, and they are not indicated explicitly in the notation of P.) Expression (11), as is explained below, also has a natural analogue ("per one particle") for a lattice system, in which the role of a finite distribution of probabilities $(p_1, \ldots, p_N)$ is played by a $\{\phi_g\}$-invariant measure μ on the phase space. It is natural to compare this analogue with $P(f)$ (that is, up to a sign, with free energy per one particle) and examine the measures for which these two values coincide (if they exist).

So far, we considered a Bernoulli topological DS (we used the Bernoulli property when taking a "piece" S_n and extending a state α to ξ_α). Now, we proceed to a general topological DS with an amenable group G and compact phase space X; we assume also that a continuous $f \colon X \to \mathbb{R}$ is given.

Let $\mathcal{U} = \{U_1, \ldots, U_l\}$ be a finite open cover of X. We set $\phi_g^{-1}\mathcal{U} := \{\phi_g^{-1}U_1, \ldots, \phi_g^{-1}U_l\}$ and

$$\mathcal{U}_{F_n} := \bigvee_{g \in F_n} \phi_g^{-1}\mathcal{U},$$

where the following notation is used: for several finite covers $\mathcal{U}_t$, where $t \in T$, $\bigvee_t \mathcal{U}_t$ denotes the cover whose elements are all nonempty intersections $\bigcap_{t \in T} U_{k_t}$ with $U_{k_t} \in \mathcal{U}_t$. We set

$$Z_{F_n, \mathcal{U}} := \sum_{U \in \mathcal{U}_{F_n}} e^{\displaystyle \sup_{x \in U} \sum_{g \in F_n} f(\phi_g x)}. \qquad (17)$$

$$P(f) := \sup_{\mathcal{U}} \lim_{n \to \infty} \frac{Z_{F_n, \mathcal{U}}}{\#F_n}, \qquad (18)$$

where $\sup_{\mathcal{U}}$ is taken over all possible finite covers of the space X. Such a definition, as opposed to (15) and (16), is no longer related to particular phase space and group action. At the same time, for a Bernoulli DS, it leads to the same number $P(f)$. (Definitions (17), (18) and (15), (16) differ mainly in that, for a Bernoulli DS, only one cover

$$\mathcal{U} = \{U_a; \ a \in A\}, \quad \text{where} \quad U_a := \{\xi; \ \pi_0(\xi) = a\} \qquad (19)$$

is used; note that

$$\mathcal{U}_{F_n} = \{U_\alpha; \ \alpha \in A^{F_n}\}, \quad \text{where} \quad U_\alpha := \{\xi; \ \xi \restriction F_n = \alpha\},$$

so the previous ξ_α belong to U_α. It turns out that, in the case under consideration, the least upper bound $\sup_{\mathcal{U}}$ involved in (18) is attained on the $\mathcal{U}$ from (19). In addition, the $\xi_\alpha \in U_\alpha$ on the right-hand side of (15) are arbitrary, and in the exponent in (17), the least upper bound over all such ξ_α is taken; it turns out that this does not affect the limit $\frac{1}{\#F_n} \ln Z_{F_n, \mathcal{U}}$ for given DS and $\mathcal{U}$.)

We have obtained an analogue of "free energy per one particle." Now, let us see what an analogue of (11) may be. The sum $\sum p_i \ln p_i$ is, of course, the negative entropy[50] of the distribution of probabilities $(p_1, \ldots, p_N)$. The analogue of this distribution for a general DS is an invariant normalized measure μ, and the analogue of entropy is the entropy h_μ mentioned in (b). In its turn, the analogue of the (average) energy per one particle is $- \int_X f \, d\mu$ (the minus sign

---

[50] In information theory, it is natural to use binary, rather than natural, logarithm. But in statistical physics, as well as in more analytical mathematical questions, natural logarithm is used; the corresponding values are only multiplied by some constant factor.

is used because it is used in the definition of f). Thus, the analogue of (11) taken with the inverse sign is

$$h_\mu + \int_X f \, d\mu \qquad (20)$$

(the entropy h_μ has already been discussed in (b)). In the special case of a Bernoulli DS, we could trace the passage from (11) to (20) in more detail, as for $P(f)$. But in the case of $P(f)$, this gave us a heuristic motivation for the definition (otherwise, it would seem that it was taken out of the blue), which is needed for neither term in (20). (In fact, something of this kind could be done for the definition of h_μ missing from (b). I have not done this because, for a DS with classical time, the definition of h_μ is known fairly well, and transferring it to the general case is straightforward.)

After everything said above, the reader, probably, will find the following theorem quite natural (still, it must be proved, and the proof is by no means easy):

$$\sup_\mu \left(h_\mu + \int_X f \, d\mu \right) = P(f), \qquad (21)$$

where the least upper bound is taken over all invariant normalized measures of the topological DS under consideration. As opposed to the simple case of a system with finitely many states, the least upper bound is not always attained at a unique measure, if attained at all. If the upper bound is attained, then the corresponding measures are called equilibrium measures (or equilibrium states, where the meaning of the term "state" is again statistical-physical rather than DS-theoretic). Relation (21) is known as the "variational principle for topological pressure." The study of related questions is the object of the so-called "thermodynamical formalism" for DSs.

I shall mention only one point. In the case of a Bernoulli DS with $G = \mathbb{Z}^m$, the existence of an equilibrium measure is proved for an arbitrary continuous function f, but there is a fundamental difference between the cases of $m = 1$ and $m > 1$. In the former case, an equilibrium measure for a Hölder function f is unique, and in the latter, it may be nonunique. For a lattice system, the nonuniqueness of an equilibrium measure corresponds to the well known (not only to humans but also to animals[51]) physical phenomenon of phase transformation.

For DSs with classical time, foundations for the thermodynamical formalism in the spirit of the approach sketched above were laid by P. Walters and

---

[51] Although, animals are familiar with phase transformation only in a continuous system (water); about phase transformation in lattice systems only humans have learned, when they began to study magnetic and electric phenomena in crystals.

D. Ruelle in the early 1970s; in particular, Walters and Ruelle proved the variational principle. This kind of activity was quite natural for Ruelle, the more so that he had done much work on mathematical questions of statistical physics before. For the same reason, he was also the first or one of the first to begin extending this theory to actions of $\mathbb{Z}^m$. Passage to general amenable groups G was performed by A. M. Stepin and A. T. Tagi-zade in [42].

There is a different approach to defining and studying interesting measures; it was suggested for DSs with classical time by Ya. G. Sinai, also about 1970 [43] (Sinai mentioned at once that this approach is suitable for nonclassical time too and, by way of example, pointed out that, for a Bernoulli DS, this approach gives measures arising in statistical physics). The corresponding measures are called Gibbs measures.[52] A Gibbs measure for a DS $\{\phi_g;\ g \in G\}$ in a compact space X is determined by a given normalized invariant measure μ_0 and a function $f \in L^\infty(X, \mu_0)$. These data are used to construct a sequence of measures μ_n absolutely continuous with respect to μ_0 and having "density" (i.e., Radon–Nikodym derivative)

$$\frac{e^{\sum\limits_{g \in F_n} f(\phi_g x)}}{\int_X e^{\sum\limits_{g \in F_n} f(\phi_g y)} d\mu_0(y)}.$$

The Gibbs measures are limit points of the sequence of measures μ_n (in the sense of weak convergence of measures). They may be noninvariant.[53] But if a Gibbs measure is a limit rather than merely a limit point, then it is invariant. The invariant Gibbs measures are equilibrium, while the converse does not generally hold. For a DS with classical time and hyperbolic behavior of trajectories, the converse assertion holds if the function f is "good."

In studying equilibrium and Gibbs measures for DSs with hyperbolic behavior of trajectories, the trick of coding is used; it allows us to represent (in a certain sense) the DS under consideration as a subsystem of a Bernoulli DS. Literally, a subsystem of a Bernoulli system is an invariant subset A in the corresponding Ω; in the topological context, it is natural to consider closed

---

[52] They indeed resemble the Gibbs distributions in classical statistical physics and, even more, the Gibbs DLR-measures for infinite systems of type $A^{\mathbb{Z}^n}$, which were introduced by Dobrushin, Lanford, and Ruelle. For a DS with classical time, the definition given below has already passed "checking by practice," but for nonclassical time, this definition is to be checked yet (except in the case known from statistical physics; but even in this case, various aspects of the relation between this definition and the DLR construction are to be clarified).

[53] In statistical physics, measures not being translation-invariant may be quite natural: it suffices to imagine that there is one phase in one half-space and a different phase in the other. This is just as possible as the situation where the entire space is filled with one phase and the corresponding measure is translation-invariant.

subsets A. I said "in a certain sense" because such an A is zero-dimensional, so a system with phase space of larger dimension cannot be topologically equivalent to the restriction $\{\phi_g \upharpoonright A\}$ of a Bernoulli DS to A. But it certainly can be a quotient system of such a DS, i.e., it can be obtained from it by a factorization modulo some equivalence relation invariant with respect to the DS. Sometimes, the set A and the factorization admit a fairly detailed description, and, although points of A are glued together under factorization, the "majority" of points are not glued together with anything. This trick (when it works) makes it possible to reduce a number of questions about the DS under examination to questions about $\{\phi_g \upharpoonright A\}$. Certainly, the success depends on whether we are able to find a "felicitous" coding. For systems with most manifest hyperbolic behavior of trajectories (such as the Anosov systems and basic hyperbolic sets), felicitous codings are related to so-called Markov partitionings, which were introduced by R. Adler and B. Weiss for hyperbolic automorphisms of the 2-torus and, almost simultaneously, by I. G. Sinai for Anosov systems; then, an improved construction suitable also for basic hyperbolic sets was suggested by D. Ruelle and R. Bowen [44].

Sinai, Ruelle, and Bowen not only constructed new invariant measures but also distinguished the cases where these measures are especially interesting. Afterwards, their approach[54] was transferred to other types of systems with somewhat (slightly) "spoiled" hyperbolicity, such as billiards, pseudo-Anosov surface homeomorphisms, or Lorentz-type attractors. As to actions of amenable groups, I am not aware of any advanced applications of this approach beyond the scope of lattice systems.

Although above we repeatedly emphasized the special position of amenable groups, in reality, works on ergodic theory (or on related questions of statistical physics) where the groups are not amenable are far from being exhausted by the few papers cited (or implied) above. There are a number of works in which some additional structures are used, unlike in works on amenable groups. For example, we can fix generators of the group[55] or some partitioning or cover of the phase space. (Even for $\mathbb{Z}^m$, we started the consideration of a Bernoulli DS with employing a fixed covering (19) (which is simultaneously a partitioning in the case under consideration); to be more precise, it was so natural at that

---

[54] Strictly, their approaches do not completely coincide formally: Bowen did not introduce a general notion of a Gibbs measure at all (this notion *per se*, as well as the notion of an equilibrium measure, is not related to coding), he simply used this name for the measures that he constructed for the systems which he considered. But altogether, from a broader point of view, this is the same approach.

[55] Even the above-mentioned modified method of averaging over a group suggested by R. I. Grigorchuk formally depends on the choice of group generators. It is not known whether the result depends on it in the general case (but in the ergodic case, it does not).

point that we did not even mentioned it first. We began to talk about other partitionings only after proceeding to more general group actions.) However, at present, it would be difficult to characterize studies of this kind by some general features.

I am not aware of new surveys on the topic of Section 1.3. The state of the art in the beginning of the last quarter century is surveyed in [45].

1.4   Bifurcations

During the past quarter century, the appearance and role of one of the fields of the theory of smooth DSs – bifurcation theory – have changed. The term "bifurcation" in its literal meaning is used in, e.g., anatomy (bronchus bifurcation). In mathematics, this term has the broader meaning of a qualitative change of the objects under consideration caused by a variation of the parameters on which these objects depend. More precise general formulations cannot be given, because the objects and their properties of interest to researchers are diverse. Precise formulations refer to particular problems.

Initially, bifurcations in mathematics were considered in relation to equilibrium figures of a rotating fluid. Namely, consider the problem: under what conditions a body consisting of a homogeneous fluid on whose particles only the mutual attraction forces implied by Newton's gravitation law act can rotate as a solid body? The corresponding figures are said to be equilibrium. The only known exact solutions to this problem are certain ellipsoidal figures (the MacLaurin and Jacobi ellipsoids) and annuli, but it is known also that there exist other figures close to those mentioned above. These figures were revealed with the use of bifurcation considerations. Namely, take an ellipsoidal equilibrium figure E_λ continuously depending on some parameter λ on which the problem under examination depends; it turns out that, when λ passes some value λ_0 (it is said to be bifurcation value), there arises a new (not necessarily ellipsoidal) equilibrium figure E'_λ, which is the closer to E_λ the closer λ to λ_0; thus, we can say that the family of figures E'_λ "branches off" from the family E_λ at $\lambda = \lambda_0$. In this case, the meaning of "bifurcation" is fairly close to its literal meaning: as the parameter increases, E_λ, so to speak, forks into E_λ and E'_λ. Analytically, the question reduces to studying a certain integral equation;[56] thus, it is natural that the "bifurcation" terminology has found common use in studying general integral equations depending on a parameter. Since a prominent figure in the theory of equilibrium figures of rotating fluids was H. Poincaré, it is no surprise that he carried over this terminology to the qualitative theory of ordinary differential equations; simultaneously, he started to apply it in a

---

[56] It is fairly unusual: the integrand is known, while the domain of integration is not.

broader sense, to any qualitative changes.

R. Thom suggested to use the term "catastrophes" instead of "bifurcations." This word is also not to be understood literally. For example, the following problems were seriously considered in catastrophe theory: violation of the stability of an elastic construction (this may be a catastrophe indeed); the formation of a bright line on the bottom of a brook by solar rays refracting in water (this can hardly excite somebody, except probably children who see this for the first time). As a possible (but not brought into the form of a mathematical model[57]) example of a catastrophe, an abrupt change in the course of a disease, after which the patient recovers very quickly, is mentioned sometimes; even if this event is a catastrophe, it is such only for bacteria.

As "catastrophe" is a synonym for "bifurcation," we may ask which term is better. As is clear from the aforesaid, neither term is to be understood literally. But the word "catastrophe" belongs to the common (literature and colloquial) language and has a definite, emotionally colored, meaning, while the initial meaning of the word "bifurcation" is known to much fewer people, and even these people hardly associate it with some emotions. Therefore, for science, the neutral word "bifurcation" is suited better; "catastrophe" is more appropriate for mass publications.

The very rich-in-content mathematical ideas of R. Thom about the singularities of smooth mappings and bifurcations of critical points of functions (Thom continued the pioneering work of H. Whitney but went significantly further) are far beyond the scope of this paper (neither the date nor contents fit). The later works on this topic do not fit either (at least in contents). Their direct applications to the theory of dynamical systems are as follows. We consider a system whose some variables (say, y; this is generally a vector rather than a number) vary rapidly and the others (say, x) vary slowly (such systems occur in reality and are fairly important); it is assumed that, when x is fixed, the fast variables y satisfy the gradient system

$$\frac{dy(t)}{dt} = -\nabla f(x, y)$$

(such a situation occurs too, but more rarely). Thus, $y(t)$ rapidly approaches a stable equilibrium point of this system, i.e., a critical point (namely, local minimum) of f regarded as a function of x. The further motion is such that $x(t)$ gradually changes, and the corresponding critical point of the function $y \mapsto f(x, y)$, which coincides with $y(t)$ almost precisely, changes too. At some

---

[57] To be more precise, mathematical models of such a phenomenon have been suggested, but, as far as I know, they were studied only numerically, without reference to catastrophe theory.

moment, the critical point bifurcates; say, it merges with another critical point and disappears, after which $y(t)$ must tend to a different critical point. Clearly, in such a situation, results about the bifurcations of critical points of functions are very important for understanding the qualitative picture, but, at this level of approximation, we shall restrict ourselves to fairly trivial and more or less direct references to these results. Next, it is claimed that, even not knowing the differential equations but only assuming that they are of the character described above and observing the changes in the real "physical" system,[58] we can make qualitative conclusions about the singularities of the function f and, thereby, understand the most essential properties of the system under consideration. In the end, some hypothetical interpretation of the experimental data is suggested. There are no other grounds (it is easy to believe that there exist fast and slow variables, but what economic, psychological, or social laws do allow us to think that the fast variables must move along the gradient of some function? if they do, then this function itself has, apparently, some economic, biological, psychological, or sociological meaning). Everything this is in a suspiciously natural-philosophic style,[59] and natural philosophy became factually obsolete even in Newton's times, although it flourished for over a century longer.[60]

---

[58] It may also be economic, biological, and even (as I heard) psychological or sociological. For systems that come from physics, their mathematical models or, at least, general features of such models are usually known to a certain extent. In these cases, the application of catastrophe theory does not cause doubts.

[59] Certainly, in the old times, the questions of psychology, economy, and sociology were not referred to natural philosophy; this is clear even from the appellation. But I am telling about the style only.

[60] As I have touched upon this matter, I shall mention the following. There is a well-known trick for illustrating interrelations between logical notions, the Euler–Venn disks. Suppose that we want to illustrate the interrelation between the notions of domestic animals, mammals, cats, and dogs. We draw two partly overlapping disks (which may be ovals, if it is more convenient for drawing) – "mammals" and "domestic animals"; inside the first disk, we draw two disks, "cats" and "dogs," which do not overlap, but each of them partly intersects "domestic animals." Nobody thinks that the set of animals is two-dimensional in a certain sense and that the four sets of interest to us are indeed disks or ovals. Could it then happen that, sometimes, the pictures drawn by catastrophers can be interpreted as conventional images of some interrelations, which are not necessarily related to the special models under consideration (in the likeness of the Euler–Venn disks, for which it is inessential that these are disks and that they are drawn in the plane)? This can be assumed if the factual aspect of the matter in economic and other questions is indeed such as described by catastrophers; on the other hand, the models in these cases are doubtful, as the critics of catastrophe theory say. Thom believed that catastrophe theory gives a new language of forms and that more complex systems and interrelations between them are, as it were, constructed from elementary blocks described according to what was said above by means of special systems with special singularities from a certain list. (*A propos*, the list has turned out to be incomplete.) But if there are no grounds for assuming that the physical system under consideration is indeed

The booklet [46] of V. I. Arnold contains several pages where this subject is described in somewhat more detail but from the same critical positions as here.[61] It also contains a bibliography. The popular paper [48] is written from different positions; the attitude of its author to the nonphysical use of the "catastrophic" ideology is positive. Finally, the authors of book [49] considered diverse applications of catastrophe theory, from those not provoking essential objections to much more "natural-philosophic" (in style) ones, on the basis of the information from the theory of singularities which is thoroughly but elementarily described in the first part of this book.

What is said above does not imply that the theory of singularities of smooth mappings and bifurcations of critical points of functions made a small contribution in DS theory and that its contribution rather directly used achievements of topologists. The conceptual influence of the former theory on the latter turned out to be significant; the role of a herald of this influence was played by V. I. Arnold about 1970. It should be mentioned that Thom claimed from the very beginning that the catastrophe theory as described above was only the first, elementary of part some more extensive theory (which would surely be universal). But he never made this general declaration more specific, and Arnold's point of departure was the concrete factual contents of the theory of singularities of smooth mappings rather than this uncertain declaration (incantation). He never proclaimed that this approach was universal (at the moment of his talk, some facts eliminating such a universality had already been known), but he reasonably pointed out that it had a fairly large, although not unlimited, domain of application, and that the boundaries of this domain could be outlined fairly precisely in some cases. A kind of the first manifesto of the new trend was his paper [50], where he explained that, to the theory of local bifurcations of DSs, a number of notions initially emerged in the theory of singularities or, more generally, in smooth topology (such as codimension, stratification, transversality, universal and versal families, moduli and their number, bifurcation diagrams, and finite definiteness) can be carried over in a natural way.

I shall briefly describe only one (probably, the simplest) idea advertised in [50]. Nonremovable degenerations of codimension k occur only in k-parameter families;[62] therefore, it is expedient to consider such equilibrium positions

---

such as described above (i.e., with slow and fast variables, the latter being partly unknown and varying along a gradient), how could these blocks and pictures be interpreted otherwise?

[61] Of course, the main contents of this booklet are the notions of the theory of singularities, its applications and history, and the earlier history of the theory of DS bifurcations, rather that a critical discussion of catastrophe theory. Arnold says even less about catastrophe theory and more about the other things in his survey [47].

[62] This means that there exist k-parameter families of DSs such that, for these families and

(i.e., those which are codim-k-degenerate) in an appropriate bifurcation context.[63] (Though, there may be other reasons for the occurrence of degenerations of fairly high codimensions; first of all, these may be the presence of symmetries. A set of equilibrium positions having a high codimension in some class of "general" DSs may be a set of equilibrium positions "with symmetries" whose codimension in the class of DSs with these symmetries is small or even zero. But can we always be confident of the exactness of these symmetries, especially if they are not consequences of laws of nature? Could not the symmetries be slightly violated in reality, and should not we be curious about what happens under such a violation?[64]) Thus, the theory of local bifurcations is not an exterior supplement to the local qualitative theory but its substantial part. Even this mere change of a point of view (although not quite new, it had never been precisely outlined on such a large scale before) substantially affected the appearance of not only bifurcation theory but also the entire local qualitative theory.

Certainly, in [50], only a part of the program could be accomplished. (Although, at the time when [50] was written, Arnold's group has already done a certain amount of work in this direction. In addition, in [50], it was shown how some earlier significant results fit into the new theory.[65]) After [50] had appeared, Arnold's group and some other mathematicians did much work on a particular implementation of the new approach. For plane flows, typical local bifurcations[66] in two- and three-parameter families were studied in almost as much detail as local bifurcation in one-parameter families were studied earlier. (Here we need to note a book of F. Dumortier, R. Roussarie, J. Sotomayor, and H. Żoladek. *Bifurcations of Planar Vector Fields* (Berlin: Springer, 1991).)

In the cases of larger dimension, the picture is far from being exhaustive

---

for all families sufficiently close to them, the DS corresponding to a certain parameter value has a degeneration of the type under consideration. (Closeness is understood as proximity in the sense of C^m, where m depends on the particular type of degenerations.) At the same time, given a $(k-1)$-parameter family, we can always obtain a family of DSs none of which has degenerations of the given type by an arbitrarily small perturbation.

[63] Undoubtedly, the authors of classical works (that appeared in the period $(-\infty, t_{[50]})$) in self-explanatory notation felt this (as they felt many things related to the new paradigm). But they gave explanations (if any) only as applied to problems of interest to them and to the cases of small codimension considered in these problems. This largely sailed past the mind of the broader community.

[64] Some work in this direction has been done both in the special context of catastrophe theory (see the information in [49]) and beyond this context.

[65] A part of this material was exposed at the end of Arnold's textbook [51] in more detail.

[66] That is, bifurcations related to objects of the local qualitative theory, such as equilibrium points and periodic trajectories of flows or fixed and periodic points of diffeomorphisms.

yet, but still, it is fairly complete, because the question about local bifurcations in such families is largely reduced to a similar question for plane flows.[67]

Along with local bifurcations, "global" bifurcations can be considered; these are bifurcations that change the phase picture as a whole and are not localized near equilibrium points or periodic trajectories. In most cases, it would be more correct to call them semilocal, because when we study such bifurcations, we pay attention not only to what is going on near some "local" object under a perturbation but also to what happens in some region of the phase space ("far away" from the local object, if a certain local object plays a role), usually in a neighborhood of an invariant set of the unperturbed system that does not reduce to an equilibrium point, and so on. However, a total control of the entire phase space is generally out of the question, so the term "global bifurcation" is, perhaps, an exaggeration. On the other hand, the classical papers by A. A. Andronov and his collaborators (the works went back to pre-war times, although some of them were published later) considered the change of the entire "phase portrait" of a flow on the entire phase plane, i.e., very true global bifurcations. Nevertheless, although the problem goes back to the old questions mentioned above, apparently, only in the last 25 years, or thereabouts, it has been eventually determined what must be added to the previous information about local or semilocal bifurcations in order to obtain a complete description of the changes of the global qualitative picture for bifurcations in typical one-parameter families for plane flows.

The results related to these (local and semilocal) questions obtained by the mid-1980s are presented in [52]; I am not aware of newer publications containing any extensive summaries of results. For typical two- and three-parameter plane flows, the passage from local and semilocal bifurcations to "truly" global ones is not studied as far as I know. For such flows, after [52] was published, bifurcations of so-called "polycycles," which generalize the bifurcations of closed separatrices, were studied in detail [53].

Another significant change in bifurcation theory is connected with the study of such semilocal (and, sometimes, global) bifurcations in dimensions > 2 for diffeomorphisms and > 3 for flows related to the complex behavior of trajectories.[68] This direction is new in principle. It is paid due attention in [52], but

---

[67] In a number of cases, the use of central manifolds automatically yields a complete reduction to plane flows, but sometimes, the central manifold has dimension 3 or 4; in these cases, the situation is more complex. Still, there is an artificial method which gives a reduction (although not so complete) to the two-dimensional case.

[68] Of course, semilocal bifurcations more or less similar to the corresponding bifurcations for plane flows but not related to the complex behavior of trajectories were studied too. Studying them is a necessary component of the theory (and they are mentioned in [52]), but they are not that interesting for us.

the further development seems to be more significant. Unfortunately, I must repeat what was said on a different occasion, namely, that there are no newer surveys comparable with [52] in completeness.

By the early 1970s, dynamical systems with complex behavior of trajectories had been more or less successfully studied in the cases where this behavior was determined by hyperbolicity. Even then (and much more now) the hyperbolic theory was complete to a certain extent, although there still remain unsolved problems, and papers on this topic appear even nowadays. The balance has definitely shifted to bifurcations. This is because of the discovery of a complex of phenomena related to nontransversal homoclinic trajectories. These phenomena are so numerous and diverse that their study is far from complete, even in the first approximation. The first to notice the fundamentally new bifurcation phenomena related in a certain way to nontransversal homoclinic trajectories was L. P. Shil'nikov, who mentioned them in his joint paper with N. G. Gavrilov [54] in 1972 (and, partly, in 1970). Specifically, in [54], four different types of bifurcations, one of which can occur on the boundaries of systems with simple dynamics (Morse–Smale systems), were discovered. For Morse–Smale systems, a symbolic description of the arising hyperbolic sets was given and secondary bifurcations under which stable periodic trajectories are born and die were specified. The closer the parameter value to the critical point, the more such trajectories can arise.[69]

After that, a possibility to study the complex behavior of trajectories in some situations beyond the scope of "pure" hyperbolic theory took shape. It is hardly an overstatement to say that, at present, bifurcation theory has become the main (although not unique) source of examples of DSs with complex behavior of trajectories that survive (in a certain sense) small perturbations (at least "many" of them). For comparison, note that, in the preceding period, hyperbolic sets were discovered independently of bifurcation theory; the same refers largely to the Lorenz attractor mentioned below. It was discovered in studying the corresponding DS with a wide range of parameter values, which revealed various bifurcations in such a system, including those related to the

---

[69] In [54], a three-dimensional flow was considered (in fact, this covers the case of a two-dimensional diffeomorphism). Multidimensional analogues of the situation studied in [54] were considered in [55–57]; in the last two papers, the study goes further than an account of analogues of the results from [54]. In [54], as in a number of other papers on this topic, it is assumed that the first-return map for the initial periodic trajectory is smoothly conjugate to a linear mapping (although conjugacy with a linear mapping was used there only for stable periodic trajectories birth, all other results were obtained without this assumption); in [55–57], this assumption is not made. Note that in studying homoclinic bifurcations, the assumption of smooth linearization simplifies technicalities, but it also often restricts the scope of problems to be considered, even if, actually, for some of these problems results do not depend upon the assumption.

Lorenz attractor; but the properties of this attractor are considered irrespective of bifurcations.[70]

At approximately the same time, S. Newhouse paid attention to nontransversal homoclinic trajectories of hyperbolic sets that do not reduce to periodic trajectories [59] (first, he considered such trajectories only for two-dimensional diffeomorphisms). Later, he called a hyperbolic set A (which is usually assumed to be locally maximal[71] and topologically transitive) which does not reduce to periodic trajectories and has a nontransversal homoclinic trajectory (thus, the unstable and stable manifolds of some trajectory from A are tangent to each other somewhere) a wild hyperbolic set (the existence of such a set in a DS is referred to as wild hyperbolicity). The term "wild" suggests that such a set is related to a number of unexpected bifurcation phenomena. First, Newhouse found an example in which wild hyperbolicity is preserved under arbitrary small perturbations[72], although the set A is Cantor-like; thus, between the stable manifolds of it trajectories, there are strips containing no such manifolds, and seemingly, all arcs of unstable manifolds of the trajectories from A that have unsuitable directions (between these manifolds, there are "gaps" too) could be moved to these strips by small perturbations. In A, the periodic trajectories are dense, and it can be shown that they have nontransversal homoclinic trajectories for an everywhere dense set of parameter values. Thus, Newhouse revealed that the space of DSs contains regions in which DSs with nontransversal homoclinic trajectories are everywhere dense. They are known as Newhouse regions now. Newhouse regions in typical finite-parameter families of DSs (the meaning of this term is evident)[73] are considered too.

---

[70] The aforesaid refers partly to the very important Henon attractor, which was introduced independently of bifurcation theory. However, later, it turned out to be closely related to this theory: first, it arises (and plays an important role) in certain bifurcations; secondly (historically, this was discovered earlier), the description of its properties is obtained at certain values of the parameters on which it depends; undoubtedly, the properties may be different at other parameter values, although (so far?) this has not been studied in detail. I do not talk about the Henon attractor here for the only reason that it is considered by Yoccoz in [1] (see also his report [58] at the Bourbaki seminar).

[71] This means that A is a maximal invariant set in some of its neighborhood. Locally maximal invariant sets (not necessarily hyperbolic) are also called isolated.

[72] In this section, I rather describe results than give precise statements. In particular, the smallness of perturbations is understood in the sense of some C^r, but I say nothing about this r (in may be different in different cases)

[73] I must warn the reader that the precise statements which refer to domains in the space of DSs and to domains in families of DSs are somewhat different. In particular, many results for regions in a space of dynamical systems are correct, or they are obtained under the assumption of smooth linearization, and to the contrary, they lose their generality for regions of families. Since I only approximately describe these results, I ignore the differences as technical.

The further studies were directed both toward discovering Newhouse regions in various situations and toward studying the properties of DSs from such regions. Both questions were stated explicitly in [60], where it was asserted that (i) under certain conditions, near a system with homoclinic tangency, there exist Newhouse regions; (ii) in such domains, there exist DSs with countably many stable periodic trajectories, and the number of such DSs is large in a certain sense. (Obviously, this assertion by itself is stronger and more surprising than that the number of such solutions increases as the parameter approaches the bifurcation value (this was discovered in [54]), but combining the result of [54] with (i) makes it not that surprising).

After Newhouse, bifurcations related to nontransversal homoclinic trajectories became popular throughout the world. However, in [60], assertion (i) was rather guessed (while (ii) could be considered more or less proved, although under assumptions that turned out to be too restrictive afterwards). A very substantial progress was made in the fundamental paper [61] of Newhouse (see also C. Robinson's paper [62], which supplements [61]). After the appearance of [61, 62], the assertions about two-dimensional diffeomorphisms stated above could be considered mainly proved and even improved thanks to the removal of redundant conditions (although, apparently, some of the later publications may be regarded as final elaborations of questions going back to the mid-1970s). Papers [63, 64] are concerned with some multidimensional analogues of the same circle of questions (the conditions imposed in [63] were weakened in [65]).

In parallel, a number of other questions about nonlocal bifurcations were studied. The qualitative picture in such problems is so complex that it is doubtful that this picture at fixed parameter values and its change under parameter variation can be completely described. In particular, systems in Newhouse regions do not admit complete description of dynamics in finite-dimensional families, since such regions contain dense sets of systems with arbitrary highly degenerate periodic trajectories and homoclinic tangencies of any order. Descriptions (somewhat incomplete) are often given only for some sets of parameter values; certainly, they are especially interesting when these sets are "significant" in some sense. In this respect, the recent papers often cannot in principle pretend to give a description of the qualitative picture as complete as that given in many cases by the theory of plane flows or of systems with uniformly hyperbolic behavior of trajectories. Naturally, in describing changes of the qualitative picture, only some essential features are often considered. In this respect, the completeness of the results on complex bifurcations is far from that inherent in the theory in Andronov's time. But this, apparently, is related to the essence of the matter. A drawback of the current state of the art is the absence of concise general specifications of what features of the qualitative picture are interesting. These features are specified in particular problems, but they are different in different contexts. It seems that the time of generalizing statements has not come yet.

The state of the art in the theory of nonlocal bifurcations in the mid-1980s is described in [52]. Yoccoz's report [1] contains several newer results and a bibliography. I draw special attention to the book of J. Palis and F. Takens [66] cited therein. In addition, I would like to mention several new papers of the group of L. P. Shil'nikov, because the works of this group (including the earlier ones) are not known well enough.[74]

The title of [67] refers to the bifurcation phenomenon when some periodic trajectory unboundedly lengthens and, in the limit, disappears; moreover, it does not leave a visible trace, as a separatrix loop or something like that (it "disappears in the blue sky"; hence the term "blue-sky catastrophe" for such bifurcations (apparently, for the first time, it was used as a joke)). For two-dimensional flows, this phenomenon was discovered by F. Fuller (who did not name it) and studied in more detail by V. S. Medvedev (under the name of "blue-sky catastrophe," which had been thought up already at that time). In [67], the same bifurcation was considered in the three-dimensional case. It had codimension 1, i.e., was typical, and occurred when passing through a certain hypersurface in the space of all systems. Simultaneously, a different bifurcation of codimension 1, maybe more interesting, was described.

It is of obvious interest to consider the question: How can an attractor[75] with complex structure arise in a smooth system of differential equations? This question is related to the following observation. So far, attractors of the type of basis sets of A axiom flows (not reducible to periodic trajectories) have arisen in no applied problems. It is therefore natural to try to construct such attractors by bifurcation methods. In [67], the following related problem is solved: How to obtain a field with a strange attractor[76] from a simple (Morse–Smale)

---

[74] In particular, [66] contains only one reference to an earlier paper of Shil'nikov, which was written before his main works on bifurcation theory were done (although, it goes beyond the scope of uniform hyperbolicity).

[75] This term is used for sets which, as it were, attract close trajectories. There are various formalizations of the property of attracting trajectories in the literature. In this paper, by an attractor, we mainly understand a compact invariant set A which is stable in the sense of Lyapunov (i.e., any neighborhood V of this set contains another neighborhood W such that all positive semitrajectories beginning in W never leave V) and such that all positive semitrajectories beginning in some neighborhood of A approach A arbitrarily closely with increasing time. (This is an almost word-for-word paraphrase of the definition of the asymptotic stability of an equilibrium point given by A. M. Lyapunov.)

[76] As far as I can judge, "strange" and "chaotic" attractors are not exact terms; rather, these are somewhat indefinite names (cf. the "fractals" of Mandelbrot), which in addition

vector field by means of a simple bifurcation of a smooth vector field? In [67], a bifurcation of codimension 1 leading to the occurrence of a hyperbolic strange attractor supported on a Smale–Williams solenoid was revealed and studied. Interestingly, when approaching the bifurcation hypersurface, the solenoid suffers no bifurcations, and the length of any closed trajectory in it tends to infinity.

Another example of the occurrence of a strange attractor under a bifurcation was considered in [68]. This attractor exists in some interval of perturbation parameter values and contains a wild hyperbolic set at all parameter values from this interval, which ensures plenty of bifurcation phenomena.[77]

In [56, 57, 69, 70], it was shown that, in a Newhouse region, DSs with saddle-node periodic trajectories of arbitrarily high multiplicity are everywhere dense. Using the results of these papers, V. Yu. Kaloshin [71] showed that, in a Newhouse region, there exist DSs where the number $N_{\mathrm{per}}(T)$ of periodic trajectories with period $\leqslant T$ increases arbitrarily rapidly as $T \to \infty$. Moreover, for any function $f(T)$, the DSs for which $N_{\mathrm{per}}(T) \geqslant f(T)$ at sufficiently large values of T form a second-category set in this domain with C^n-topology (for any positive integer n); thus, the phenomenon under consideration is by no means exceptional. Actually, the result follows from the above-mentioned results on density in the C^r-topology in Newhouse regions of systems with countable sets of periodic trajectories of undetermined order of degeneracy, i.e., periodic trajectories of period T for which the corresponding Poincaré map looks like $\bar{x} = x + o(\|x\|^r)$. Thus, by perturbing the identity map in the C^r-topology, we can obtain any large $(> f(T))$ number of fixed points of the Poincaré map and, accordingly, periodic trajectories of period $\leqslant T$; these trajectories are structurally stable, and consequently they survive under small perturbations. This answers a question raised about 30 years ago. Although, for analytic systems, the question about the possibility of a superexponential (in T) growth of the number of peri-

---

carry a shadow of emotion (the initial surprise). As a rule, the term "strange attractors" is used for attractors (i.e., sets which, so to speak, attract trajectories) that are "complex" and "strange" in some sense; they are not manifolds or something like that (say, they are not composed of several "pieces" of manifolds). Chaotic attractors are characterized by the quasi-random behavior of their trajectories. Already in the early 1970s, V. M. Alekseev suggested to formalize "quasi-randomness" as the positivity of topological entropy. (But there are different (nonequivalent) formalizations (see [66]) which make it possible to treat some attractors with zero entropy as chaotic.) The chaoticity and strangeness understood in this sense are not synonyms: an attractor may be a manifold but have positive entropy; it may have zero entropy but be not a manifold and, in general, have more or less complex structure. That the majority of known strange attractors are chaotic is a different matter (this is because they are somewhat hyperbolic, although to a lesser extent than the true hyperbolic sets).

[77] Paper [66] contains references to earlier works of western authors, who also mentioned the emergence of strange attractors under certain bifurcations. Usually these attractors were Lorenz-type attractors (see below) rather than uniformly hyperbolic attractors.

odic trajectories with period $\leqslant T$ remains open. Subsequently, Kaloshin proved that, nevertheless, in a sense it is "typical" that the growth of $N_{\mathrm{per}}(T)$ is not much faster than exponential.

In [72, 73], it was shown that sometimes, in the presence of a nonrough heteroclinic contour, bifurcation phenomena similar to those related to non-transversal homoclinic trajectories may occur. But there are also significant differences. Thus, if the contour contains saddles with different "divergences," then Newhouse regions can exist arbitrarily close to the system considered such that these regions contain dense sets of systems having simultaneously count-able sets of saddle, stable, and unstable periodic trajectories.

The recent (and partly semipopular) survey of Shil'nikov [74] contains an additional information (with references) on bifurcations related to nontransver-sal homoclinic trajectories.

Yet another significant achievement related to semilocal bifurcations is the study of the bifurcations of the Lorenz attractor. This attractor, as its name suggest, was discovered by E. Lorenz in the course of a numerical experiment on a special third-order system, which was interesting because of hydrodynamic considerations. Lorenz's discovery had attracted attention only about 10 years later; moreover, applied (in the broad sense of the word) and pure mathematicians had become interested in it for different reasons. Applied mathematicians had become convinced of the real existence of strange (or chaotic) attractors; most of them were not aware of the hyperbolic strange attractors discovered by theoreticians, because such attractors did not arise in problems which come from the natural sciences. For mathematicians, the Lorenz attractor was inter-esting not merely as yet another manifestation of the already known possibility of weird behavior of trajectories but, on the contrary, as an object which, al-though having some properties close to those of hyperbolic attractors (applied mathematicians were most impressed by these properties), differ from them in other properties (it is these niceties that were interesting for mathemati-cians). The mathematical interpretation of Lorenz's results was initiated by J. Guckenheimer in a paper entitled "A strange, strange attractor." The word "strange" is repeated not only because the movie of S. Cramer suggests so, but also because (i) the attractor under consideration is strange in the sense that its trajectories have dynamically complex structure (they have countably many saddle periodic trajectories, continuum many Poisson stable trajectories, etc.), while being preserved under small perturbations (this resembles the uniformly hyperbolic attractors, which had already been known at that time but still were regarded as something strange by many); (ii) it is not uniformly hyperbolic (al-though it has certain hyperbolic properties); (iii) the intrinsic structure of this attractor, as opposed to that of uniformly hyperbolic sets, does not remain invariable under small perturbations; to be more precise, it continuously changes in any one-parameter family of generic systems with such an attractor. Thus, if we treat bifurcations as arbitrary qualitative changes, bifurcations occur for all values of the parameter of such a family, and this phenomenon is nonremovable in the sense that it is inherent in all sufficiently close families of DSs. Such a phenomenon *per se* is not new: it takes place for the old example of the family of flows on the torus that has the form $\dot{x} = 1$, $\dot{y} = \lambda$ in coordinates x, y mod 1. But, arbitrarily slightly varying this family, we can ensure that the bifurcations occur only when λ belongs to the complement to some open dense set (thus, from the topological point view, the bifurcation parameter values are exceptional[78]). At the time when the paper of Guckenheimer appeared, the nonremovability was not quite new either: a similar phenomenon was revealed before by S. Smale and R. Abraham for a different example,[79] but Guckenheimer described this phenomenon in a different situation.[80]

Because of what is said above, as applied to the Lorenz attractor, bifurcations are understood not as mere changes of its intrinsic structure but as changes that are substantial in a certain sense. Certainly, such are the bifurcations under which the Lorenz attractor appears or disappears, but there are also some other bifurcations of this kind. The particular specification of what qualitative changes are substantial is determined by a particular description of the Lorenz attractor and is somewhat tied to a particular situation in this sense.

A model of the Lorenz attractor was suggested by R. Williams, who developed Guckengeimer's approach. His model is often called the "geometric Lorenz attractor." In this model, we can study the intrinsic structure of a given attractor (including the behavior of the trajectories contained in it); it can also be proved that there do indeed exist third-order systems with such an attractor, and that the attractor survives under small perturbations. However, the system under consideration was subject to a number of constraints, which

---

[78] At this point, a collision between the topological and metric points of view, which was mentioned by A. N. Kolmogorov on a different occasion, occurs: if the families under consideration are sufficiently smooth, then the set of bifurcation values of the parameter always has positive measure. Similar questions refer to KAM theory and, as applied to the example under consideration, to the more special theory of M. Herman and J.-C. Yoccoz; I mentioned his report at the very beginning.

[79] Yet earlier, H.-C. Andersen described the princess on a pea, who tried to reduce herself to general position all night long in order to make the feelings that bothered her exceptional, but she did not succeed.

[80] It is clear from what has been said about Newhouse regions that this phenomenon is observed for these regions too. But, as mentioned, the results on Newhouse regions obtained by the mid-1970 were rather guessed than rigorously proved.

might hold in some systems but did not hold in the initial Lorenz system; thus, strictly speaking, it remained unclear whether Lorenz had actually dealt with the attractor named after him (i.e., whether the attractor arising in his system is the geometric Lorenz attractor as described by Guckenheimer and Williams).

These questions were answered in [75], where the conditions ensuring the existence of a Lorenz attractor were relaxed to geometric condition on a certain first-return map. The verification of these conditions for the Lorenz system itself still relies in part on a numerical experiment, but such an experiment has been performed many times by various groups of mathematicians with the use of different programs, so the results do not cause doubts.[81] The intrinsic structure of the attractor considered in [75] is the same as that of the attractor considered by Williams (only some changes under certain bifurcations are unlike), but the properties of the system near the attractor are different (the attraction of trajectories to the attractor may be weaker). In subsequent works of other authors (largely concerned with the ergodic properties of the Lorenz attractor; the other problems have mainly been solved[82]), the Lorenz attractor was understood in the sense of [75]. In addition, in [75], its intrinsic structure was studied in more detail (in particular, its symbolic description well suited for bifurcation problems was given).

For us, it is important that, in [75], not only the preservation of the Lorenz attractor under small perturbations and the constant variation of its structure was confirmed for the model under consideration, but also the scenario of the birth of such an attractor, i.e., a sequence of bifurcations leading to its birth, was described, and bifurcations occurring after its birth were considered; in particular, the scenario of the destruction of the Lorenz attractor was presented.

A later article by Tucker (W. Tucker, A rigorous ODE solver and Smale's 14th problem. *Found. Comput. Math.*, **2** (1) (2002), 53–117) describes a numerical experiment for the Lorenz system with rational arithmetic and granted error estimate.

However, strangely enough, this (automatically irreproachable, it would seem) paper raises some doubts. According to Tucker, he succeeded in verifying the original conditions from Guckenheimer's paper for the Lorenz system (with appropriate parameter values), and, accordingly, the attractor turned out

---

[81] I emphasize again that the truth of the fundamental fact of the existence of a DS with a Lorenz attractor had became clear as soon as the first publications of J. Guckenheimer and R. Williams appeared; simple special systems of differential equations for which a Lorenz attractor is born under some bifurcation are given in the papers cited in [66]; neither result is related to numerical experiments. Numerical experiments are needed "only" to verify the presence of a Lorenz attractor in the system considered by Lorenz.

[82] See [76] for details.

to be not only qualitatively, but even quantitatively the same as in the Williams model; however, Tucker did not check the later and less restrictive condition of Shil'nikov (which he does not even mention), although, according to previous numerical experiments, it is accepted that the Lorenz system satisfies only the Shil'nikov conditions.

Investigations of the Lorenz attractor were also carried out by means of reliable computations in the following papers:

K. Mischaikow and M. Mrozek. Chaos in the Lorenz equations: A computer assisted proof. *Bull. Amer. Math. Soc.*, **32** (1) (1995), 66–72.

K. Mischaikow and M. Mrozek. Chaos in the Lorenz equations: A computer assisted proof. Part II. The details. *Math. Comput.*, **67** (223) (1998), 1023–1046.

K. Mischaikow, M. Mrozek, and A. Szymczak. Chaos in the Lorenz equations: A computer assisted proof. Part III. Classical parameter values. *J. Diff. Equations*, **169** (1) (2001), 17–56.

The results of the last three papers are not as complete as those of Tucker, in the sense that (as far as I can judge) they do not discuss the coincidence of this attractor with the geometric attractor in the sense of Guckenheimer–Williams, but only prove that several properties (actually the most interesting ones) of the latter coincide with those of the former.

2 "Named" problems

2.1 Structurally stable systems

Even the simplest examples show that the qualitative properties of DSs may change or remain intact under arbitrarily small perturbations. A trivial example is a DS in which no motion occurs, when the identity diffeomorphism or a flow with zero phase velocity field is considered. Certainly, arbitrarily small perturbations may then give different qualitative pictures. A less degenerate example is when the diffeomorphism has a fixed point or the flow has an equilibrium point (at which the corresponding vector field vanishes) that is asymptotically but not exponentially stable. Then, under an arbitrarily small (in the sense of C^1) perturbation, the fixed (equilibrium) point may disappear at all, or it may happen that the point does not disappear but stability is violated. If the fixed point is exponentially stable, then it does not disappear and remains asymptotically stable under a C^1-small perturbation; the same refers to an exponentially stable equilibrium point. In this case, the local qualitative picture (near the fixed or equilibrium point) is preserved. The cases where the qualitative picture in the entire phase space survives small perturbations deserve special attention. In this connection, the following definition was introduced (the idea was due to A. A. Andronov and L. S. Pontryagin (1937)).

A diffeomorphism f of a smooth closed manifold M is said to be *structurally stable* (the literal translation of the original Russian term is "rough") if, for any diffeomorphism g sufficiently C^1-close to f, there exists a homeomorphism $\chi \colon M \to M$ that conjugates f and g in the sense that

$$\chi \circ f = g \circ \chi. \tag{22}$$

A flow $\{\phi_t\}$ on a smooth closed manifold M determined by a smooth phase velocity field $\mathbf{v}$ (so that $\frac{d}{dt}\phi_t(x) = \mathbf{v}(\phi_t(x))$) is said to be structurally stable if any vector field $\mathbf{w}$ sufficiently C^1-close to $\mathbf{v}$ determines a flow $\{\psi_t\}$ equivalent to the flow $\{\phi_t\}$ in the sense that there exists a homeomorphism $\chi \colon M \to M$ that maps the trajectories of the first flow to trajectories of the second flow and preserves the direction of motion on the trajectories. It is worthwhile to mention some special features of this definition (see [77] for more details). In the case of continuous time, it is not required that the homeomorphism χ conjugates the unperturbed and perturbed flows in the sense that

$$\chi \circ \phi_t = \psi_t \circ \chi \quad \text{for all } t \tag{23}$$

126

(which may seem to be the natural analogue of (22)). The point is that, if a flow has a closed trajectory, then the period of this trajectory may change under perturbation, and (23) is then impossible; at the same time, we do not regard a change in the period as a change of the qualitative picture. It is not required that χ is a diffeomorphism either, because, if the diffeomorphism f has a fixed point (which must be preserved under small perturbations for a structurally stable diffeomorphism), then the eigenvalues of the corresponding matrix of linear approximation may change under a perturbation, while if χ in (22) were a diffeomorphism, this could not occur. For flows, similar considerations are somewhat harder to formulate (because we require equivalence rather than the fulfillment of (23)), but the conclusion is the same – χ cannot generally be a diffeomorphism. The definition of Andronov and Pontryagin required in addition that the homeomorphism χ be C^0-close to the identity homeomorphism whenever g is sufficiently close to f or $\mathbf{w}$ to $\mathbf{v}$. Later, M. M. Peixoto suggested to remove this condition; thus, there exist two logically different versions of structural stability, in the sense of Andronov and Pontryagin and in the sense of Peixoto. The former is formally more restrictive than the latter, but it has been proved that, actually, these versions are equivalent. Thus, I shall not distinguish between them, although their equivalence is a very nontrivial fact (I shall talk about this further on). Instead of "structural stability," the term "roughness" is used sometimes (especially in Russia).

After A. A. Andronov and L. S. Pontryagin had introduced the notion of a structurally stable system and characterized structurally stable flows on the plane (to be more precise, on the two-dimensional sphere) in their classical work, the problem of a qualitative characterization of the behavior of trajectories in structurally stable systems in other cases naturally arose. The definition said what happens under perturbations; the characterization which I mean was concerned only with the behavior of the trajectories of the unperturbed system. M. M. Peixoto transferred the Andronov–Pontryagin theorem to flows on closed surfaces; the statement remained almost unchanged. In all these cases, structurally stable flows form an open everywhere dense set in the space of all flows with the C^1 topology. Naturally, there arose the question of what systems are structurally stable in multidimensional cases (for systems with discrete time, this question was interesting even in the two-dimensional case[83]). The very direct generalization of the structural stability conditions for two-dimensional flows leads to the so-called Morse–Smale systems; by analogy with this case, we might conjecture that the structurally stable systems form an open everywhere

---

[83] The structurally stable diffeomorphisms of the circle were described by A. A. Mayer shortly after the appearance of the paper of Andronov and Pontryagin. They also form an open everywhere dense set in the space of all C^1-diffeomorphisms.

dense set in the space of all DSs on the manifold under consideration with C^1-topology. But these conjectures turned out to be false, except the conjecture for the Morse–Smale systems. The fairly dramatic story of the study in this domain was described by many authors, including myself (see below); therefore, I shall proceed to the answer at once. In the course of the "hyperbolic revolution," S. Smale conjectured that structural stability is equivalent to the hyperbolicity[84] of the set of nonwandering points (this is the main condition), the density of the periodic points in this set (the combination of these two conditions is called the Smale A axiom), and the transversality of the intersections of the corresponding stable and unstable manifolds (this is the strong transversality condition). The sufficiency of these conditions was proved (by R. C. Robinson in complete generality) in the end of the preceding two decades, and their necessity was proved only now, although a very important step was made long ago by C. Pugh, who proved a seemingly simple but actually difficult closure lemma. It asserts that, if a given smooth system has a nonwandering point x, then x can be made periodic by an arbitrarily C^1-small perturbation. A somewhat simplified and refined (in comparison with the initial) proof is given in [80]; in the same paper, it is shown that a similar lemma is valid in the class of Hamiltonian systems. (In this class, as well and in the class of volume-preserving DSs, the C^k-analogue of this lemma for a sufficiently large k is not valid! The references to the paper of M. Herman and Z. Xia on this topic are given in [1].) But even after this lemma was proved, the proof of the necessity of the structural stability conditions conjectured by Smale took much time. For dynamical systems with discrete time, the necessity was proved by R. Mañé [82, 83], and for flows, by S. Hayashi [85]. (In [77], some "intermediate" works are cited, which also played a role.) It is worth mentioning that Mañé and Hayashi had to supplement the closure lemma by some similar assertions (as "obvious" as this lemma). Toyoshiba indicated that Hayashi's paper requires certain modifications and indicated which (H. Toyoshiba. A property of vector fields without singularity on $\mathcal{G}^1(M)$. *Ergodic Theory and Dynam. Syst.*, **1** (2001), 303–314).

The above-mentioned equivalence of Andronov–Pontryagin structural stability and Peixoto structural stability follows from the sufficiency for the former of conditions necessary for the later. No simpler proof of this equivalence is known, although, seemingly, this fact should be more elementary.

A property somewhat weaker than structural stability is Ω-stability; roughly

---

[84] The definitions of the Morse–Smale systems and hyperbolic sets are given in textbooks on smooth dynamical systems, in surveys [78, 79], and in the relevant articles in *Mathematical Encyclopaedia*. It is worth mentioning that, as applied to equilibrium points, the terminology commonly used in the theory of smooth dynamical systems starting with 1960s differs from the previous one: since then, exponentially stable (unstable) foci and nodes are also classified with hyperbolic equilibrium points, while earlier, only saddles were said to be hyperbolic.

speaking, this is the preservation under small perturbations of the set of non-wandering points together with the dynamics on it. A precise definition can be found in any textbook on hyperbolic dynamics or in survey [78] on hyperbolic sets. The term is because the set of nonwandering points is often denoted by Ω. We also use this notation. A necessary and sufficient condition for Ω-stability was stated as a conjecture by S. Smale and J. Palis. Its main component is the hyperbolicity of the set of nonwandering points; in addition, it involves the density of the periodic points in this set and the so-called acyclicity. There are equivalent conditions, which require either the hyperbolicity and acyclicity of some other set, which is formally smaller than Ω *a priori*, or only the hyper-bolicity of a certain set (the set of chain recurrent points), which is formally larger than Ω; in reality, if the above-mentioned conditions holds, then all these sets coincide with Ω. The equivalence of these conditions and their sufficiency were proved by S. Smale and his collaborators; by modern standards, the proof is comparatively easy (sufficiency is easier to prove than the sufficiency of the corresponding condition for structural stability). The proof of necessity turned out to be as difficult as for structural stability. For systems with discrete time, it was obtained by J. Palis [84] right after the appearance of Mañé's paper; for flows, it was given by S. Hayashi.

Survey [77] describes the state of the art in the early 1980s. (The historical account is supplemented by M. Peixoto reminiscences [81].) For more recent advances, see [82–85].

The definition of structurally stable systems involves C^1-small perturba-tions. Considering diffeomorphisms $f, g, \ldots$ (vector fields $\mathbf{v}, \mathbf{w}, \ldots$) of class C^k and understanding the closeness of g to f (of $\mathbf{w}$ to $\mathbf{v}$) in the definition of the structural stability of the initial DS (determined by f or $\mathbf{v}$) in the sense of C^k-closeness, we obtain the definition of the property which is natural to call C^k-structural stability. (As previously, the homeomorphism χ is not required to be smooth, and we may require or not the closeness of χ to the identity transformation, so there are two versions of C^k-structural stability, in the sense of Andronov and Pontryagin and in the sense of Peixoto.) In the new termi-nology, structural stability is C^1-structural stability. What can be said about C^k-structural stability for $k > 1$?

So far, nothing contradicts the conjecture that C^k-structural stability is equivalent to C^1-structural stability (except, certainly, that the notion of C^k-structural stability applies only to DSs of class C^k). However, there are only two positive results concerning this conjecture: (i) in dimension 1 (both for flows and for diffeomorphisms); (ii) for flows on orientable two-dimensional closed mani-folds and on the three simplest nonorientable two-dimensional closed manifolds, namely, for those with Euler characteristics 1 (projective plane), 0 (torus), and -1. In all these cases, C^k-structurally stable systems form an open everywhere dense set in the space of all DSs of class C^k on the given manifold. In other cases, the efforts to find necessary conditions for C^k-structural stability by using the approach that has led to success for $k = 1$ depend on the validity of the C^k-version of the closure lemma. In light of the above considerations, the prospects are uncertain. It can be added that the somewhat (seemingly, quite naturally) strengthened assertion of the closure lemma is not valid even in the C^2 case; see the reference to a paper by C. Gutierrez cited in [1] and [86] (the strengthening of the lemma which is shown in [86] to be false for $k = 2$ is close to the assertions proved and used by Mañé and Hayashi for $k = 1$).

The notion of structural stability can be defined for systems with noncompact phase manifolds and for flows on compact manifolds with boundary. The situation in these cases is far from being completely clarified. It is known that it differs in some respect from that described above.

I am not aware of works related to the topic of the last three paragraphs that are more recent than those cited in [77].

Finally, I would like to dwell on the structural stability of smooth self-mappings of the interval $[0, 1]$ that are not one-to-one (for diffeomorphisms of the interval, the question is trivial). The definition of structural stability is carried over to this case word for word (with "diffeomorphism" replaced by "smooth mapping"). One new circumstance is evident: in the presence of critical points (that is, points x where $f'(x) = 0$), the mapping f cannot be C^1-structurally stable. Indeed, by an arbitrarily C^1-small perturbation, we can change the character of a critical point so strongly that the change is "sensed" even by our fairly crude approach, when everything is considered up to topological conjugacy. For example, we can ensure that the perturbed mapping take some interval to one point, or, *vice versa*, that each point have only finitely many preimages. Therefore, in the presence of critical points, a rich-in-content theory must consider C^k-structural stability for $k > 1$. What is said above makes it clear that this involves significant difficulties.

Using the specifics of the one-dimensional case and transferring consideration to the complex domain (where, fortunately, some useful results had been obtained), O. S. Kozlovskii was able to overcome these difficulties in the simplest nontrivial case of the so-called unimodal mappings, that is, smooth self-mappings of the interval with precisely one critical point [87]. The result is that, for any $k > 1$, a unimodal C^k-smooth self-mapping of the interval is C^k-structurally stable if and only if it satisfies the A axiom (suitably reformulated for one-dimensional mappings with critical points) and its critical point is nondegenerate (its second derivative does not vanish). Kozlovskii proved also that the unimodal mappings satisfying the A axiom (to which the condition that the critical point must be nondegenerate is trivially added) are everywhere dense in the space of all unimodal mappings of class C^k. Properly speaking, this is the main result, which comparatively easily implies the necessity of the structural stability condition stated above; its sufficiency had been known earlier.

2.2 Hilbert's 21st problem

In this section, we shall deal with the linear system of ordinary differential equations in the complex domain

$$\frac{dy}{dx} = C(x)y, \qquad y = (y^1, \ldots, y^p) \in \mathbb{C}^p. \tag{24}$$

Except in one case specified below, it is assumed to be holomorphic in the entire extended complex plane (the Riemann sphere) $\overline{\mathbb{C}}$, except at several singular points $a_1, \ldots, a_n$. Denote $S = \overline{\mathbb{C}} \setminus \{a_1, \ldots, a_n\}$. The holomorphy of the system at a point $x \in \mathbb{C}$ means simply the holomorphy at this point of the coefficients of the matrix $C(x)$. If ∞ is not among the points a_i, then system (24) must be holomorphic at the point ∞; this means that, if we rewrite (24) in terms of the new independent variable $\zeta := 1/z$, then the coefficients of the resulting system

$$\frac{dy}{d\zeta} = -\frac{1}{\zeta^2} C\left(\frac{1}{\zeta}\right) y \tag{25}$$

must have a removable singularity at the point $\zeta = 0$. Generally, solutions to system (24) branch (the simplest example is

$$p = 1, \quad n = 2, \quad S := \overline{\mathbb{C}} \setminus \{0, \infty\}, \quad \frac{dy}{dx} = \frac{\alpha}{x} y; \tag{26}$$

the solutions are $y = Cx^\alpha$). But they can be treated as functions on the universal covering surface $\widetilde{S}$ of the domain S, on which they are single-valued holomorphic functions. (Any solution to a linear system can indeed be extended to the entire surface $\widetilde{S}$; this is proved in approximately the same way as the assertion (probably familiar to the reader) that, in the real domain, the solutions to a linear system can be extended over the entire interval (or half-line, or line $\mathbb{R}$) where the coefficients of this system are defined and continuous. For nonlinear systems, neither the complex nor real versions of the extension theorem are generally valid.) I shall denote the points of $\widetilde{S}$ by tilde, assuming that $\widetilde{x}$ lies over $x \in S$; thus, it is better to denote solutions to (24) by $y(\widetilde{x})$. It may happen that some solution to a particular system of form (24) becomes single-valued already when lifted to a covering surface of the domain S which is lower than $\widetilde{S}$ (sometimes, a solution is unique even on the very domain S). But nothing prevents us from lifting it to $\widetilde{S}$ anyway.

We denote the projection $\widetilde{S} \to S$ that maps each point $\widetilde{x} \in \widetilde{S}$ to the point $x \in S$ covered by it as π. The homeomorphisms $\sigma \colon \widetilde{S} \to \widetilde{S}$ for which $\pi \circ \sigma = \sigma \circ \pi$ (i.e., which permute points $\widetilde{x}$ over the same x) are called deck transformations. The group of all deck transformations is denoted by Δ. It is isomorphic to a more concrete object, namely, to the fundamental group $\pi_1(S, x_0)$; in the case under consideration, this is the free group with $n-1$ generators corresponding to passages around some $n-1$ points among the points a_i. However, the groups $\pi_1(S, x_0)$, as opposed to Δ, partly depend on x_0 (these groups with different x_0 are isomorphic, but there is no "standard" isomorphism between them, as there is no "standard" isomorphism $\pi_1(S, x_0) \to \Delta$).

A sum of two solutions to a linear system is a again a solution, and a solution multiplied by a constant scalar is a solution; therefore, the solutions to (24) form a vector space $\mathfrak{Y}$, which is complex in the case under consideration. Since a solution is uniquely determined by its value at some $\widetilde{x}_0$, the space $\mathfrak{Y}$ is p-dimensional. The basis in $\mathfrak{Y}$ is what is called a fundamental system of solutions in the theory of differential equations. In addition to the vector-matrix system (24), we consider the matrix system

$$\frac{dY}{dx} = C(x)Y, \tag{27}$$

where Y is a square matrix of the pth order. Its columns are solutions to (24), and their linear independence is equivalent to the nondegeneracy of the matrix $Y(\widetilde{x})$ (for any $\widetilde{x} = \widetilde{x}_0$; to prove nondegeneracy for any $\widetilde{x}_0$, it is sufficient to prove it for some $\widetilde{x}_0$); if the matrix $Y(\widetilde{x})$ is nondegenerate, then it is called a fundamental matrix of system (24) (thus, a fundamental matrix is a matrix whose columns form a fundamental system of solutions). A general solution to (24) has the form $y = Y(\widetilde{x})c$ with a constant $c \in \mathbb{C}^p$. When the basis in $\mathfrak{Y}$ formed by the columns $Y(\widetilde{x})$ is used, the coordinates of this y form the column vector c.

Everything said above is analogous to definitions and statements of the theory of linear systems in the real domain. But the following assertion is new: If $y(\widetilde{x})$ is a solution to (24) or $Y(\widetilde{x})$ is a solution to (27) and $\sigma \in \Delta$, then $\widetilde{x} \mapsto y(\sigma\widetilde{x})$ or $\widetilde{x} \mapsto Y(\sigma\widetilde{x})$ is also a solution to (24) or (27), respectively. Indeed, any point $\widetilde{x}_0 \in \widetilde{S}$ has a neighborhood $\widetilde{U}$ in $\widetilde{S}$ such that π homeomorphically maps it to a neighborhood $U := \pi(\widetilde{U})$ of the point $x_0 = \pi\widetilde{x}_0$ in S; clearly, $\sigma\widetilde{U}$ is a neighborhood of $\sigma\widetilde{x}_0$ in $\widetilde{S}$ which is also homeomorphically mapped to U under π, and

$$\sigma(\pi \restriction \widetilde{U})^{-1} = (\pi \restriction \sigma\widetilde{U})^{-1}. \tag{28}$$

That $y(\widetilde{x})$ is a solution to (24) in $\widetilde{U}$ means that $y_1(x) := y((\pi \restriction \widetilde{U})^{-1}x)$ is a solution to (24) in U. That $\widetilde{x} \mapsto y(\sigma\widetilde{x})$ is a solution to (24) in $\widetilde{U}$ means that

$y_2(x) := y(\sigma(\pi \upharpoonright \widetilde{U})^{-1}x)$ is a solution to (24). But, by virtue of (28), we have $y_2(x) = y((\pi \upharpoonright \sigma\widetilde{U})^{-1}x)$, and this is indeed a solution to (24), because $y(\widetilde{x})$ is a solution in $\sigma\widetilde{U}$ (and in the entire domain $\widetilde{S}$). Thus, $y(\sigma\widetilde{x})$ is indeed a solution to (24) near any point $\widetilde{x}_0 \in \widetilde{S}$. Using the isomorphism $\Delta \approx \pi_1(S, x_0)$, we can describe the passage from $y(\widetilde{x})$ to $y(\sigma\widetilde{x})$ as a change of a solution to (24) under its analytic continuation over a chain of disks along the closed path corresponding to σ. Such an interpretation involves several identifications of different objects: Δ is identified with $\pi_1(S, x_0)$ and holomorphic functions on $\widetilde{S}$, with the Weierstrass sets of their elements.

In his way, there arises the transformation

$$\sigma_*: \mathfrak{Y} \to \mathfrak{Y}, \qquad (\sigma_* y)(\widetilde{x}) = y(\sigma^{-1}\widetilde{x}),$$

which is obviously linear and nondegenerate. We have $(\sigma_* \tau_*) = \sigma_* \tau_*$ (if we defined $(\sigma_* y)(\widetilde{x})$ as $y(\sigma\widetilde{x})$, then we would obtain $(\sigma_* \tau_*) = \tau_* \sigma_*$, as in 1.3, (a)). Denoting the group of nondegenerate linear transformations of the space $\mathfrak{Y}$ by $\mathrm{GL}(\mathfrak{Y})$, we obtain a representation

$$\Delta \to \mathrm{GL}(\mathfrak{Y}), \qquad \sigma \mapsto \sigma_*; \tag{29}$$

it is called a *monodromy representation* (for (24)). Taking some basis in $\mathfrak{Y}$, i.e., a fundamental system of solutions united into a fundamental matrix $Y(\widetilde{x})$, we can pass to a matrix representation $\chi: \Delta \to \mathrm{GL}(p, \mathbb{C})$, for which $\chi(\sigma)$ describes the change of coordinates in this basis of an element $y \in \mathfrak{Y}$ under the passage to $\sigma_* y$. It is easy to see that $Y(\widetilde{x}) = Y(\sigma\widetilde{x})\chi(\sigma)$, which can be taken as an equivalent definition of χ. The representation χ is also called a monodromy representation. It is determined by $n - 1$ nondegenerate matrices being the images of generators of Δ (this means that these matrices describe an analytic continuation of the solutions under going around $n - 1$ singular points); so it is a more concrete object than (29). But system (24) itself determines χ not quite uniquely: when a basis in $\mathfrak{Y}$ is changed, the representation χ is replaced by the conjugate representation $C\chi C^{-1}$, where C is some constant matrix. In this connection, the term "monodromy" is used for the whole class of conjugate representations $\{C\chi C^{-1}; C \in \mathrm{GL}(p, \mathbb{C})\}$ (it is indeed determined by system (24)).

We are interested in systems (24) having singularities of a comparatively "weak" type. A singular point $a_i \neq \infty$ is said to be *Fuchsian* if $C(z)$ has a pole of the first order at this point; the singular point ∞ is said to be Fuchsian if the coefficients of (25) have a pole of the first order at the point $\zeta = 0$. A system whose all singular points are Fuchsian is called a Fuchsian system. It can be proved that, for Fuchsian systems with singular points $a_1, \ldots, a_n$ and, possibly, ∞, the matrix $C(x)$ has the form

$$C(x) = \sum_{i=1}^{n} \frac{B_i}{x - a_i},$$

where B_i are constant matrices. The point ∞ is singular if and only if $\sum B_i \neq 0$.

The "weakness" of the Fuchsian singularities manifests itself in the behavior of solutions near singular points. If x tends to a Fuchsian special point a, then $|y(x)|$ may increase or decrease not faster that some power of $|x-a|$. This needs be refined, because even for (26) with $\operatorname{Im} \alpha \neq 0$, we can ensure an arbitrarily rapid growth of $|y|$ with decreasing $|x|$ if 0 is approached on a spiral along which $|y|$ decreases slowly in comparison with the number of turns around 0 (then $|y|$ changes mainly at the expense of these turns). The required refinement is very simple: $|y|$ must increase or decrease not faster than some power of $|x - a|$ as x tends to a remaining inside a fixed angle with vertex a.

An isolated singular point a of system (24) is said to be *regular* if, for this point, the same condition of no more than power increase or decrease of solutions as x tends to a holds (with the same refinement concerning an angle). It turns out that, at a regular singular point, the coefficients on the right-hand side of the system have a pole (not necessarily of the first order), so there exist regular singular points that are not Fuchsian. But even for a pole of the second order, the singular point may be not regular. Only about 10 years ago, an algorithm that makes it possible (if it is possible to perform all the computation) to determine whether a singular point is regular was constructed. It turned out to be very burdensome. It would be imprudent to assert that it cannot be simplified but, apparently, the arrangement of the regular systems among all systems whose coefficients have a pole at a is rather complex, so an algorithm "recognizing" them cannot be very simple.

If all singular points of a system are regular, then the system itself is said to be regular.

Now, we can proceed to the topic of this section, which is Hilbert's 21st problem (it is also known as the Riemann–Hilbert problem). The problem is: Show that there always exists a linear differential equation (actually, a system of equations rather that one equation is meant) of Fuchsian type with given singular points and given monodromy group (the modern statement of the problem refers of a monodromy representation, which is more precise). Hilbert might have meant not only the systems which are called Fuchsian today but also regular (in modern terminology) systems. Irrespective of the obscurity related to the terminology of the beginning of the century, there exist two problems, "Fuchsian" and "regular." The latter had shortly been solved positively by J. Plemelj (*a propos*, actually, that was the first felicitous use of the theory of singular integral equations; the foundations of this theory were laid by Plemelj at precisely that moment, although formally Plemelj did not mention them in this connection). Plemelj tried also to derive a positive solution to the problem in the Fuchsian setting, but in reality, his reduction works only under a certain additional condition on the monodromy representation; thus, what he obtained was a sufficient condition for the positive solvability of the Fuchsian version of the 21st problem. The condition is that at least one of the monodromy matrices corresponding to going around the singular points $a_1, \ldots, a_n$ reduces to a diagonal form. But this was recognized much later. Yu. S. Il'yashenko said that he noticed a gap in Plemelj's proof in 1975, when talking about Fuchsian systems in the course of his lectures. In the literature, the incompleteness of Plemelj's argument was mentioned, apparently, only in the 1985 survey [88]. But at that time, there still remained the hope that the answer was always positive for Fuchsian systems. The discovery by A. A. Bolibrukh of a contradicting example [89][85] was quite a surprise. Continuing his work in this direction, Bolibrukh, on the one hand, constructed a series of counterexamples of various types and, on the other had, found new sufficient conditions of various degrees of generality for the problem to have a positive solution; in this work, other authors participated too. One of the new sufficient conditions, which is due to V. P. Kostov and A. A. Bolibrukh, is that the representation χ must be irreducible.

The new methods found application in some other problems of the analytic theory (such as the Birkhoff problem about the standard form of a system in a neighborhood of an irregular special point). It should be mentioned that Bolibrukh usually considered vector bundles instead of integral equations; such a geometric approach was used for the first time by H. Rohrl exactly in the beginning of the past two decades (his paper contains, in particular, a different proof of the Plemelj theorem); another persistent "ingredient" of his papers is A. Levelt's improvement (1961) of the classical local theory constructed in the past century (largely by L. Fuchs and H. Poincaré). A presentation of this topic (although it has become already not quite complete) is given in [90].

A nonlinear analogue of the Riemann–Hilbert problem can be stated (and investigated) [91].

A. A. Bolibruch's posthumous paper "Differential equations with meromorphic coefficients" in *Contemporary Problems of Mathematics*, issue 1 (Moscow: Mat. Inst. im. V. A. Steklova, 2003), pp. 29–83, contains a survey of some of his

---

[85] This, undoubtedly, was one of the best papers among those published in *Matematicheskie Zametki* (if not the best one). Although *Zametki* are translated into English (as *Mathematical Notes*), this paper was not translated, because, at that time, the rubric of brief communications (where the most recent novelties are published) was not translated at all!

results. Note also his lectures *Fuchsian Differential Equations and Holomorphic Bundles* (Moscow: MTsNMO, 2000) [in Russian].

2.3 The Dulac conjecture

The conjecture was that the system of two equations

$$\dot{x} = f(x, y), \quad \dot{y} = g(x, y) \tag{30}$$

in $\mathbb{R}^2$, where f and g are polynomials (say, of the nth degree), can have only finitely many limit cycles. Below, we denote the number of these cycles by $L(f, g)$. H. Dulac himself regarded the statement "$L(f, g) < \infty$" as a theorem which he proved (1923) rather than as a conjecture. In reality, Dulac correctly understood that the proof reduces to analyzing the first-return map along a polycycle[86] and that the difficulty of the problem is related to the nonanalyticity of this mapping; he established some properties of the first-return map, but they were insufficient for making the required conclusion that it could have only finitely many fixed points. In 1977, F. Dumortier cast doubts on the completeness of Dulac's argument. In the summer of 1981, R. Moussu made these doubts a subject of an extensive discussion by writing to several colleagues about them. Independently, in the same summer, Yu. S. Il'yashenko found a mistake in Dulac's memoir (his position was more categorical: he directly indicated a mistake at a particular place), after which the Dulac theorem was renamed the Dulac conjecture. In the literature, renaming was accomplished, apparently, in the preprint of Yu. S. Il'yashenko cited in the survey [88] already mentioned and, then, in this survey. Paper [92] summarizes this stage of critical familiarization with Dulac's memoir (to which, certainly, many other things were added).

Now, the Dulac conjecture is proved completely. For $n = 2$, this was done by R. Bamon (R. Bamon. Quadratic vector fields in the plane have a finite number of limit cycles. *Inst. Hautes Etudes Sci. Publ. Math.*, **64** (1987), 111–142) and in the general case, by Yu. S. Il'yashenko and J. Ecalle [93, 94]. The methods used by Il'yashenko and Ecalle are different; both of them were applied to other problems too. The theory of Il'yashenko is called geometric,

---

[86] A polycycle is a closed curve L formed by separatrices joining some equilibrium points and by these points themselves, where the directions of motion (with increasing t) along the separatrices correspond to the same traversal direction of L. If L has points of self-intersection (they can be only equilibrium points), it is additionally required in fact that a small deformation of the oriented curve L could yield a closed curve without self-intersections (otherwise, the birth of a limit cycle from the oriented curve L is out of the question).

The term "polycycle" has become popular in recent years, although it is not quite standard yet, as far as I know. Earlier, the terms "separatrix contour (polygon)," "complex cycle," etc. were often used.

while the approach of Ecalle is related to a new process of summation of divergent series, which is well suited for the analytic theory of ordinary differential equations. Il'yashenko's paper "Finiteness theorem for limit cycles" (*Usp. Mat. Nauk*, **45** (2) (1990), 143–200) is an introduction to [91], while a brief exposition of Ecalle's approach is contained in [95].[87] The predecessors of Ecalle (as far as his summation process is concerned, but not with regard to the Dulac problem) were J.-P. Ramis and W. Balser. Books [96] and *Formal Power Series and Linear Systems of Meromorphic Ordinary Differential Equations* (New York: Springer, 2000) by Balser is a simplified exposition of this process (from positions slightly different from Ecalle's) and a number of its applications (to problems interesting for Balser himself). Book [96] ends with the proof of a theorem of B. Braaksma, according to which the formal solutions to nonlinear meromorphic ordinary differential equations are always summable by the new process, so there is nothing better to desire!

In this item, it is pertinent to recall Hilbert's 16th problem. Its first half refers to real algebraic geometry and the second, to the theory of differential equations. Each of the halves is an aptly stated large-scale scientific problem, but, as far as I can judge, they not only formally refer to different fields of mathematics but also are far from each other in essence, although Hilbert, apparently, did not completely realize this. We are interested in the second part of the 16th problem. It consists in obtaining an upper bound for the number of limit cycles of system (30), where f and g are polynomials of the nth degree, in the form of an explicit function of n. The very setting of the problem implies that

$$H(n) := \max \left\{ L(f,g); \; f \text{ and } g \text{ are } n\text{th-degree polynomials} \right\} < \infty \qquad (31)$$

for any n. So far, it is unknown whether this is true.

At one time, I. G. Petrovskii and E. M. Landis [97–99] believed that they had managed to prove (31) and obtain a bound for $H(n)$; in particular, they claimed that $H(2) = 3$. However, it is known now that system (30) in which f and g are second-degree polynomials can have four limit cycles (the example is due to Shi Songlin; it is presented in [92, 100]).

Certainly, it often happens that somebody finds a correct approach to a problem but makes a mistake in implementing it. Sometimes, this is a mere unfortunate oversight, and sometimes, something essential is missing. But in

---

[87] Concerning [93, 94], Smale remarked in his article mentioned in the preface (S. Smale. Mathematical problems for the next century. In *Mathematics. Frontiers and Perspectives*, eds. V. Arnold, M. Atiyah, P. Lax, and B. Mazur (Providence, RI: Amer. Math. Soc., 2000), pp. 221–244) that "these two papers have yet to be thoroughly digested by the mathematical community."

the case under consideration, the situation turned out to be more complicated. Landis and Petrovskii's papers contain valuable ideas which influenced the further development of the theory of ordinary differential equations in the complex domain, but as to Hilbert's 16th problem, the approach suggested in [97, 98] is blocked by obstacles that have not been overcome so far.

In [97, 98], considerations were transferred to the complex domain, where the geometric object corresponding to system (30) is a two-dimensional foliation (foliation with two-dimensional leaves) with singularities (these are the points where $f = g = 0$) rather than a system of curves (trajectories). Landis and Petrovskii discussed some properties of this foliation; for this purpose, they introduced a system of relevant notions having many things in common with the general theory of foliations, which arose somewhat earlier and began to successfully develop at approximately the same time. In other places of [97, 98], special features of the foliations corresponding to (30) with polynomial f and g were mentioned; in particular, a discussion of the typical properties of such foliations was begun. All these considerations are very rich in content. But it turns out that, in the complex domain, system (30) may have infinitely many limit cycles (apparently, Petrovskii and Landis allowed such a possibility. Its careful examination was performed by Yu. S. Il'yashenko; a reference is given in [88]). Thus, Petrovskii and Landis tried to prove that the number of those complex limit cycles that can fall in the real domain is comparatively small and can be estimated explicitly. A systematic verification organized in the late 1960s by S. P. Novikov at a seminar specially organized for this purpose (in which I actively participated as "devil's advocate") showed that this part of [97, 98] was groundless (the same conclusion was independently made by Yu. S. Il'yashenko). But for Hilbert's 16th problem, this part is crucial.[88]

Some progress was made in studying two local (with respect to the parameter) versions of Hilbert's 16th problem. The first version, the study of which was initiated by Yu. S. Il'yashenko, is concerned with estimating (in terms of the degree of the corresponding polynomials) the maximum number of limit cycles which can be born under a polynomial perturbation of a polynomial Hamiltonian system (as a rule, of a system with polynomial Hamiltonian). The state of the art in the mid-1980s is surveyed in [88] (in the section about "weakened Hilbert's problem"); after this survey had appeared, the study continued, but no more recent summary of results was published, as far as I know. Another

---

[88] The part of [97, 98] where the general properties of the foliations under consideration are discussed contains gaps and inaccuracies (this was mentioned at once by several people), but the defects of this part can largely be corrected by refining formulations. Landis and Petrovskii corrected most of them in [99]. But they agreed with the criticism concerning the passage from general questions to Hilbert's 16th problem proper only later [101] (after the seminar mentioned above was held).

version was suggested by V. I. Arnold. Its initial formulation turned out to be too optimistic [102]. Now, a weak version is studied; it is known as "Hilbert–Arnold's problem," although formally neither Hilbert nor Arnold stated such a problem: Prove that, in a typical k-parametric family of vector fields in the plane $\mathbb{R}^2$ (or, after natural extensions, on the projective plane $\mathbb{R}P^2$), a polycycle gives birth to only a finite number of limit cycles, which is estimated from above by a constant depending only on k. This conjecture is proved for polycycles containing only elementary singular points (i.e., such that at least one eigenvalue of the matrix of linear approximation is nonzero). V. Yu. Kaloshin obtained a bound of the form e^{ck^2} (c is a constant) for the number of possible limit cycles born in such a situation. (Supposedly, a bound which is polynomial in k must exist.)

2.4 Homogeneous flows and the Raghunathan conjecture

A significant advance in the theory of DSs of algebraic origin – homogeneous flows – was M. Ratner's proof of M. Raghunathan's conjecture and its metric analogue (because of this analogy, references to the metric Raghunathan conjecture are sometimes encountered; strictly speaking, this is incorrect. Raghunathan himself stated his conjecture only for a special case; the general statement and the metric analogue were suggested by S. Dani). I shall give the necessary definitions and, simultaneously, some information about homogeneous flows.

A homogeneous flow is a DS on a homogeneous space G/D, where G is a Lie group and D is its closed subgroup; this system is determined by the left[89] action on G/D of some subgroup $H \subset G$: under the action of an element $h \in H$, a coset gD (an element of G/D) is mapped to hgD. Strictly, the entire group G acts on G/D on the left, but we are (ultimately) interested only in the restriction of this action to H. In what follows, we always assume that G/D has finite volume; this means that there is a finite measure μ on G/D invariant with respect to the given action of G. (The facts known about more general cases are largely examples.) In the most popular (and old) examples, the subgroup H is one-parameter, i.e., $H = \{h_t; \ t \in \mathbb{R}\}$; thus, a flow in the usual sense (with classical time) is obtained; however, multidimensional subgroups H can be considered too.

We denote the Lie algebra of the group G by $\mathfrak{g}$ and the Lie group of all linear transformations $\mathfrak{g}$ treated as a vector space by $\mathrm{GL}(\mathfrak{g})$. The theory of Lie

---

[89] It is assumed that G/D consists of left cosets gD. If we considered right cosets Dg (it would be better to denote the set of right cosets by $D\backslash G$, but the standard notation is G/D), then the homogeneous flow would be determined by the right action.

groups considered the so-called *adjoint representation* Ad, which is a certain homomorphism of Lie groups

$$\text{Ad}\colon\ G \to \text{GL}(\mathfrak{g}), \qquad g \mapsto \text{Ad}_g$$

(and determines an action of G on $\mathfrak{g}$ according to the rule $(g, X) \mapsto \text{Ad}_g x$). Its definition in the general case requires knowing a number of basic facts from the theory of Lie groups; citing them would take space, while a reader familiar with these facts should be familiar with the definition of Ad too. But it is easy to give the definition of Ad in the important special case where G is a matrix Lie group, i.e., a subgroup of the group of nondegenerate matrices $\text{GL}(n, \mathbb{R})$ or $\text{GL}(n, \mathbb{C})$ (with some n) being a smooth submanifold of this group. The group $\text{GL}(n, \mathbb{R})$ or $\text{GL}(n, \mathbb{C})$ is an open subset in the space $\text{Mat}(n, \mathbb{R})$ or $\text{Mat}(n, \mathbb{C})$ of all matrices (including degenerate ones) of order n. As a vector space (i.e., with matrix multiplication disregarded), the matrix space is isomorphic to $\mathbb{R}^{n^2}$ or $\mathbb{C}^{n^2}$ (we consider the matrix coefficients as usual coordinates in $\mathbb{R}^{n^2}$ or $\mathbb{C}^{n^2}$ enumerated differently). Thus, in the case under consideration, G is a subset of a vector space, and it is clear what G being a smooth submanifold means. It turns out that $\mathfrak{g}$ can be regarded merely as the tangent space to G at the point being the identity matrix I, but it is more convenient to move this space in parallel to itself from I to 0 (the zero matrix). (Thus, the usual tangent lines to G at I are the lines $I + tA$, where $t \in \mathbb{R}$ or $\mathbb{C}$ and $A \in \mathfrak{g}$.) In this case, the action of G on $\mathfrak{g}$ reduces to a mere conjugation of matrices:

$$\text{Ad}_g X = gXg^{-1} \qquad (g \in G \subset \text{GL}(n, \mathbb{R})\ \text{or}\ \text{GL}(n, \mathbb{C}),\ X \in \mathfrak{g}).$$

An element $g \in G$ is said to be *unipotent* if all the eigenvalues of the transformation Ad_g are equal to 1. A Lie subgroup $U \subset G$ is unipotent if all of its elements are unipotent. A special case is the so-called *horospherical* subgroups. A subgroup $H \subset G$ is said to be *horospherical* if there exists an element $g \in G$ such that $g^n h g^{-n} \to e$ as $n \to \infty$ for all $h \in H$ (e is the identity element of the group G). The term is because, in one special case, such subgroups are closely related to the so-called horospheres in Lobachevskii geometry. (The explanation of this relation would be a too lengthy digression, since it would require describing the group-theoretic interpretation of geodesic flows on manifolds (or, at least, on surfaces) of constant negative curvature. The reader familiar with this interpretation, probably, knows about the relation.) A one-parameter horospherical subgroup is said to be *horocycle* (because of the same relation).

A homogeneous flow (generally, with multidimensional time) on a quotient space G/D of a Lie group G which is obtained under the action of a unipotent subgroup $U \subset G$ on the cosets by left translations is said to be *unipotent*.

If the subgroup is horospherical (horocycle), then the flow is also said to be horospherical (horocycle).

The Raghunathan–Dani conjectures refer to unipotent flows (possibly, with multidimensional "time") on quotient spaces G/D of finite volume. The first conjecture says that the closure of each trajectory of such a flow is a homogeneous subspace of finite volume and the second, that if an ergodic measure for this flow is finite on the compact sets, then is concentrated on a homogeneous subspace of finite volume and has a simple algebraic description there; to be more precise, it originates from the Haar measure on the subgroup determining this homogeneous subspace. The second conjecture was proved by Ratner without the assumption that G/D has finite volume; only the finiteness of the ergodic measure under consideration was required.

An important role in the proof is played by a property of one-parameter unipotent subgroups defined by Ratner (it is called the Ratner property). According to Ratner, this property was suggested to her by her previous works on homogeneous flows, which were formally concerned with quite different questions. (In these works, the rigidity phenomenon for horocycle flows was discovered: in certain cases, the existence of a metric isomorphism between two such flows implies that this isomorphism has an algebraic origin, being obtained from some inner automorphism of G in combination with a translation by a constant time along trajectories.) On the other hand, there are a number of papers where special cases of both conjectures are proved. Most of them are concerned with horospherical subgroups; the first result of this kind was obtained as early as 1936 by G. A. Hedlund. However, by modern standards, the case of horospherical flows is too simple, and these papers not only considered the conjecture of Raghunathan and Dani but also (and to a larger extent) contained far-reaching studies of the topological and metric properties of such flows. G. A. Margulis established the validity of the topological Raghunathan conjecture in a significantly more complicated case, namely, for certain one-parameter[90] unipotent (but not horocycle) flows on $SL(3, \mathbb{R})/SL(3, \mathbb{Z})$. This allowed him to prove the Oppenheim–Davenport conjecture about quadratic forms, which is well known in number theory (the idea to use unipotent flows in proving this conjecture dates back to Raghunathan). Then, Margulis and Dani proved the topological Raghunathan conjecture for all "generic" unipotent flows in the same space and obtained some results supplementing the preceding number-theoretic results of Margulis. (In general, it seems that, in works of Margulis *et al.*, a new area of geometric number theory emerges, where the "scene" is Lie groups and their homogeneous spaces rather than Euclidean space and the torus. The case of nilpotent groups, which is closest to the Euclidean case, was considered in this

---

[90] That is, obtained under actions of one-parameter subgroups.

context before the work of Margulis, but those considerations were aimed to "group and dynamical" interpretation of known facts rather than to obtaining new number-theoretic results.)

After both conjectures (topological and metric) were proved, the validity of some of their generalizations was established. Although, they were not quite generalizations, because they required the connectedness of U (the question to what degree this additional requirement is necessary is being discussed), but in other respects, the conditions on U were relaxed to the following requirements: (i) U is generated by unipotent elements (Ratner); (ii) U is generated by quasi-unipotent elements, i.e., by elements g for which the eigenvalues of Ad_g are of modulus 1 (A. N. Starkov). In case (i), the conclusions remain the same as above, and in case (ii), they are somewhat modified (e.g., the closure of a trajectory is still a manifold, but it is not necessarily a homogeneous space; it is only related to a homogeneous space in a certain way).

In relation to the Ratner theorem, it is worth mentioning that cases where the closures of some trajectories are not manifolds had been known for a long time. G. A. Margulis pointed out that this phenomenon certainly has place if a one-parameter homogeneous flow has the property of uniform partial hyperbol-icity, which is well known in the theory of smooth dynamical systems (in the case under consideration, it is equivalent to the flow being non-quasi-unipotent).

For completeness, I shall mention several older results on homogeneous flows. In a certain sense, the theory of homogeneous flows reduces to the the-ory of ergodic homogeneous flows. Namely, in the nonergodic case, Starkov described a partitioning of G/D into ergodic components which are finite-fold covered by homogeneous spaces (although the components themselves are not necessarily homogeneous spaces); moreover, the restriction of the initial flow to each component can be lifted to a homogeneous flow on the covering. For this reason, ergodic flows deserve paramount attention; much information about their properties related to ergodic theory is gathered. There is a criterion for the ergodicity of a homogeneous flow in terms of algebraic conditions on the "input data" that determine the flow. Its main ingredients were obtained by L. Auslander (the solvable case), C. Moore (the semisimple case), and S. Dani (the general case); some finishing touches were put independently by Starkov and D. Witte. Spectra of homogeneous flows were described. For homogeneous flows, the conjecture of V. A. Rokhlin mentioned in Section 1.3, (a) was proved (by Starkov on the basis of the Ratner metric theorem). For the special case where $G = \mathrm{SL}(2, \mathbb{R})$, fairly many results about the properties of homogeneous flows (i.e., geodesic and horocycle flows on surfaces G/D of constant negative curvature) were obtained even without the finiteness assumption on the volume (which is area in the case under consideration) of the surface G/D (although the question is far from exhausted). This is closely related to actions of Fuchsian groups on the Poincaré disk, and these actions attract attention for more than 100 years already. The summary of results given in [103] makes it possible to compare properties of flows and actions.

Homogeneous flows are the subject-matter of Ratner's report [104], surveys [105–107], and book [108].

2.5   The Seifert conjecture

This conjecture was that a smooth flow without equilibrium points on the three-dimensional sphere $\mathbb{S}^3$ must have a closed trajectory. It was bases on the theorem of G. Seifert according to which all flows obtained by small perturbations of the Hopf flow, which is described below, have closed trajectories.

Let us represent the three-dimensional sphere $\mathbb{S}^3$ as the set of those points (z, w) of the two-dimensional complex plane $\mathbb{C}^2$ (which is four-dimensional from the real point of view) for which $|z|^2 + |w|^2 = 1$. The phase velocity of the Hopf flow is the vector field that assigns the vector (iz, iw) to a point (z, w). The trajectories of the Hopf flow are the circles $\{e^{it}z, e^{it}w\}$; the partitioning of $\mathbb{S}^3$ into these circles is the Hopf fibration,[91] which is well known in topology; hence the term "flow." In addition to the proof given by Seifert himself, there exist at least two other proofs of his theorem due to F. B. Fuller and M. Bottkol (some references and an exposition of Fuller's idea are given in [79]). Fuller used the Fuller index, which is a topological characteristic (introduced by Fuller) of the behavior of trajectories near a closed trajectory,[92] while Bottkol used an ingenious version of perturbation theory (suggested by Jürgen Moser in one paper about periodic solutions near an equilibrium point). Thus, the methods of these works are more general than the results (which can hardly be said about the proof of Seifert himself). But they also apply only to small perturbations of the Hopf flow.

The Seifert conjecture is related to the following torus conjecture: if a phase velocity field on the boundary of the solid torus $\mathbb{D}^2 \times \mathbb{S}^1$ is directed everywhere inside (or everywhere outside) the solid torus and has no equilibrium points, then there is a closed trajectory. Intuitively, this trajectory must make one turn around the solid torus; thus, the example constructed by Fuller, in which

---

[91] To be more precise, the Hopf fibration is the mapping $\mathbb{S}^3 \to \mathbb{S}^2$ obtained by identifying each of the circles specified into a point. It is related to H. Hopf's discovery (unexpected at that time) that the homotopy group $\pi_3(\mathbb{S}^2)$ is nontrivial.

[92] This index has nothing in common (except the term "index") with the Conley index mentioned in Section 1.1. As to classical roots, the Fuller index is related to the Kronecker–Poincaré index rather than to the Morse index.

a closed trajectory exists but is homotopic to zero in the solid torus, was a "warning bell" against too much confidence in naïve intuition.

Both conjectures have been disproved even for analytic flows. This service was mainly rendered by K. Kuperberg [109, 110], who constructed C^∞-counterexamples, after which W. Thurston and E. Ghys specified an analytic modification of the construction. It is worthwhile to mention the contributions made by previous authors. F. W. Wilson constructed counterexamples to multidimensional analogues of the Seifert conjecture. That was not a great surprise, since it was rather obvious that, in the multidimensional case, quasiperiodic trajectories can completely "replace" periodic trajectories (as in Wilson's examples), but part of technical tricks proved useful afterwards. A surprise was the 1974 counterexample of P. Schweitzer[93] (which was "genuine," three-dimensional) with flow of smoothness C^1. After that, smooth versions of both conjectures did not inspire great confidence, but it required much effort to improve the smoothness of counterexamples even to C^2 (J. Harrison).

Interestingly, the Seifert conjecture is true for the so-called contact flows (H. Hofer [112]). Contact flows live on odd-dimensional manifolds M^{2n+1}. Such a flow is defined with the use of a so-called contact form λ, that is, a Pfaffian form for which the $(2n+1)$-dimensional form

$$\lambda \wedge \underbrace{d\lambda \wedge \ldots \wedge d\lambda}_{n \text{ times}} \tag{32}$$

is everywhere nonvanishing. (I omit the smoothness specification.) This definition resembles the definition of a symplectic structure on an even-dimensional manifold,[94] but there is at least one essential distinction. A symplectic form does not distinguish any directions at points of M, while a contact form determines a one-dimensional direction at every point x; this is the degeneracy direction of the form $d\lambda$: a vector $X \in T_x M$ has this direction if

$$d\lambda(X, Y) = 0 \quad \text{for any } Y \in T_x M.$$

---

[93] The example of Schweitzer is described in Tamura's book [111].

[94] The similarity becomes even more manifest when the following theorem of G. Darboux is taken into account: In a neighborhood of any point x, there are local coordinates $(x_1, y_1, \ldots, x_n, y_n, z)$ such that $\lambda = dz + \sum y_i\, dx_i$. However, a contact structure is usually defined not as a mere pair (M, λ) with contact form λ but as the field of $2n$-dimensional tangent subspaces $E_x \subset T_x M$, where

$$E_x = \{X \in T_x M; \ \lambda(X) = 0\},$$

determined by this form on M. Let us specify that only a field E_x of $2n$-dimensional tangent subspaces which is obtained in this way, with the use of some contact form, is called a contact structure. Equivalently, all contact forms $f\lambda$, where f are scalar functions, determine the same contact structure as λ.

In addition, a vector $X_x \in T_x M$ with this direction for which $\lambda(X_x) = 1$ is fixed. The vector field X is sometimes called the field of G. Reeb. A contact flow is a flow with phase velocity field X.

In reality, the Hofer theorem is more general; it asserts that, if the one-dimensional cohomology group $H^1(M, \mathbb{R})$ for a closed three-dimensional manifold M is trivial, then any contact flow on M has a closed trajectory. This is a special case of A. Weinstein's conjecture, in which the manifold is not assumed to be three-dimensional.

Form (32) can be taken for a volume form on M. (Those who relate the notion of volume to Riemann geometry should take into account that, if a nowhere vanishing exterior m-form Ω on a smooth m-manifold M is given, then M admits a Riemannian metric for which the volume element is expressed by the form Ω.) It is easy to prove that the contact flows preserve the volume determined by this form. There arises the question as to whether the Seifert conjecture is true for any flows preserving volume. G. Kuperberg (K. Kuperberg's son) constructed a counterexample to this conjecture (see a reference in [110]).

3 Some other achievements

I repeat that the style of this section is often even briefer than that of the preceding sections; frequently, instead of explaining results, I only cite them. But I still give references to the literature where part of important results are presented and references to other papers are made.

3.1

First, there are domains which attracted attention much more than a quarter century ago and which still remain domains of extensive studies; the directions of study are more or less the same as before, although these domains were enriched by important new ideas, notions, methods, etc. The report of Yoccoz [1] considers two such directions, which differ in the character of the motions under examination. One of them is concerned with quasiperiodic trajectories and trajectories close to quasiperiodic in some respect (such as "cantori"), and the other deals with the hyperbolic behavior of trajectories; again, not only "pure hyperbolicity" as it was shaped in the 1960s but also similar cases are considered. As I said, I am not going to duplicate Yoccoz's report. I shall make only several small literature additions.

(*a*) A specific part of the "hyperbolic" theory is formed by the results (obtained largely by French mathematicians, although the initiative was W. Thurston's) about three-dimensional Anosov flows and related geometric questions. The 1991 situation is described in [78]. One of the most recent works in this area is [113].

(*b*) In the 1960s, in studying a classical geometric object – geodesic flows on manifolds of negative curvature – the role played by geometry was definitely second to the role played by DS theory. During the past 20 years, the role of geometry noticeably increased, especially when the curvature is assumed to be only nonnegative rather than strictly negative; see the survey [114] and papers [115, 116],[95] where newer examples (and more references) are given.

---

[95] In [115], for one class of closed manifolds of nonpositive curvature (the so-called manifolds of rank 1), a partitioning of the unit tangent bundles into two invariant sets (introduced earlier) such that the behavior of trajectories in infinitesimal terms (in terms of variational equations) on one of them is, so to speak, more hyperbolic than on the other was studied. If we can rely on the infinitesimal characteristics, then we can expect that the behavior of trajectories on the first set is "more stochastic" than on the second. It is shown in [115] that this is so in two respects (which are closely related), in respect of topological entropy and in respect of the asymptotic behavior of the number of closed geodesics of length $\leqslant T$ with increasing T.

(*c*) Yoccoz paid comparatively little attention to special properties of invariant measures in the case where the trajectories have hyperbolic behavior. Relatively old results (part of which are supposed to be known in [1]) are given in [2]. Next, I mention the recent paper [117] (where some references to earlier works are cited), which has completed the many-year efforts in studying some old well-known questions about hyperbolic measures. A compactly supported normalized measure μ invariant with respect to a diffeomorphism $f: M \to M$ (which is assumed to be of class $C^{1+\epsilon}$ in what follows) is said to be *hyperbolic* if the iterations of the mapping f (to be more precise, of its "tangent extension," "differential," or "derivative" Tf) "exponentially" change the tangent vectors almost everywhere (in the sense of this measure), i.e., if

$$\varlimsup_{n \to \infty} \frac{\ln \|T_x f^n X\|}{n} \neq 0$$

for all $X \in T_x M$ at almost all $x \in M$.

The main result of [117] is a metric analogue of the well-known topological fact about the local structure of hyperbolic sets, which is mentioned in any textbook or survey on hyperbolic theory. If A is a locally maximal hyperbolic set, then locally (in a neighborhood of any point $x \in A$) is has the structure of the direct product of some set B in the local unstable fiber $W^u_{\mathrm{loc}}(x)$ and some set C in the locally stable fiber $W^s_{\mathrm{loc}}(x)$; a point $(y, z) \in B \times C$ corresponds to the point of the set A that lies in the intersection of the local stable fiber $W^s_{\mathrm{loc}}(y)$ passing through y with the locally unstable fiber $W^u_{\mathrm{loc}}(z)$ passing through z. (I omit the details.) This fact is frequently used, and its weakened version holds for nonuniform hyperbolicity. Having the local structure of direct product, it is natural to ask whether each hyperbolic invariant measure can be represented as the direct product of a measure on $W^u_{\mathrm{loc}}(x)$ and a measure on $W^s_{\mathrm{loc}}(x)$. For nonhyperbolic measures, it would be unnatural to expect that such a representation holds, and even for hyperbolic measures, there is no hope to obtain a decomposition of this kind in the general case. However, in [117], a weakened analogue of a local representation of hyperbolic measures in the form of such a direct products is obtained; roughly speaking, the representation holds with an error, which is an additional multiplier that varies slower than an arbitrarily small power of ϵ with decreasing the size of the spherical ϵ-neighborhood of a point x. An additional information about the properties of the measure-"factors" is also obtained.

Although [115] is outwardly as analytical as the 1960s works, it uses, in addition to the notion of rank 1 and the two sets mentioned above, the following objects of geometric origin: the Buseman function, the absolute, and some measures on the absolute constructed with the use of a trick similar to the Poincaré series in the theory of automorphic forms.

In relation to this question, it is also proved in [117] that hyperbolic measures have some "good" properties, which do not generally hold for arbitrary invariant measures of dynamical systems. For example, as is known, there exist different notions of the dimension of a set; each of them reflects, in some reasonable way, properties that are natural to regard as "dimensional." In the general case, different dimensions do not coincide, but they do coincide for "good" sets. It is less well-known that, similarly, there exist different notions of dimension for measures. It turns out that, for hyperbolic measures, a number of dimensions coincide.

The very statements of these results may seem "technical" (in other places, I cited results whose statements sounded more "directly"), but since I have once played a certain role in the creation of hyperbolic theory, I believe, I may permit myself to rely on my impression of their importance. Yet, I have already mentioned that the selection of material for Section 3 was more subjective than for Sections 1 and 2.

(*d*) In addition to [1], where cantori were considered, there are earlier papers concerning similar objects for certain flows [118–120] and papers in which three approaches different from that of [1] were developed. A. B. Katok [121] mentioned the possibility of employing old ideas of G. Birkhoff to study the corresponding invariant sets for two-dimensional diffeomorphisms. A. Fathi [122] treats cantori as a kind of generalized solutions (of the type of their most recent version, the so-called viscous solutions) to the Hamilton–Jacobi equation. Finally, R. Mañé made new observations related to superlinear Lagrangians; they concern not only cantori (see [123, 124]).

Another example of a successful continuation of the activity begun earlier is the works of A. D. Bruno and his collaborators in local theory (see [88, 125]). From the time of Poincaré (or even earlier), the most important method of local theory consists in constructing a formal (i.e., represented by formal power series) "normalizing" transformation which reduces a (local) flow near an equilibrium point or a diffeomorphism near a fixed point to a certain simpler form, called the normal form. The members of Bruno's group have studied a wide circle of related questions with maximal completeness and considered many applications. The questions concerning normal forms include constructing normalizing transformations (one of the services rendered by A. D. Bruno earlier is a geometric trick, which applies to the multidimensional case and generalizes the Newton polygon[96]), studying the structure of normal forms and the possibilities suggested by the ambiguity of the construction (when it takes place),

---

[96] This method applies also to locally studying systems of algebraic equations; as is known, the polygon was invented by Newton for precisely this purpose in the special case of one equation with two unknowns.

and analyzing the convergence of the constructed transformations. The applications refer not only to ordinary differential equations proper (including those depending on parameters, which is important for bifurcation theory), but also to some partial differential equations.

When some method is well elaborated, the "nontriviality" of a problem can be estimated by the degree of completeness to which it can be studied by this method and the amount of additional considerations needed to solve it. Some people do not understand this; hence the epigraph (quotation from Mark Twain) to Bruno's book alluding to an uncle who wondered what happens to failed mechanics, gunsmith, shoemakers, metalworkers, engineers... Mark Twain believed that they become watch-makers, while Bruno hinted that their "abilities" can find a different application...

3.2 The theory of singular perturbations

This is one more example of a successful continuation of studies begun earlier. The theory deals with systems of ordinary differential equations of the form

$$\epsilon\dot{x} = f(x, y, \epsilon), \quad \dot{y} = g(x, y, \epsilon), \tag{33}$$

where ϵ is a small parameter. "Singular" means that the small parameter is included as a multiplier of the derivative rather than as a mere parameter on which the right-hand side of the system depends (the right-hand side of (33) may depend on ϵ too, but this is not so essential). Certainly, we can introduce a new ("slow") time $s := t/\epsilon$ and, denoting differentiation with respect to it by prime, rewrite (33) as

$$\dot{x} = f(x, y, \epsilon), \quad \dot{y} = \epsilon g(x, y, \epsilon). \tag{34}$$

Now, ϵ is contained only in the right-hand side, but in the new terms, to an interval of time t having finite length T, there corresponds an interval of slow time s having length T/ϵ, which unboundedly increases as $\epsilon \to 0$ (whereas we, naturally, want to study the behavior of solutions to (33) at least on a finite time interval not decreasing as $\epsilon \to 0$); thus, anyway, the obtained perturbation problem is not quite usual.

In any case, clearly, x varies much faster than y, and therefore, the properties of the solutions depend mainly on the properties of the system of "fast motions"

$$\dot{x} = f(x, y, 0), \tag{35}$$

in which y plays the role of a constant parameter. Under additional assumptions about (35), an attempt to describe slower variation of y can be made.

The simplest case is in which the solutions to (35) tend (as $t \to \infty$) to an exponentially stable equilibrium point $x_0(y)$, which, in general, depends on the "parameter" y. In this case, it is natural to assume that this y in system (33) changes with time approximately as a solution to the system

$$\dot{y} = g(x_0(y), y, 0). \tag{36}$$

If the solutions to (36) tend to an exponentially stable equilibrium point y_0 or to an exponentially stable closed trajectory l_0 with increasing time, then the solutions to (33) also tend to an equilibrium point (namely, to $(x_0(y_0), y_0)$) or, respectively, to a closed trajectory L_ϵ near the curve $L_0 := \{(x_0(y), y); \; y \in l_0\}$. Results of this kind (certainly, under appropriately refined conditions on the system) were obtained about 50 years ago. It is possible not only to prove the assertion about the limit of trajectories as $\epsilon \to 0$ but also to study their dependence on ϵ in more detail, namely, to obtain an asymptotic expansion up to an arbitrary power of ϵ for these trajectories. For the periodic trajectory L_ϵ, an asymptotic series in the powers of ϵ is obtained, while for trajectories with a fixed initial value (x', y'), where $x' \neq x_0(y')$, the corresponding expansion contains also terms including $\ln \epsilon$. The uniqueness of a closed trajectory L_ϵ near L_0 can be proved as well. This result somewhat differs in character from that on the asymptotic behavior of L_ϵ. Indeed, *a priori*, there might exist two closed trajectories L'_ϵ and L''_ϵ at a "distance" of higher order of smallness than any power of ϵ apart, say, of exponential order $O(e^{-1/\epsilon})$ or $O(e^{-1/\sqrt{\epsilon}})$. The above-mentioned asymptotic expansion up to an arbitrary ϵ^n could not "feel" this.

A more complicated situation occurs when, at some ("bifurcation") values of y, two equilibrium points of system (35) (the stable point $x_0(y)$ and some unstable point) merge but (35) has a stable equilibrium point $x_1(y)$ somewhere away from this "junction" point to which the trajectory under consideration can be "attracted" after passing the bifurcation value y. Suppose that those y for which such a junction occurs form a smooth hypersurface M in the y space (to be more precise, analytical condition on M under which this is so are imposed) and that trajectories of system (36) reach M having direction transversal to M (this can be stated in the form of explicit conditions on f and g). If this is so for a trajectory of system (36) to which the y-component of the trajectory of system (33) under consideration is close, then it is natural to suppose that, after this moment, the latter trajectory quickly passes to $x_1(y)$ and, then, its y-component is close to the trajectory $y(t)$ of system (36) in which $x_0(y)$ is replaced by $x_1(y)$ and the x-component is close to $x_1(y(t))$. This process may repeat itself. Then, we may expect that the trajectories of system (33) consist of arcs of two types: some of them are similar to those described above in the simpler situation and close to arcs of the form $\{(x_i(y_i(t)), y_i(t))\}$, where the y_i are solutions to systems of form (36) (with some $x_i(y)$ instead of $x_0(y)$), and the others are close to some arcs each going from two merged equilibrium points of system (35) to some stable equilibrium point. The arcs of the first type can be traversed in finite time, and the arcs of the second type can be traversed very quickly. It is easy to understand that such a system of arcs may form a closed curve L_0. In this case, we may expect that (again, under appropriately refined conditions), for small ϵ, (33) has a closed trajectory L_ϵ near L_0. It is worth mentioning that such objects are known in physics, where they provide a mathematical description for some types of relaxation oscillations.[97]

The study of these questions began in the 1940s with a consideration of the case where x and y are "one-dimensional" (by J. Haag and A. A. Dorodnitsyn). In the 1950s, L. S. Pontryagin and E. F. Mishchenko obtained important results in the multidimensional case. They determined the first terms of asymptotic expansions for various arcs of trajectories and for the periodic trajectory L_ϵ and its period. These asymptotic expansions turned out to be significantly more complex than in the case considered above (they involve fractional powers of ϵ), and they could hardly be predicted. Somewhat later, N. Kh. Rozov clarified the structure of the entire asymptotic expansion in the case of one-dimensional x and y. In this case, the uniqueness of a closed trajectory L_ϵ is proved very easily, so the theory had acquired a certain completeness, while in the multidimensional case, the questions about the uniqueness of L_ϵ and about the structure of the whole asymptotic expansion remained open. (The state of

---

[97] The original meaning of this term was related to the physical nature of oscillatory systems. Relaxation systems were opposed to systems of different (more usual) character (which were called Thomson type systems at one time); an example of the latter is the usual radiogenerator. It includes an oscillation contour, supplied with energy; on the other hand, the contour "loses" energy, mainly because of generator's radiation (which is the generator is created for) and partly because of resistance. The stable periodic trajectory describing the generated oscillations is close to one of the trajectories describing the free oscillations of the contour, and its amplitude is determined by the balance between the supplied and lost energy. In a relaxation oscillatory system, oscillations arise otherwise: first, in some part of the system, energy is somehow accumulated, and then, the accumulated energy is "discharged" in a different part of the system (hence the term, which comes from *relaxation* – relief, discharge). Depending on the structure of the system, this process can be "balanced" in such a way that the oscillations be similar to harmonic oscillations (although there is no oscillatory contour); but the discharge may also be very fast in comparison with the slower accumulation of energy, and the oscillation is then "close to discontinuous" ("discontinuity" corresponds to discharge). In the latter case, a mathematical description of the system is often given by a system of form (33), in which, as the "slow" variable y varies, the phenomena briefly described above (junction of two equilibrium points, etc.) occur; the small parameter ϵ is "responsible" for the rate of discharge. In the mathematical literature, there is the tradition to use the term "relaxation oscillations" only in this latter case.

the art in the beginning of the period under consideration is described in detail in [126].)

Answers to these questions were found about 10 years ago. The uniqueness of L_ϵ was proved independently by C. Bonet and four Russian mathematicians, E. F. Mishchenko, Yu. S. Kolesov, A. Yu. Kolesov, and N. Kh. Rozov (see [127]).

Recall that one of the conditions above was that, when a trajectory of system (36) reaches the hypersurface M (where stable and unstable equilibrium points of system (35) merge), the vector $g(x_0(y), y, 0)$ (determining the direction of this trajectory) is transversal to M. The question arises: How do the trajectories of system (33) near the point where $g(x_0(y), y, 0)$ is tangent to M behave? The existence of such points is a phenomenon fairly "typical" in the obvious sense, except in the case where y is "one-dimensional," so the question cannot be brushed aside as referring to some exceptional situation. If y is one-dimensional (so that M is merely a point), then the question does refer to an exceptional situation, but if (33) depends on a parameter a, then such a situation may arise for some $a = a_0$, and it becomes sufficiently "typical." In such a case, the natural problem of studying the behavior of solutions to (33) not only for $a = a_0$ but also for values of a close to a_0 arises. Replacing ϵ by 0 on the right-hand side of (33), we obtain a somewhat simplified system for which the point M is equilibrium. When a varies on some interval, in system (33) (not simplified), an equilibrium point close to M experiences a bifurcation. Of special interest is the case where this is a Hopf bifurcation; it is interesting to watch how the newborn limit cycle, first small and having almost elliptic form, grows with varying a into a quite different "almost discontinuous" limit cycle considered above.

This last question was investigated first. The study was performed in the beginning of the period under consideration by a group of French mathematicians related to Strasbourg (E. Benoit, F. and M. Dieners, J.-L. Callot, A. Troesch, E. Urlacher, and others); the initiative was due to G. Reeb (renowned for his contribution to foliation theory); see [128–131]. Somebody found the "growing" (with changing a) limit cycle to resemble a flying duck at some values of a; its different parts were given the corresponding names, from "bill" to "tail." Soon, all the trajectories involved in these problems (including those where the parameter a was absent but y was "non-one-dimensional") came to be called "duck trajectories" (even nonclosed ones, let alone the absence of similarity with ducks – flying, swimming, walking, or fried), and the whole research area was called "hunting ducks." A distinguishing feature of the French works was a systematic use of nonstandard analysis (see [128, 129]). Apparently, the French authors, as well as the authors of [129], believed that the language of nonstandard analysis is more convenient for arguing in problems of this kind (including constructing various asymptotic expansions). The statements of final assertions, even when given in the nonstandard language, are easy to translate into the standard language; moreover, the authors of [128, 129] sometimes expressly explain how to do this. Though, the paper [130] initiated the employment of only standard mathematics in "hunting ducks." In [127], a number of questions about "ducks" for "non-one-dimensional" y was studied within the framework of standard analysis.

Obviously, as y varies, an equilibrium $x_0(y)$ of system (35) may suffer a bifurcation of a different kind, when a stable limit cycle branches off this equilibrium and the equilibrium itself persists but becomes unstable. Seemingly, we might then expect that the trajectory of system (33) rapidly approaches the limit cycle. But in reality, the situation is more complex. The phenomenon of "stability loss delay," when the x-component of the trajectory remains near $x_0(y)$ for a long time after the bifurcation (while the y-component changes approximately according to (36)), is possible (and even typical in a certain sense). The study of this phenomenon was initiated by L. S. Pontryagin and M. A. Shishkova in 1973 for one special (but quite "representative") case. Fairly complete results in the general case were obtained over 10 years later by A. I. Neishtadt. References are given in [52]; see also [131–133]. In [134], a similar phenomenon related to the loss of stability of a cycle is discussed.

Furthermore, for all the y under consideration, the trajectories of (35) may tend to a stable closed trajectory $C(y)$ rather than to an equilibrium. In this case, it is natural to expect that the x-component of the solution to (33) with initial value (x', y') rapidly approaches $C(y')$ and, then, always remains near $C(y)$, where y is the y-component of the trajectory under consideration, and the evolution of y is approximately described by the equation $\dot{y} = g(x, y, 0)$ averaged along $C(y)$. Again, we can consider the case where the averaged system has an exponentially stable equilibrium or an exponentially stable closed trajectory. The main work has essentially been done by N. M. Krylov and N. N. Bogolyubov (and by a number of other authors in special cases) long before the beginning of the period under consideration, but they considered the problem in a somewhat different setting; in the spirit of the approach presented here, the problem has been studied by L. S. Pontryagin and L. V. Rodygin in 1960 (see [126]).

In the case considered above, in system (35), a one-frequency oscillation mode is established. It is also quite possible that multifrequency oscillations occur from the very beginning (without a transition process). If they do not depend on y, then the question reduces to studying the behavior of the solutions to the equation $\dot{y} = g(t, y, \phi)$ with "multifrequency" dependence of g on t. Important results on problems of this type were obtained (also before the period we are interested in) by N. N. Bogolyubov and his collaborators (first of all, by Yu. A. Mitropol'skii). Of a different character is the situation where the multifrequency oscillations in (35) depend on y. Seemingly, in this case, general results can hardly be obtained, because in realistic problems, the character of the corresponding trajectory of (35) may change substantially under an arbitrarily small change of the initial values (x', y'). For example, it may be dense on a torus of large dimension or be a closed trajectory for very close initial values. Over what should the equation for y be averaged in such a situation?

In 1960, I pointed out that it is possible to obtain a completely satisfactory answer in a sufficiently general case[98] by considering the behavior of only a "majority" of trajectories rather than that of all of them. The "exceptional" trajectories, for which the averaging method in its natural (or "naïve"?) setting does not work, correspond to a set of initial values whose measure tends to zero as $\epsilon \to 0$. To be more precise, for fixed $\delta > 0$ and $T > 0$, the measure $\mu_{\delta,T}(\epsilon)$ of the set of those (x', y') for which the error of the averaging method on the interval $[0, T]$ exceeds δ tends to zero. What we consider is, so to speak, the convergence of solutions (as $\epsilon \to 0$) in the measure of initial values. Simultaneously, related but less general (at least, formally) results were obtained by T. Kasuga.

I did not estimate $\mu_{\delta,T}(\epsilon)$ but only proved that this value tends to 0 as $\epsilon \to 0$, because the theorem was so general that I could not hope for such an estimate (at least, for any satisfactory estimate). About 15 years later, i.e., in the beginning of the period under consideration, it turned out that such an estimate can be obtained under reasonable restrictions, namely, when there are "true" multifrequency oscillations in (35) (solutions are quasiperiodic). The most essential step was made by A. I. Neishtadt, who considered the important case where the dependence of the basis frequencies of these oscillations on y is nondegenerate. He showed that $\mu_{\delta,T}(\epsilon)$ is at most proportional to $\sqrt{\epsilon}/\delta$, and in the class of power estimates this result is final: there exist examples where $\mu_{\delta,T}(\epsilon)$ differs from $\sqrt{\epsilon}/\delta$ only by a not very substantial logarithmic factor. Later, by the same methods, results of this kind were extended (with suitable modifications) to some systems with degenerate dependence of frequencies on y; most of the progress was made by V. I. Bakhtin [135]. References to earlier works are given in [136].

---

[98] Properly speaking, we do not even assume the presence of multifrequency oscillations in system (35); we suppose that, for each y, the system has a "good" invariant measure and a "good" system of first integrals $I_1(x, y), \ldots, I_k(x, y)$, and it is ergodic on almost all surfaces $I_j = \text{const}$.

3.3   Exponentially small effects in perturbation theory

Properly speaking, even many of the results mentioned in Section 3.2 go beyond the "power" perturbation theory, because even for analytic systems, the corresponding series in powers of the small parameter usually diverge, and therefore, any assertion about the existence and uniqueness of a closed trajectory (much more, of a quasiperiodic solution) are related to something beyond this theory. But these are assertions of qualitative character. During the past 25 years, a number of quantitative results which refer to exponentially small effects were obtained. The results on stability loss mentioned in Section 3.2 are partly of such a character. Some other directions in which such results were obtained are

  (a) the problem of separating motions;
  (b) the problem of splitting separatrices;
  (c) the problem about the preservation of adiabatic invariants;
  (d) Nekhoroshev's theory;
  (e) the problem about Arnold diffusion.

Some information and references can be found in [52, 136, 137]. Additional references are (only works of the past 25-years are included, but they contain references to those comparatively few earlier works that are not mentioned in the books cited above):

  (a) [132, 134, 138];
  (b) [139–142];
  (d) [143–145];
  (e) [144].

3.4   The entropy formula

Even before the period under consideration began, the conjecture that, for any compact smooth manifold M and smooth mapping $f \colon M \to M$, the topological entropy $h_{\text{top}}(f)$ is not smaller than the spectral radius of the induced mapping f_* in the "complete" homology group $H_*(M, \mathbb{R}) := \bigoplus H_i(M, \mathbb{R})$ with real coefficients was stated. The situation in the beginning of this period is described in [146]. This conjecture has been proved by Yomdin [147–149].

3.5   Integrable and nonintegrable systems

The problem of integrating differential equations (i.e., solving them) is as old as the theory of differential equations itself. Some time (before Cauchy), the theory of differential equations (which was not formally regarded as a separate section of analysis at that time) consisted mainly of tricks developed for integrating certain equations or classes of equations. (Though, almost simultaneously, methods of perturbation theory arose in celestial mechanics. But apparently, at that time, they were regarded as a part of celestial mechanics rather than of the theory of differential equations.) Sometimes, first integrals instead of solutions are found. (The authors who say "solution" instead of "integral" often abbreviate "first integral" to "integral.")

It is pertinent here to explain what is the origin of the terminology. The term "integration" is used because solving a differential equation can be regarded as a generalization of usual integration ($\int f(x)\, dx$ is a solution to the simplest differential equation $\frac{dy}{dx} = f(x)$). For this reason, the functions which are solutions to differential equations (and tuples of functions which are solutions to systems of differential equations) are called integrals of these equations (systems). Nowadays, they are most often called simply solutions (as in this paper), although the same term then refers to a function satisfying an equation and to the process of finding this function. (Still, Bourbaki retained the archaic terminology.) As opposed to an integral, a first integral is a function constant along a solution. The term is used because the first stage in seeking a solution ("integral") is finding a first integral or, in the case of a system, a "complete" system of first integrals, i.e., a system of functions $F_1, \ldots, F_k$ such that any solution satisfies the system of equations

$$F_1 = c_1 = \text{const}, \quad \ldots, \quad F_k = c_k = \text{const} \tag{37}$$

for some $c_1, \ldots, c_k$. When such functions F_i are known, solving a system of differential equations reduces to solving the "finite" system (37). It was believed that the latter system is simpler and, anyway, solving it is beyond the scope of the theory of differential equations. In practice, when the first integrals were any complicated, this second stage of solution turned out be far from easy. In problems that were solved in practice (except in very simple ones), at this stage, complicated elliptic-type functions were employed. It is easy to believe that, for more complex first integrals, system (37) can be solved only on a computer, which is, generally, not at all easier (it is often even harder) than numerical integration of the initial system of differential equations. Even so, the knowledge of the first integrals may allow us to make qualitative conclusions about the behavior of solutions that by no means can be immediately seen from the system of differential equations itself.

Naturally, at the early stage of the development of the theory, a large number of relatively simple problems were integrated; integrating them was often a matter of luck. Any handbook contains many examples of this kind, which are, so to speak, "odd" in the sense that each of them was integrated on its own, out of any relation to other problems. Many of the integrated problems refer to analytical mechanics, which had long been the main "consumer" of the theory of differential equations and of analysis in general.

Integrability has somewhat different meanings in the "general" theory of ordinary differential equations and in analytical mechanics (which deals with Euler–Lagrange or Hamilton equations). In the general theory, integrability usually means that the function f of interest to us (which is a solution or a first integral; in the case of a system of ordinary differential equations, several first integrals forming a complete system may be considered) is obtained from the simplest functions, polynomials, by applying certain operations. These operations are the algebraic operations, differentiation, (indefinite) integration, and exponentiation (passage from g to e^g). Obviously, such an f has an explicit expression, which generally contains integrals ("quadratures") that may not always be "taken"; for this reason, f is said to be represented by quadratures, and the differential equation (system) is said to be integrated by quadratures. The list of operations does not include a number of "elementary functions," such as logarithm and direct and inverse trigonometric functions. However, it is easy to see that, in constructing f, they can be replaced by the operations specified above. For example, $\ln g = \int \frac{g'}{g}\, dx$. Sometimes, one more operation is added to the list, namely, that of solving an algebraic equation whose coefficients are already constructed functions. In this case, generally, f has no explicit expression, and it is said to be represented by generalized quadratures.

In analytical mechanics, integrability usually means the existence of a complete system of first integrals which are either comparatively simple (algebraic) or arbitrary analytic functions of coordinates in the phase space. The case where the first integrals are represented by quadratures is not distinguished. In the problems that have been integrated, both classical and new, the first integrals are often algebraic (or even rational) functions of suitably introduced coordinates. Negative results (that a certain system has no first integrals or has no complete system of first integrals) are usually much easier to obtain for algebraic integrals than for analytic ones. At the same time, the answer to the question whether or not a given function on the phase space is algebraic depends on the coordinates; thus, negative results on algebraic first integrals also depend on coordinates, and we cannot be sure that the system would not become integrable if some other phase variables were used.

There is yet another version of the notion of integrability, which goes back to S. V. Kovalevskaya. It is generally known that she found a new case where the equations of motion of a heavy solid body admit a fourth integral[99] and, eventually, a complete system of first integrals. It is less well-known that her work was largely related to the problem: In what cases all solutions, when

---

[99] To pedantically observe the distinction between "integrals" and "first integrals," we should say "a fourth first integral," strange as it sounds.

considered in the complex domain, are meromorphic functions of time?[100] (The existence of the fourth integral in the new case which she revealed (as well as in the cases known earlier) was obtained as a byproduct.) In this connection, the term "Kovalevskaya integrability" has come into use.

There exist no general integration methods as universal as the differentiation rules. Still, a number of important problems of analytical mechanics (which are stated mathematically in the form of Euler–Lagrange or Hamilton equations) was solved on the basis of only two (partially overlapping) methods. First, in the presence of continuous symmetry groups, these problems have the corresponding first integrals (this is a theorem of E. Noether). Secondly, we can pass to a certain partial differential equation (the Hamilton–Jacobi equation) and try to choose coordinates in which the variables are separated. There is no guarantee that one of these methods applies in a particular case; in return, a number of integrable problems of analytical mechanics were found by proceeding in the reverse direction and trying to find problems to which these methods do apply. Note that solving problems of this kind (I mean those of them that have been solved) is usually especially distinctly divided into the two stages specified above, finding a sufficiently complete system of first integrals and using these integrals to obtain explicit expressions for the time dependences of coordinates and momenta. At the first stage, an important role is played by a theorem of J. Liouville,[101] according to which, for a "mechanical" DS with n degrees of freedom, it is sufficient to find n functionally independent first integrals $F_1, \ldots, F_n$ such that all the Poisson brackets $\{F_i, F_j\}$ vanish (in this case, we say that these integrals are in involution). Even at this stage, far-reaching qualitative conclusions about the behavior of trajectories can be made; people were making such conclusions in special cases for a long time, but the general observation is due to V. I. Arnold (for this reason, the corresponding theorem is often referred to as the Liouville–Arnold theorem). In principle, when the variables are separated, the first stage can be obviated, at least, the first integrals can be not involved explicitly. However, in practice, these integrals do

---

[100] Kovalevskaya pointed out that her results apply also to the more general question: When are all solutions single-valued functions of complex time? The validity of this assertion was confirmed by A. M. Lyapunov.

[101] As far as I understand, this theorem for the most important case of an autonomous (i.e., not including time explicitly) n-degree-of-freedom system of equations of mechanics was first published by little-known mathematician E. Bour, while Liouville generalized it to the nonautonomous case, which does not play a noticeable role. (Although, Liouville referred to his earlier oral report. In addition, he proved earlier the special case of this theorem for $n = 2$, but this case was essentially known to C. Jacobi and S.-D. Poisson before Liouville.) This confirms the saying that the affluent gain and the indigent lose. When the conditions of the Liouville theorem hold, the system is said to be Liouville integrable.

arise, and they are given special attention.

Shortly before the beginning of the period under consideration, P. Lax suggested a new, third, integration method (in the sense of finding a system of first integrals), which is now known as the (L, A)-pair method. To be more precise, this is a third method for finding integrable problems. It is not related in some special way to analytic dynamics, but most of its applications deal with Hamiltonian systems and provide Liouville integrable systems. The method applies to partial differential equations; actually, it originates from a study of one of such equations (the Korteweg–de Vries equation), which was previously considered from different positions. (First, this equation was studied by numerical methods, and this study was related to earlier works concerned with other equations and different questions; this is an instructive and dramatic story, but it would lead us too far away from our theme. The analytical study initiated in this connection has led to a discovery of another method for integrating this equation, which is related to the inverse problem of scattering theory. Apparently, Lax wanted to comprehend the obtained results from a different standpoint.) But Lax had understood at once that his method had a wider domain of applicability.

Probably, the majority of modern applications of the (L, A)-pair method and its modifications, including the most interesting applications, refer to partial differential equations. However, the method has produced fairly many results on ordinary differential equations too.

The method of (L, A)-pairs *per se* refers to the first stage (finding first integrals). But very soon after it was discovered, P. Lax and S. P. Novikov added very important considerations related to the second stage.

If we try to directly apply the proof of the Liouville theorem in order to obtain explicit expressions for coordinates and momenta as functions of time, we will find that, for a number of classical problems of mechanics, it first seems that the analytic functions involved are fairly complicated; however, the answer can be expressed in terms of comparatively simpler functions by employing various special considerations. Such simplifications look usually as results of incidental coincidences. It would be desirable to have some general statements more directly specifying the functions which are indeed necessary to use. Such statements can hardly be formulated in a satisfactory and natural way in the same terms in which the Liouville theorem is stated (if they could, they would be found approximately a century ago). But it turned out that they can be formulated with the use of a certain modification of (L, A)-pairs, namely, of (L, A)-pairs with spectral parameter. Such pairs do not always exist, and when they do exist, they are harder to find than mere (L, A)-pairs. But in return, they give much more.

As well as the "usual" method of (L, A)-pairs, the method of spectral (L, A)-pairs was extensively applied not only to ordinary differential equations but also to partial differential equations. This led to a deep study of the algebro-geometric aspects of integrable systems – so deep that it provoked a discovery of new facts even in algebraic geometry proper. Although, this discovery was somewhat related to partial differential equations; in general, in a number of most important publications on the algebro-geometric aspects of integrable systems, most attention was given to partial differential equations.[102] But recently, a book has appeared in which the exposition is entirely oriented to Hamiltonian systems from the very beginning [150]. I also draw reader's attention to the recent collection [151] of Russian translations of older papers by J. Moser on integrable systems.

During the past 25 years, in addition to the discovery of new integrable systems and study of their properties, some "opposite" results were obtained; namely, the nonintegrability[103] of a number of special systems was proved. The first results of this kind were obtained about 1900, but the results of that period were mostly of fairly bounded character. It was usually proved that a system of mechanical origin had no complete system of algebraic first integrals. Obviously, the algebraicity of a first integral depends on a coordinate system. Certainly, when a mechanical problem is said to have no algebraic first integrals except those already known in mechanics, the coordinates which naturally arise in the very statement of the problem are meant. Nevertheless, such a result makes an impression of incompleteness, because it is still possible that "additional" (not yet known) first integrals exist but are expressed by transcendental functions in the "natural" coordinates.

To prove "true" nonintegrability, geometric rather algebraic considerations are needed. Although even Poincaré used them, a systematic interpretation of these questions was initiated in works of V. M. Alekseev, which appeared a little less than 10 years before the period under consideration. Then, a significant contribution was made by V. V. Kozlov and S. L. Ziglin; see Kozlov's survey [152] and [136]. The more recent book [153] contains newer nonintegrability results (in addition to a "positive" information on the integration of a number of special systems[104]). Important works of I. A. Taimanov and G. Paternain are devoted to geodesic flows (see [154]).

---

[102] If I would write a survey on partial differential equations, I should include this subsection in the section "New or "renewed" directions." But I write about DS theory, and, as you see, I included it in Section 3.

[103] Here and in what follows, by nonintegrability we mean the lack of Liouville integrability.

[104] The method of (L, A)-pairs is mentioned in [153] under the name "Heisenberg representation." Spectral (L, A)-pairs are not considered.

In some cases, the nonexistence of analytic first integrals is considered. To what degree is analyticity essential? Sometimes it is essential, and sometimes it is not. An extreme example of the latter situation is Anosov systems with "good" invariant measure; they do not even have measurable first integrals (because of ergodicity). But this, as I said, is an extreme example. For systems with two degrees of freedom, there are fairly good results that ensure (under certain assumptions) the absence of "additional" first integrals with any decent smoothness. The main role (as Poincaré anticipated) is played by homoclinic trajectories (and by the related hyperbolic sets); the other conditions imposed in nonintegrability theorems are, so to speak, additional. When the number of degrees of freedom is larger than two, the situation becomes substantially more complicated: no sufficiently convenient additional conditions are known yet (leaving aside analytic first integrals).

L. Butler [155] constructed a series of (related to each other) examples of analytic Riemannian metrics with integrable geodesic flows for which the "additional" (to the energy integral) first integrals are of class C^∞ but not analytic and the partitioning of the phase space into regions filled with invariant tori (which are involved in the Liouville–Arnold theorem) do not possess the geometric properties which would be implied by the analyticity of the first integrals and which are used to derive constraints on the geometry of the phase space of an integrable system. Accordingly, these constraints themselves do not hold. Using the idea of Butler, A. V. Bolsinov and I. A. Taimanov [156] constructed "improved" examples in which integrable analytic geodesic flows have positive topological entropies. Moreover, the restrictions of these flows to certain "exceptional" invariant submanifolds are Anosov flows (notwithstanding that the properties of Anosov flows and those of integrable flows are, as it were, two opposite extremes).

3.6 The Conley theory

Up to about 1970, any far-reaching applications of topological methods were only concerned with systems of a few special types (such as those of variational nature). The applications of topology to systems of, so to speak, general character were fairly primitive (this, of course, does not mean that they were not important). They were largely somehow related to rotation of vector fields on the boundaries of domains and to the Poincaré–Kronecker index[105] of the zeros

---

[105] As is known, in mathematics, there are many objects of various nature called indices (not counting subscripts and superscripts); for this reason, to avoid confusion, it is necessary to add some identifying words to the term "index"; these are often the names of the authors of the corresponding notions. When it is clear from context which index is meant, we say simply "index."

of a vector field. In essence, purely topological (and fairly simple) notions and assertions were used; they bore no specific relation to dynamical systems.[106] There was also the Wazewski principle (to which Conley's ideas were closely related). But this principle had long occupied a special, isolated position.

The new notions introduced by Conley are essentially related to dynamical systems (as well as the Wazewski principle, though), and they require no *a priori* assumptions about the special features of the system under consideration.[107] Conley emphasizes another aspect of the matter, namely, a kind of "roughness" (in the same sense as in the phrase "roughly but reliably") of the corresponding objects. His theory involves a far-reaching generalization of the classical notion of Morse index. The point of departure of M. Morse was variational problems, and he considered indices of nondegenerate critical points of functions; but it is known for a long time that the Morse index can be considered from a different angle – from the point of view of the theory of dynamical systems. Namely, a critical point of a function is an equilibrium point of the corresponding gradient flow, and its Morse index can be interpreted naturally in terms of the properties of the flow near this equilibrium point. The Conley theory considers not only equilibrium points but also a large class of so-called isolated (or locally maximal) compact invariant (i.e., consisting of trajectories) subsets in the phase space of the flow, which is not necessarily gradient. For these sets, some topological characteristics are introduced, which retain the name "indices." To be more precise, these indices characterize not only the intrinsic structure of this set proper; rather, they characterize some special features of the behavior of trajectories near this set. The new indices are more general and complex objects than the original Morse indices, but they essentially reduce to the Morse indices in the cases considered by Morse; this justifies the use of the term.

As soon as the new theory emerged, papers on its applications to mathematical questions of celestial mechanics (such as the three-body problem and related questions) and to problems related to traveling waves started to appear; later, it was applied to the theory of partial differential equations. Unexpected was the use of index considerations in the works of C. Conley and E. Zehnder mentioned in Section 1.1, which, as I said, played a noticeable role in the formation of symplectic geometry as a separate discipline of high taxonomy rank. Although, now the role of "index" considerations in this discipline has

---

[106] Although the work which has to be done to fit a problem under examination into the corresponding topological framework may be quite nontrivial and essentially depend on the special features of the dynamical system under consideration.

[107] Although, the application of the Conley theory to a given system may be meaningless or nonimplementable in practice. Such a possibility should be taken into account whenever and whatever general theory is applied.

apparently decreased.

The term "Conley theory" is justified by the role played by C. Conley in the creation and development of this theory. Conley himself pointed out that R. Easton also played an important role in its creation. Later, a whole group of authors worked on its development and applications.

A brief exposition of the basic notions of the Conley theory is contained in [79]. To this theory, the lectures of Conley [157] and Mischaikow [158] are devoted.

3.7   Singularities in the n-body problem

The problem is to describe the motion of material points (particles) attracting each other according to Newton's law. A mathematical description of such a system is given by a Hamiltonian system in the $6n$-dimensional phase space with variables p_i and q_i, where $i = 1, \ldots, 3n$. Here $q_{3i-2}, q_{3i-1}, q_{3i}$ are the usual coordinates of the ith particle in the usual (physical) space $\mathbb{R}^3$ and $p_{3i-2}, p_{3i-1}, p_{3i}$ are the projections of its momenta to the coordinate axes in $\mathbb{R}^3$. Even not writing out the system, we can easily understand that it has singularities at those points where the coordinates of the ith particle coincide with those of the jth particle for some $i \neq j$. Let us denote the set of such points in the configuration space (the space of q_i variables) by Σ; then the singular points of the system are the points of the set $\Sigma \times \mathbb{R}^{3n}$. If the initial values $(p(0), q(0))$ are taken outside this set, then, locally, there exists a solution $(p(t), q(t))$. It is not necessarily extendable over the entire positive semi-axis of t values. If the maximal interval of existence of the solution is finite, say, it is an interval $[0, T)$, where $T < \infty$ (we consider only positive values of t, although this is inessential), then we say that the solution has a singularity at $t = T$. (The vector-function $(p(t), q(t))$ indeed has a singularity at $t = T$.) It is easy to prove that $q(t)$ unboundedly approaches Σ as $t \to T$. If the limit $\lim q(t)$ as $t \to T$ exists (it unavoidably belongs Σ), then we say that a collision occurs in the system at $t = T$. (The etymology is obvious: in the limit of $t \to T$, two or more particles are at the same point of the physical space $\mathbb{R}^3$). Until recently, the question whether there exist singularities different from collisions remained open. It is known that there are no such singularities for $n \leqslant 3$. For $n = 2$, this assertion is obvious, and for $n = 3$, it was proved about 100 years ago.

If a collision is double, then the further motion of the system (during some time) can be defined in a reasonable (and quite evident) way; this observation was made very long ago, by L. Euler.[108] Euler's definition of motion after a

---

[108] Voltaire criticized Euler on this account; he said (certainly, he was right) that Euler's description of motion after a collision was physically unreal. However, the motion of particles

double collision does increase the maximal interval of existence of the solution, but (except in the case of $n = 2$) not necessarily to infinity. Thus, for $n = 3$, the motion may end up with a triple collision (before that, double collisions may occur, but during a finite time interval ending strictly before the triple collision).[109]

Just before the beginning of the past quarter century, J. Mather and R. McGehee [159] discovered that, even at $n = 4$, singularities of a different type are possible. They found an example where infinitely many double collisions occur during a finite time T. All four bodies move along the same stright line in the same direction. In the end, as $t \to T$, the three bodies move away to infinity; one of them move in one direction and the two others move in another direction, unboundedly approaching each other (this gives energy for the whole process). The fourth body oscillates between the two latter bodies, alternately colliding with each of them. The question arises: Do there exist singularities which are not collisions and which occur at the end of some collision-free interval $[0, T)$? This question cannot be answered by considering only the one-dimensional case, which complicates the problem. In 1992, Z. Xia [160] showed that such a phenomenon can occur at $n \geqslant 5$.[110] The question remains open only for $n = 4$.[111]

3.8 Stable ergodicity

The C^2-smooth Anosov systems with "good" invariant measure are ergodic and remain ergodic under small (only C^1-small) perturbations that break neither the C^2-smoothness of the system nor the presence of a "good" invariant measure. This property of systems (that they remain ergodic under small (in the sense of some C^r) perturbations that does not break the presence of a "good" invariant measure) may be called "stable ergodicity." The question is: Do there exist other stably ergodic systems?

---

which fly very close to each other but still do not collide is real. Euler's motion after collision describes the limit of such a motion as the distance between the flying particles under consideration tends to zero. Though, Euler himself motivated his definition of motion for $t > T$ otherwise, on the basis of its analytical properties.

[109] In addition to this qualitative picture, there is an analytical theory, which deals with the character of functions describing singularities of various collision types. Thus, the situation of singularities at $n = 3$ as a whole cannot be summarized in a few sentences. An approximately fifteen-years-old survey is contained in [136].

[110] Xia points out that J. Gerver has established the possibility of singularities which do not reduce to collisions by a different method in the n-body problem with some large n.

[111] The idea to consider motions close to the motions described by Mather and McGehee but not "fitting" in a straight line but going on in the plane or in $\mathbb{R}^3$ suggests itself. However, so far, this approach has not led to success.

A positive answer to this question is due to C. Pugh and M. Shub (whom other authors joined). First, only separate examples were considered, then some (although fairly special) classes of stably ergodic systems were revealed. Certainly, their other ergodic properties (and the behavior of these properties under perturbations) are interesting too. For some classes of systems, as well as for the Anosov systems, not only stable ergodicity but also "stable K-property" and even "stable Bernoulli property" (defined similarly) hold. Undoubtedly, a new chapter in smooth ergodic theory has begun.

At the time where I was writing this paper, many works in this domain existed only as preprints. The publications known to me are [161–164]. Close questions are considered in [165, 166]. These papers consider DSs of special form (certain skew products) and the preservation of ergodicity and stronger properties under small perturbations in classes of such systems.

3.9 "Abstract" (purely metric) ergodic theory

My interests are related largely to smooth dynamical systems, which unavoidably determines my attitude to this topic in a very general aspect. During the preceding period, the main achievements in purely metric ergodic theory were the development of "entropy" theory and, then, of the Ornstein theory, which borders on entropy theory; the influence of these achievements on "smooth" theory is hard to overrate. The achievements of the past 25 years, as far as I can judge, do not have such an effect. References to works of the preceding period are given in [4, 45, 167].

On the structure of systems with invariant measure

In what follows, by a DS we understand a dynamical system $\{\phi^n\}$ with discrete time in a "good" space X with normalized invariant measure μ (defined on a σ-algebra $\mathfrak{B}$ of subsets of X). H. Furstenberg and R. Zimmer [168–171] developed a theory in which a DS is represented in the form of the "inverse limit" of a sequence (possibly, transfinite) of DSs which starts with the trivial DS in a space reducible to one point and in which every DS is a so-called "primitive extension" of the preceding one (each DS whose number is a transfinite ordinal α being the limit of an increasing sequence of ordinals is the inverse limit of the sequence of DSs with numbers $\beta < \alpha$). Here, three new notions arise: inverse (or projective) limit, extension, and primitive extension. I shall not give a precise definition of any of these notions, but I shall try to explain what they are, at least in part, neglecting some details and employing analogies.

First, I must tell about extensions of spaces with measure. These are a kind of analogue of bundles in topology. A space with measure $(X, \mathfrak{B}, \lambda)$ is an extension of a space with measure $(Y, \mathfrak{C}, \mu)$ (which might be called a base in the spirit of topology but is more frequently referred to as a factor of $(X, \mathfrak{B}, \lambda)$) if there is fixed a surjection ("onto" mapping) $\pi \colon X \to Y$ ("projection" to the base (factor)) which is measurable (if $A \in \mathfrak{C}$, then $\pi^{-1} A \in \mathfrak{B}$) and preserves measure (under the same assumption, $\lambda(\pi^{-1} A) = \mu(A)$). By analogy with topology, the sets $\pi^{-1}(y)$, where $y \in Y$, are called fibers. An example is the projection of the direct product $Y \times Z$ of spaces with measure $(Y, \mathfrak{C}, \mu)$ and $(Z, \mathfrak{D}, \nu)$, which is endowed with the corresponding σ-algebra of measurable sets and the measure $\lambda := \mu \times \nu$, onto one of the "factors" Y or Z (say, on Z). In this example, the fiber is $\pi^{-1}(z) = Y \times \{z\}$, and on the fiber, the natural measure λ_z is defined; this is simply the measure μ of Y "transferred" to the fiber $\big(\lambda_z(A \times \{z\}) := \mu(A)\big)$. By the Fubini theorem, for a measurable set $A \subset Y \times Z$, we have

$$\lambda(A) = \int_Z \lambda_z(A \cap \pi^{-1}(z)) \, d\nu(z). \tag{38}$$

Interpreting measures as probabilities, we can (in the spirit of probability theory) interpret $\lambda_z(A \cap \pi^{-1}(z))$ as the conditional probability of the "event" A under the condition z. (Note that, generally, the set of these "conditions" (i.e., of points $z \in Z$) is uncountable; therefore, in the case under consideration, the notion of conditional probability is fairly delicate.) Alternatively, in "pure" measure theory, a "conditional measure" is considered. It turns out that, for "good" spaces with measure, a system of "conditional measures" arises on the fibers of the extension $\pi \colon X \to Y$; each fiber $\pi^{-1}(y)$ is endowed with its own measure λ_y and

$$\lambda(A) = \int_Y \lambda_y(A \cap \pi^{-1}(y)) \, d\mu(y),$$

by analogy with (38). Here and in what follows, I omit the necessary stipulations on the neglect of sets of measure zero and measurability of various objects. For a function f on X, such carelessness is admissible when we consider the L_2-norms of its restrictions to various fibers, i.e.,

$$\|f\|_y := \int_{\pi^{-1}(y)} |f(x)|^2 \, d\lambda_y(x).$$

In what follows, an analogue of the "layerwise product" of two bundles is also encountered (it is largely considered as applied to the case of vector bundles, in which it is called the Whitney sum). Let $(X, \mathfrak{B}, \lambda)$ and $(Y, \mathfrak{C}, \nu)$ be extensions of $(Z, \mathfrak{D}, \nu)$ with projections π and ρ. Their fiberwise product is

$$X \times_Z Y := \{(x, y); \ x \in X, \ y \in Y, \ \pi(x) = \rho(y)\}.$$

It is assumed to be endowed with the measure $\lambda \times_Z \mu$, where

$$\lambda \times_Z \mu(A) := \int_Z \lambda_z(A \cap \pi^{-1}z)\mu_z(A \cap \rho^{-1}(z)) \, d\nu(z).$$

The measurable subsets of the space $X \times_Z Y$ are defined in such a way that the last definition makes sense for them; I shall not explain this in more detail. Note that $X \times_Z Y$ can be regarded as a natural extension of Z.

A DS $\{\psi^n\}$ in a phase space $(X, \mathfrak{B}, \lambda)$ is called an extension of a DS $\{\phi^n\}$ in $(Y, \mathfrak{C}, \mu)$ if the former space is represented as an extension of the latter with projection π and this projection commutes with the corresponding transformations, i.e., $\phi \circ \pi = \pi \circ \psi$. If DSs $\{\chi^n\}$ in X and $\{\psi^n\}$ in Y are extensions of a DS $\{\phi^n\}$ in Z, then their fiberwise product $\{(\chi \times_Z \psi)^n\}$ is the DS in $X \times_Z Y$ for which

$$(\chi \times_Z \psi)(x, y) := (\chi(x), \psi(y)).$$

It preserves the fiberwise products of the corresponding measures and is an extension of the DS $\{\phi^n\}$ in Z.

Some classes of DS extensions are introduced. We need two classes, "weakly mixing" and "compact." In the case where the base reduces to one point, the corresponding DSs are weakly mixing or metrically isomorphic to group translations of compact commutative groups. (The DSs of the latter type can also be characterized as follows: the orbits of the corresponding unitary operator U_ϕ on the Hilbert space $L_2(X, \lambda)$ (here $(U_\phi f)(x) := f(\phi x)$) are conditionally compact. Hence the term. Applying the general definition of a compact extension to the special case under consideration, we obtain immediately this property.) As is known, weak mixing is equivalent to the continuity of the DS spectrum (i.e., of the operator U_ϕ considered on the orthogonal complement to the constants); in addition, if the DS is ergodic, then compactness is equivalent to the discreteness of the spectrum. Thus, for ergodic DSs, the two classes introduced above correspond to DSs with continuous and discrete, respectively, spectra, but now this is not done in spectral terms.

In the general case, when the base is nontrivial, a weakly mixing extension of a DS $\{\phi^n\}$ in Y is defined as an extension $\{\psi^n\}$ in X such that its "fiberwise square" $\{(\psi \times_Y \psi)^n\}$ has no invariant sets except the full preimages of the invariant sets of the DS $\{\phi^n\}$ (treated as a quotient DS of $\{(\psi \times_Y \psi)^n\}$). When Y reduces to one point, we obtain the well-known definition of spectrum continuity: the Cartesian square $\psi \times \psi$ acting on $X \times X$ is ergodic. One of the definitions of a compact extension is as follows. In $L_2(X, \mathfrak{B}, \lambda)$, the functions f such that the restrictions

$$\{U_{\phi^n} f \restriction \pi^{-1}(y); \ n \text{ is arbitrary}\} \subset L_2(\pi^{-1}(y), \lambda_y) \tag{39}$$

of the functions $U_{\phi^n} f$ with all possible n to the fibers π^{-1} are conditionally compact subsets are dense; moreover, the ϵ-nets for such sets with different y can be chosen so that they are compatible in a certain sense; namely, for any $\epsilon > 0$, there exist $g_1, \ldots, g_k \in L_2(X, \mathfrak{B}, \lambda)$ such that their restrictions to the fibers are ϵ-networks for sets (39). Thus, both definitions (of weakly mixing and compact extensions) are, as it were, "fiberwise" modifications of the definitions of weakly mixing and conditionally compact orbits $\{U_{\phi^n} f\}$.

A primitive extension is an extension which is either weakly mixing or compact. Finally, the limit of an inverse system is defined in virtually the same way as in algebra and topology with the only difference that, in the case under consideration, we deal with different structures (spaces with measure on which transformations act), and we must take care of the corresponding structure on the limit space.

The above definitions and main structural theorem can be generalized to DSs with nonclassical time ranging over a commutative group G of finite rank. The most significant change is in the definition of primitive extensions. In the nonclassical case, an extension is called primitive if G can be represented as a direct product $G_1 \times G_2$ in such a way that the extension is compact when treated as an extension of a DS with time ranging over G_1 and weakly mixing when treated as an extension of a DS with time ranging over G_2. Thus, we cannot ensure that each term of our sequence of extensions is of one of the two simplest types, but we can ensure that, in each term, these two types are combined in a simple way.

An application of the ideas of ergodic theory or topological dynamics (such as the structural theorem [168, 169] or the much simpler considerations from [172, 169], depending on the situation) to Bernoulli DSs made it possible to obtain comparatively simple and uniform proofs of a number of theorems of number theory, including the well-known van der Waerden theorem (which does not require applying the structural theorem) and the theorem of E. Szemeredi, which is the far-reaching development of the van der Waerden theorem conjectured by P. Erdös and P. Turan. (These two theorems are stated below.) Many of these theorems were known earlier but, e.g., the n-parameter analogue of the Szemeredi theorem turned out to be new; its proof in the spirit of original Szemeredi's argument would be, apparently, very cumbersome (if possible at all).

I shall explain the character of such applications of DS theory for the example of the van der Waerden theorem, which asserts that, if $\mathbb{Z}_+$ (it is a little more convenient to start with zero) is divided into m disjoint subsets A_i, then, for any $l \in \mathbb{N}$, there exists an i such that A_i contains a "segment" of an arithmetic progression of length l. It is fairly easy to prove the following assertion:

If $\phi\colon X \to X$ is a continuous self-mapping of a metric compact space, then, for any $l \in \mathbb{N}$, there exist a sequence $n_k \to \infty$ and a point $x \in X$ such that $\phi^{in_k}x \to x$ for all $i = 1,\ldots,l$. Take a point $\xi = \{\xi_n\}$ in $\{1,\ldots,r\}^{\mathbb{Z}_+}$ for which ξ_n is equal to number of the set A_i that contains n. Let us apply the above-mentioned assertion to the restriction of the topological Bernoulli shift σ to the closure X of the trajectory $\{\sigma^n\xi\}$. For the corresponding point $x = \{x_n\}$ and sufficiently large values of k, the zeroth coordinate of each of the points $\sigma^{in_k}x$ with $i = 1,\ldots,l$ coincides with the zeroth coordinate of the point x, i.e.,

$$x_0 = x_{n_k} = x_{2n_k} = x_{3n_k} = \ldots = x_{ln_k}.$$

Since x belongs to the closure of the set $\{\sigma^n\xi\}$, the first ln_k coordinates of the point $\sigma^j\xi$ coincide with the first ln_k coordinates of the point x for some j. Therefore,

$$\xi_j = \xi_{j+n_k} = \xi_{j+2n_k} = \xi_{j+3n_k} = \ldots = \xi_{j+ln_k},$$

which means that the numbers $j, j + n_k, j + 2n_k, j + 3n_k, \ldots, j + ln_k$ (which form a segment of an arithmetic progression of length l) belong to the same A_i.

The Szemeredi theorem asserts that, if a subset $A \subset \mathbb{Z}_+$ has "positive upper density," i.e., if

$$\varliminf_{n\to\infty} \frac{\text{the number of elements in } A \cap [a_n, b_n]}{b_n - a_n} > 0$$

for some $a_n, b_n \in \mathbb{Z}_+$ such that $b_n - a_n \to \infty$, then A contains arbitrarily long segments of geometric progressions. The idea that this theorem can be proved with the use of some metric (in the sense of measure) analogue of the topological assertion which so easily implies the van der Waerden theorem suggests itself. This guess is correct, but the proof of the needed metric assertion[112] is much harder than that of the topological assertion. It is obtained by applying the structural theorem when "moving" step by step along the corresponding "sequence" of extensions (this sequence is generally transfinite, so the aforesaid should be understood *cum grano salis*).

Multiplicities of spectra

One of the versions of the spectral theorem for a unitary operator U in a separable Hilbert space H is that U has a model, which is a fairly precisely determined operator V in an as precisely determined Hilbert space K. The

---

[112] This assertion is that, if ϕ is an endomorphism of a the Lebesgue space, then, for any $l > 0$ and any measurable A with $\mu(A) > 0$, there exists an $i > 0$ such that

$$\mu(A \cap \phi^i A \cap \ldots \cap \phi^{li} A) > 0.$$

operator V is a model for U in the sense that it is conjugate to U by a unitary isomorphism $W\colon H \to K$. In this model, K is constructed from a finite or infinite number of mutually orthogonal "blocks," which are the spaces $L^2(\mathbb{S}^1, \nu)$, where $\mathbb{S}^1 := \{\lambda \in \mathbb{C};\ |\lambda| = 1\}$ and ν is a measure on $\mathbb{S}^1$. The measures ν on different blocks either are orthogonal or coincide; n_1 blocks correspond to the measure ν_1, $n_2 > n_1$ blocks correspond to the measure ν_2, etc.; there may be an infinite number of blocks corresponding to some measure ν_∞. Finally, the operator V is such that the blocks are invariant with respect to it and, if g belongs to one of the blocks, say, if $g \in L^2(\mathbb{S}^1, \nu)$, then $(Vg)(\lambda) = \lambda g(\lambda)$. The set $\{n_1, n_2, \dots\}$ (it may be finite or infinite; it may end with the symbol ∞, or it may not contain this symbol) is called the set of spectral multiplicities for U. It is uniquely determined by the operator, and all operators unitary conjugate to U have the same set of spectral multiplicities (while each measure ν_i is determined up to passage to an equivalent measure).

As soon as an automorphism ϕ of a Lebesgue space was assigned a unitary operator U_ϕ (by B. Koopman in the late 1920s), the question arose as to what operators can be obtained in this way, at least in the case of an ergodic ϕ (which is the most important case for "abstract" ergodic theory). In particular, what sets of spectral multiplicities can these operators have? The question about the corresponding measures (more precisely, about the corresponding classes of equivalent measures) is not posed at this point, but I shall distinguish between the cases of discrete, continuous, and mixed spectra.

In the case of a discrete spectrum, the answer is almost obvious: only $\{1\}$. Examples of continuous (even Lebesgue[113]) spectra for which the sets of multiplicities are $\{\infty\}$ have been known for a long time. After World War II, fairly many examples with other sets of spectral multiplicities, mostly $\{1\}$ (which is called the simple spectrum, or the spectrum of multiplicity 1), were gradually collected; very significant progress has been made over the past decade. It is easy to show that, for a mixed spectrum, the set of multiplicities must start with 1 (because of the discrete component); it turns out that any such set is realized for some ϕ. This was proved by different methods in the paper of J. Kwiatkowski and M. Lemanczyk cited in [173] and in the paper [174] of O. N. Ageev. In the case of a continuous spectrum, the problem has not yet been completely solved, but it was proved that any set beginning with 1 can be realized [175]. For sets beginning with $n_1 > 1$, the situation is not quite clear; among various examples the following one deserves special mention. When examples with $\{1\}$ appeared, V. A. Rokhlin asked whether there exists a ϕ with "double" continuous spectrum (with set of multiplicities $\{2\}$); recently, Ageev and V. V. Ryzhikov gave a positive answer. There is a recent survey [173] of

---

[113] Recall of the Banach problem mentioned in Section 1.3 (a).

this topic. I have written up a detailed and somewhat simplified version of the results of Ageev and Ryzhikov (D. V. Anosov. On spectral multiplicities in ergodic theory. In *Problems of Comtemporary Mathematics*, issue 3 (Moscow: MTsNMO, 2003)). Ageev has announced a new result: For any natural number n, there is an ergodic automorphism ϕ of the Lebesgue space such that the spectrum of U_ϕ is continuous, homogeneous, and of multiplicity n.

Approximations by periodic transformations and joinings

The best-known ideas and notions of ergodic theory refer to spectra and entropy. The simplest prototypes of the former arise in quasiperiodic oscillations and of the latter, in sequences of independent random trials. The prototypes *per se* bear no special relation to ergodic theory; they were known long before this theory emerged. The far-reaching development of the corresponding ideas has led to wider applications of these prototypes, not only to the "abstract" (purely metric) theory but also to "applied" ergodic theory, which should not surprise us if we recall where these ideas have come from. (The use of ideas of probabilistic origin in studying DSs with hyperbolic behavior of trajectories might seem somewhat unexpected for an outside observer, but such a use can also be regarded as a realization of the prophetic sentences of H. Poincaré (in his book on the probability theory[114]) that instability generates stochasticity.)

During the past three decades or thereabout, a new direction – a new system of notions and ideas – was formed, and its origin is indeed intrinsic. Possibly, this is the reason why the applications of the new direction to smooth DSs are scarce; the examples are mainly of combinatorial character (say, in spaces of sequences). It seems that the only well-known example of a smooth DS that refers to this direction is a horocycle flow. (There also exist artificial smooth examples.) But, as is known, most frequently, it takes half century for new ideas to find extensive applications outside their native domains; by now, only half of this term has passed.

I shall define only two notions from this area (actually, there are much more of them).

Let ϕ be an automorphism of a Lebesgue space (X, μ), and let $A \subset X$ be a measurable subset such that the sets

$$A, \ \phi A, \ \ldots, \ \phi^h A \tag{40}$$

are pairwise disjoint. Such sets are said to form a Rokhlin tower (for ϕ) of

---

[114] My impression is that, at the very end of the nineteenth–beginning of the twentieth century, several physicists expressed the same point of view (I remember two names: W. Kelvin and M. Smoluchowski). But I cannot give references.

height h. If $\mu\left(X \setminus \bigcup_{i=0}^{h} \phi^i A\right) < \epsilon$, I shall call such a tower a Rokhlin ϵ-tower. To understand the meaning of this notion, let us artificially close the chain of mappings

$$A \xrightarrow{\phi} \phi A \xrightarrow{\phi} \ldots \xrightarrow{\phi} \phi^h A$$

for a while by the mapping that takes each point of the form $\phi^h x$, where $x \in A$, to x. We obtain a periodic transformation ψ of the space $\bigcup_{i=0}^{h} \phi^i A$, which coincides with ϕ at least on the set $\bigcup_{i=0}^{h-1} \phi^i A$, whose measure is $> 1 - (\epsilon + 1/n)$. If h is large and ϵ is small, then ψ, as it were, approximates ϕ with a high accuracy.

It turns out that, for virtually any ϕ (with some trivial exceptions), there exist Rokhlin ϵ-towers of height h, where h is arbitrarily large and ϵ is arbitrarily small. V. A. Rokhlin used this fact to prove that the set of weakly mixing ϕ is "massive" in the space of all automorphisms ϕ with uniform (much more, with weak) topology. In the late 1960s, A. B. Katok, V. I. Oseledets, and A. M. Stepin used approximation considerations to study particular automorphisms ϕ and to construct ϕ with certain properties. Their statements were not merely direct applications of Rokhlin towers; they involved a quantitative condition on the rate of approximation by periodic transformations (see the article "Approximation by periodic transformations" in *Mathematical Encyclopaedia* or [167]). If no conditions were imposed, ϕ could be virtually arbitrary, and we could say nothing particular about its properties. Then, D. Ornstein suggested a modification of the same idea in which the qualitative condition was replaced by a requirement of qualitative character. The Ornstein condition is as follows: For any $\epsilon > 0$ and any measurable set B, there exists a Rokhlin ϵ-tower (40) such that the set B can be approximated by a union A' of some of the sets from (40) up to ϵ, i.e., so that the measure of the symmetric difference between B and A' is $\mu(B \triangle A') < \epsilon$. When this condition holds, ϕ (and the DS $\{\phi^n\}$) is said to have rank 1.

The automorphisms ϕ of rank 1 have a number of common properties. Thus, all of them are LB-automorphisms[115] and have simple spectra; an automorphism ψ commuting with ϕ is the weak limit of some sequence ϕ^{n_i}, which readily implies that the centralizer of ϕ either reduces to $\{\phi^n\}$ (this is always the case if ϕ is mixing) or is uncountable (the ϕ with uncountable centralizers have been classified). At the same time, in many respects, the properties of

---

[115] *LB* is an abbreviation for *loosely Bernoulli*. Initially, such automorphisms were also called standard, but now, this term is abandoned. See the section about the equivalence of DSs in the sense of Kakutani in [176].

DSs of rank 1 are very diverse (already the preceding sentence describes an instance of this diversity). For example, an ergodic automorphism with discrete spectrum is of rank 1, but the converse is not necessarily true. An automorphism ϕ of rank 1 may have square root[116] (it may even have continuum many pairwise nonisomorphic roots), or it may have no square roots at all; it may even happen that ϕ^2 has roots of all orders. Two automorphisms of rank 1 may be weakly isomorphic (i.e., each of them may be isomorphic to a factor of the other) not being isomorphic. Many unexpected examples of this kind (some of which contradict assumptions that might seem natural otherwise) are interesting, if only because they are examples of automorphisms of the Lebesgue space, irrespective of the fact that their constructions give automorphisms of rank 1. It should be mentioned that the constructions rarely use directly the definition of a DS of rank 1 given above; instead, some equivalent, more constructive, definitions are used as a rule. These definitions are longer, but they involve explicitly the parameters of certain constructions providing all automorphisms ϕ of rank 1. The success of a construction of an example depends on whether we are able to control the influence of these parameters on the properties of a ϕ to be constructed.

As there exist automorphisms of rank 1, there must be automorphisms of other ranks. The idea is that they should be well approximated in terms of several Rokhlin towers.

A detailed exposition of the theory of DSs of finite rank is contained in [177]. I shall only mention three facts. For a $k \in \mathbb{N}$, we have

$$\operatorname{rank} \phi^k = k \cdot \operatorname{rank} \phi.$$

The elements of the set of spectral multiplicities do not exceed the rank, but even the simplicity of the spectrum does not ensure the finiteness of the rank. For automorphisms of finite rank, the Rokhlin conjecture (mentioned in Section 1.3) that the mixing property implies mixing of all degrees is proved.

The second notion which I shall dwell on is that of joinings, which goes back to works of H. Furstenberg and D. Ornstein; a systematic use of joinings was initiated by D. Rudolf.

A joining of an automorphism ϕ of a measure space (X, μ) is an invariant normalized ergodic measure ν on the Cartesian power

$$\phi^{\times n} \colon X^n \to X^n, \quad (x_1, \ldots, x_n) \mapsto (\phi x_1, \ldots, \phi x_n)$$

which is projected to μ under the natural projection of X^n onto any of the factors X. (Note that, strictly speaking, we deviate from the orthodox concept

---

[116] That is, there may exist an automorphism ψ such that, at the same time, $\psi^2 = \phi$. Other roots are defined similarly.

of purely metric theory according to which everything should take place in a Lebesgue space. In the theory of Lebesgue spaces, X^n is considered only with the measure

$$\mu^{\times n} = \underbrace{\mu \times \ldots \times \mu}_{n \text{ times}};$$

it is a joining, but the other joinings of a weakly mixing ϕ are concentrated on sets with zero measure $\mu^{\times n}$, which must be neglected from the orthodox point of view. Thus, we have to represent the space (X, μ) not as a Lebesgue space but as a space of a different type, usually as a "standard Mackey" space, and verify afterwards that the result (information about joinings) does not depend on the choice of a particular realization.)

The "joining" distinctions between DSs refer to the amount of joinings: some DSs have "few" joinings, while other DSs have "many" of them. A DS $\{\phi^n\}$ always has joinings which are direct products of measures of the form

$$A_1 \times \ldots \times A_k \mapsto \mu(\psi_1 A_1 \cap \ldots \cap \psi_k A_k),$$

where ψ_i are automorphisms commuting with ϕ. If there are no other joinings (i.e., the joining are "few" in this sense), then the automorphism ϕ is said to be simple. M. Ratner proved that, for a horocycle flow $\{\phi_t\}$ on a closed surface of constant negative curvature (and in some other cases), all the ϕ_t with $t \neq 0$ are simple. The joinings are "particularly few" if ϕ is simple and its centralizer reduces to $\{\phi^n\}$; in this case, we say that ϕ has minimal self-joinings.

The notion of joinings *per se* bears no relation to approximation by periodic transformations. However, in almost all cases in which joinings have been found or any important information on joinings has been obtained, ϕ admits a rapid (in some sense) approximation. The work of Ratner is exceptional in this respect, as well as the proof of the Rokhlin conjecture for ϕ with singular spectrum obtained by B. Host through the use of joinings.

Joinings are the subject matter of the survey [178]; the paper [179] supplements it.

3.10   New periodic trajectories in the three-body problem

After the Russian text of this paper was published, I became aware of the discovery of principally new closed trajectories in the three-body problem:

A. Chenciner and R. Montgomery. A remarkable periodic solution of the three body problem in the case of equal masses. *Ann. Math.*, **152** (3) (2000), 881–901.

References

[1] J.-C. Yoccoz. Recent developments in dynamics. In *Proc. Int. Congr. Math., Zürich, Switzerland, 1994* (Basel: Birkhäuser, 1995), pp. 246–265.

[2] A. B. Katok and B. Hasselblatt. *Introduction to the Modern Theory of Dynamical Systems* (Cambridge: Cambridge University Press, 1995).

[3] J. Moser. Dynamical systems – past and present. In *Proc. Int. Congr. Math., Berlin, 1998*, vol. I: *Plenary Lectures and Ceremonies* (Bielefeld: University of Bielefled, 1998), pp. 381–402.

[4] I. P. Kornfel'd, Ya. G. Sinai, and S. V. Fomin. *Ergodic Theory* (Moscow: Nauka, 1980) [in Russian].

[5] V. I. Arnold and A. B. Givental'. Symplectic geometry. *Itogi Nauki i Tekhn. Sovrem. Probl. Mat. Fund. Napravleniya*, **4** (1985), 5–139.

[6] M. Gromov. Pseudo-holomorphic curves on symplectic manifolds. *Invent. Math.*, **82** (2) (1985), 307–347.

[7] H. Hofer and E. Zehnder. *Symplectic Invariants and Hamiltonian Dynamics* (Basel: Birkhäuser, 1994).

[8] *Holomorphic Curves in Symplectic Geometry*, eds. M. Audin and J. Lafontaine (Basel: Birkhäuser, 1994).

[9] M. Bialy and L. Polterovich. Hamiltonian diffeomorphisms and Lagrangian distributions. *Geom. Funct. Anal.*, **2** (2) (1992), 173–210.

[10] M. Bialy and L. Polterovich. Invariant tori and symplectic topology. In *Sinai's Moscow Seminar on Dynamical Systems. Amer. Math. Soc. Transl.*, ser. 2, vol. 171 (Providence, RI: Amer. Math. Soc., 1996), pp. 23–33.

[11] D. McDuff and D. Salamon. *Introduction to Symplectic Topology* (New York: Clarendon Press, 1995).

[12] V. I. Arnold. Topological problems of the theory of wave propagation. *Uspekhi Mat. Nauk*, **51** (1) (1996), 3–50.

[13] W. Ballmann. Der Satz von Lyusternik und Schnirelmann. In *Beitrage zur Differentialgeometrie. Bonner Math. Schriften*, vol. 102 (Bonn: University of Bonn, 1978), 1–25.

[14] I. A. Taimanov. Closed extremals on two-dimensional manifolds. *Uspekhi Mat. Nauk*, **47** (2) (1992), 143–185.

[15] V. Bangert. On the existence of closed geodesics on two-spheres. *Int. J. Math.*, **4** (1) (1993), 1–10.

[16] J. Franks. Geodesics on S^2 and periodic points of annulus homeomorphisms. *Invent. Math.*, **108** (2) (1992), 403–418.

[17] J. Franks. Rotation vectors and fixed points of area preserving surface diffeomorphisms. *Trans. Amer. Math. Soc.*, **348** (7) (1996), 2637–2662.

[18] S. Matsumoto. Arnold conjecture for surface homeomorphisms. Preprint.

[19] N. Hingston. On the growth of the number of closed geodesics on the two-sphere. *Internat. Math. Res. Notices*, **9** (1993), 253–262.

[20] A. V. Bolsinov and A. T. Fomenko. *Introduction to the Topology of Integrable Hamiltonian Systems* (Moscow: Nauka, 1997) [in Russian].

[21] H. O. Peitgen and P. H. Richter. *The Beauty of Fractals* (Berlin: Springer, 1986).

[22] M. Yu. Lyubich. Dynamics of rational transformations: Topological picture. *Uspekhi Mat. Nauk*, **41** (4) (1986), 35–95.

[23] A. E. Eremenko and M. Yu. Lyubich. Dynamics of analytic transformations. *Algebra i Analiz*, **1** (3) (1989), 1–70.

[24] L. Carleson and T. W. Gamelin. *Complex Dynamics* (New York: Springer, 1993).

[25] M. Lyubich. Dynamics of quadratic polynomials. I, II. *Acta Math.*, **178** (2) (1997), 185–297.

[26] W. de Melo and S. van Strien. *One-Dimensional Dynamics* (Berlin: Springer, 1993).

[27] M. Feigenbaum. Universality in the behavior of nonlinear systems. *Uspekhi Fiz. Nauk*, **141** (2) (1983), 343–374.

[28] F. P. Greenleaf. *Invariant Means on Topological Groups and Their Applications* (New York: Van Nostrand Reinhold, 1969).

[29] U. Krengel. *Ergodic Theorems* (Berlin–New York: De Gruyter, 1985).

[30] A. A. Tempel'man. *Ergodic Theorems on Groups* (Vilnius: Mokslas, 1986) [in Russian].

[31] R. I. Grigorchuk. An individual ergodic theorem for actions of free groups. In *XII Workshop on the Theory of Operators in Function Spaces, Tambov, 1987*, (Tambov: Tambovskii Gosudarstvennyi Pedagogicheskii Institut, 1987) [in Russian], vol. 1 p. 57.

[32] A. Nevo and E. M. Stein. A generalization of Birkhoff's pointwise ergodic theorem. *Acta Math.*, **173** (1) (1994), 135–154.

[33] A. Nevo and E. M. Stein. Analog of Wiener's ergodic theorems for semisimple Lie groups I. *Ann. Math.*, **145** (3) (1997), 565–595.

[34] G. M. Margulis, A. Nevo, and E. M. Stein. Analog of Wiener's ergodic theorems for semi-simple Lie groups. II. Preprint.

[35] V. I. Arnold and A. L. Krylov. A uniform distribution of points on the sphere and some ergodic properties of solutions to linear ordinary differential equations. *Dokl. Akad. Nauk SSSR*, **148** (1) (1963), 9–12.

[36] D. A. Kazhdan. Uniform distribution in the plane, *Trudy Mosk. Mat. O-va*, **14** (1965), 299–305.

[37] Y. Guivarc'h. Généralisation d'un théorème de von Neumann. *C. R. Acad. Sci. Paris, Sér A*, **268** (18) (1969), 1020–1023.

[38] O. N. Ageev. Dynamical systems having Lebesgue components of countable multiplicity in their spectra. *Mat. Sb.*, **136** (3) (1988), 307–319.

[39] D. V. Anosov. On the contribution of N. N. Bogolyubov to the theory of dynamical systems. *Uspekhi Mat. Nauk*, **49** (5) (1994), 5–20.

[40] D. Ornstein and B. Weiss. Ergodic theory of amenable group actions. I. The Rohlin lemma. *Bull. Amer. Math. Soc.*, **2** (1) (1980), 161–164.

[41] D. Ornstein and B. Weiss. Entropy and isomorphism theorems for actions of amenable groups. *J. d'Analyse Math.*, **48** (1987), 1–141.

[42] A. M. Stepin and A. T. Tagi-zade. A variational characterization of the topological pressure of amenable transformation groups. *Dokl. Akad. Nauk SSSR*, **254** (3) (1980), 545–549.

[43] Ya. G. Sinai. Gibbs measures in ergodic theory. *Uspekhi Mat. Nauk*, **27** (4) (1972), 21–64.

[44] R. Bowen. *Methods of Symbolic Dynamics (Collection of Works)* (Moscow: Mir, 1979) [Russian translation].

[45] A. M. Vershik and I. P. Kornfel'd. Periodic approximations and their applications. Ergodic theorems, spectral and entropy theories for general group actions. *Itogi Nauki i Tekhn. Sovrem. Probl. Mat. Fund. Napravleniya*, **2** (1985), 70–89.

[46] V. I. Arnold. *Catastrophe Theory* (Moscow: Nauka, 1990).

[47] V. I. Arnold. Catastrophe theory. *Itogi Nauki i Tekhn. Sovrem. Probl. Mat. Fund. Napravleniya*, **5** (1986), 219–277.

[48] A. V. Chernavskii. Applications of catastrophe theory to psychology. In *The Number and Thought*, no. 2 (Moscow: Znanie, 1979).

[49] T. Poston and I. Stewart. *Catastrophe Theory and Its Applications* (London–San Francisco–Melbourne: Pitman, 1978).

[50] V. I. Arnold. Lectures on bifurcations and versal families. *Uspekhi Mat. Nauk*, **27** (5) (1972), 119–184.

[51] V. I. Arnold. *Additional Chapters of the Theory of Ordinary Differential Equations* (Moscow: Nauka, 1978) [in Russian].

[52] V. I. Arnold, V. S. Afraimovich, Yu. S. Il'yashenko, and L. P. Shil'nikov. Bifurcation theory. *Itogi Nauki i Tekhn. Sovrem. Probl. Mat. Fund. Napravleniya*, **5** (1986), 5–219.

[53] S. I. Trifonov. The cyclicity of elementary polycycles of typical smooth vector fields. *Trudy Mat. Inst. Steklov*, **213** (1997), 152–212.

[54] N. K. Gavrilov and L. P. Shil'nikov. On trigonometric dynamical systems close to systems with nonrough homoclinic curve. I, II. *Mat. Sb.*, **88** (4) (1972), pp. 475–492; **90** (1) (1973), pp. 139–156.

[55] S. V. Gonchenko. On stable periodic motions of systems close to a system with nonrough homoclinic curve. *Mat. Zametki*, **33** (5) (1983), 745–755.

[56] S. V. Gonchenko, D. V. Turaev, and L. P. Shil'nikov. Dynamical phenomena in multidimensional systems with nonrough homoclinic Poincaré curve. *Dokl. Ross. Akad. Nauk*, **330** (2) (1993), 144–147.

[57] S. V. Gonchenko, L. P. Shilnikov, and D. V. Turaev. Dynamical phenomena in systems with structurally unstable Poincaré homoclinic orbits. *Chaos*, **6** (1) (1996), 15–31.

[58] J.-C. Yoccoz. Quadratic polynomials and the Henon attractor *Asterisque*, no. 201–203 (1991), exp. no. 734 (1992), 143–165.

[59] S. Newhouse. Non-density of axiom A(a) on S^2. *Proc. A.M.S. Symp. Pure Math.*, **14** (1970), 191–202.

[60] S. Newhouse. Diffeomorphisms with infinitely many sinks. *Topology*, **13** (1) (1974), 9–18.

[61] S. Newhouse. The abundance of wild hyperbolic sets and nonsmooth stable sets for diffeomorphisms. *Publ. Math. IHES*, **50** (1979), 101–151.

[62] C. Robinson. Bifurcations to infinitely many sinks. *Comm. Math. Phys.*, **90** (3) (1983), 433–459.

[63] J. Palis and M. Viana. High dimension diffeomorphisms displaying infinitely many periodic attractors. *Ann. Math.*, **140** (1) (1994), 207–250.

[64] S. V. Gonchenko, D. V. Turaev, and L. P. Shil'nikov. On the existence of Newhouse regions near systems with nonrough homoclinic Poincaré curves (the multidimensional case). *Dokl. Ross. Akad. Nauk*, **329** (4) (1993), 404–407.

[65] N. Romero. Persistence of homoclinic tangencies in higher dimensions. *Ergod. Theory and Dyn. Dystems*, **15** (4) (1995), 735–757.

[66] J. Palis and F. Takens. *Hyperbolicity and Sensitive Chaotic Dynamics at Homoclinic Bifurcations* (Cambridge: Cambridge University Press, 1993).

[67] L. P. Shil'nikov and D. V. Turaev. On blue-sky catastrophes. *Dokl. Ross. Akad. Nauk*, **342** (5) (1995), 596–599.

[68] D. V. Turaev and L. P. Shil'nikov. An example of a wild strange attractor. *Mat. Sb.*, **189** (2) (1998), 137–160.

[69] S. V. Gonchenko, L. P. Shilnikov, and D. V. Turaev. On models with nonrough Poincaré homoclinic curves. *Physica D*, **62** (1–4) (1993), 1–14.

[70] S. V. Gonchenko, D. V. Turaev, and L. P. Shilnikov. On models with non structurally stable Poincaré homoclinic curve. *Dokl. Akad. Nauk SSSR*, **320** (2) (1991), 269–272.

[71] V. Kaloshin. Generic diffeomorphisms with superexponential growth of number of periodic orbits. Preprint no. 1999/2, SUNY Stony Brook Inst. for Math. Sc. (1999).

[72] S. V. Gonchenko, D. V. Turaev, and L. P. Shil'nikov. On Newhouse regions for two-dimensional diffeomorfisms close to a diffeomorfism with a nonrough homoclinic contour. *Trudy Mat. Inst. Steklov*, **216** (1997), 76–125.

[73] S. V. Gonchenko and L. P. Shil'nikov. On two-dimensional analytic area-preserving diffeomorfisms with a countable set of elliptic stable periodic points. *Regulyarnaya i Khaoticheskaya Dinamika*, **2** (3/4) (1997), 106–123.

[74] L. P. Shil'nikov. Homoclinic orbits: Since Poincaré till today. Preprint no. 571, Weierstrass-Inst. für Angew. Analysis und Stochastik, Berlin (2000).

[75] L. P. Shil'nikov, V. S. Afraimovich, and V. V. Bykov. On attracting nonrough limit sets of Lorentz attractor type. *Trudy Mat. Inst. Steklov*, **44** (1982), 150–212.

[76] R. V. Plykin, E. A. Sataev, and S. V. Shlyachkov. Strange attractors. *Itogi Nauki i Tekhn. Sovrem. Probl. Mat. Fund. Napravleniya*, **66** (1991), 100–148.

[77] D. V. Anosov. Structurally stable systems. *Trudy Mat. Inst. Steklov*, **169** (1985), 59–93.

[78] D. V. Anosov and V. V. Solodov. Hyperbolic sets. *Itogi Nauki i Tekhn. Sovrem. Probl. Mat. Fund. Napravleniya*, **66** (1991), 12–9.

[79] D. V. Anosov, I. U. Bronshtein, S. Kh. Aranson, and V. Z. Grines. Smooth dynamical systems. *Itogi Nauki i Tekhn. Sovrem. Probl. Mat. Fund. Napravleniya*, **1** (1985), 151–242.

[80] C. C. Pugh and R. C. Robinson. The C^1 closing lemma, including Hamiltonians. *Ergod. Theory and Dyn. Systems*, **3** (2) (1983), 261–313.

[81] M. M. Peixoto. Acceptance speech for the TWAS 1986 award in mathematics. In *The Future of Science in China and the Third World. Proc. of the Second General Conf. Organized by the Third World Acad. Sci*, eds. A. M. Faruqui and M. H. A. Hassan (Singapore: World Scientific, 1989), 600–614.

[82] R. Mañé. On the creation of homoclinic points. *Publ. Math. IHES*, **66** (1988), 139–159.

[83] R. Mañé. A proof of the C^1 stability conjecture. *Publ. Math. IHES*, **66** (1988), 161–210.

[84] J. Palis. On the C^1 Ω-stability conjecture. *Publ. Math. IHES*, **66** (1988), 211–215.

[85] S. Hayashi. Connecting invariant manifolds and the solution of the C^1 stability and Ω-stability conjectures for flows. *Ann. Math.*, **145** (1) (1997), P. 81–137; Correction. *Ann. Math.*, **150** (1) (1999), 353–356.

[86] C. C. Pugh. Against the C^2-closing lemma. *J. of Diff. Equations*, **17** (2) (1975), 435–443.

[87] O. S. Kozlovski. Structural stability in one-dimensional dynamics. Preprint, University of Amsterdam (1997).

[88] V. I. Arnold and Yu. S. Il'yashenko. Ordinary differential equations. *Itogi Nauki i Tekhn. Sovrem. Probl. Mat. Fund. Napravleniya*, **1** (1985), 7–149.

[89] A. A. Bolibrukh. The Riemann–Hilbert problem on the complex projective line. *Mat. Zametki*, **46** (3) (1989), 118–120.

[90] D. V. Anosov and A. A. Bolibruch. *The Riemann–Hilbert problem* (Wiesbaden–Braunschweig: Vieweg, 1994).

[91] *Trudy Mat. Inst. Steklov*, **213** (1997).

[92] Yu. S. Il'yashenko. Dulac's memoir "On limit cycles" and related questions of the local theory of differential equations. *Uspekhi Mat. Nauk*, **40** (6) (1985), 41–78.

[93] Yu. S. Ilyashenko. *Finiteness Theorems for Limit Cycles* (Providence, R.I.: Amer. Math. Soc., 1991).

[94] J. Ecalle. *Fonctions analysable et solution constructive du probleme de Dulac* (Paris: Hermann, 1992).

[95] J.-C. Yoccoz. Nonaccumulation of limit cycles. *Asterisque*, no. 161–162 (1988), **3**, exp. no. 690, (1989), 87–103.

[96] W. Balser. *From Divergent Power Series to Analytic Functions* (Berlin: Springer, 1994).

[97] I. G. Petrovskii and E. M. Landis. On the number of limit cycles for the equation $\frac{dy}{dx} = \frac{P(x,y)}{Q(x,y)}$, where P and Q are polynomials of degree 2. *Mat. Sb.*, **37** (2) (1955), 209–250.

[98] E. M. Landis and I. G. Petrovskii. On the number of limit cycles for the equation $\frac{dy}{dx} = \frac{P(x,y)}{Q(x,y)}$, where P and Q are polynomials. *Mat. Sb.*, **43** (2) (1957), 149–168.

[99] I. G. Petrovskii and E. M. Landis. Corrections to the papers "On the number of limit cycles for the equation $\frac{dy}{dx} = \frac{P(x,y)}{Q(x,y)}$, where P and Q are polynomials of degree 2" and "On the number of limit cycles for the

equation $\dfrac{dy}{dx} = \dfrac{P(x,y)}{Q(x,y)}$, where P and Q are polynomials." *Mat. Sb.*, **48** (2) (1959), 253–255.

[100] V. V. Amel'kin, N. A. Lukashevich, and A. P. Sadovskii. *Nonlinear Ocsillations in Second-Order Systems* (Minsk: Izd. BGU, 1982).

[101] E. M. Landis and I. G. Petrovskii. A letter to the Editorial Board. *Mat. Sb.*, **73** (1) (1967), 160.

[102] A. Kotova and V. Stantso. On few-parameter generic families of vector fields on the two-dimensional sphere. In *Concerning Hilbert 16th problem*, eds. Yu. Ilyashenko and S. Yakovenko (Providence, RI: Amer. Math. Soc., 1995), pp. 155–201.

[103] A. Starkov. Fuchsian groups from the dynamical viewpoint. *J. of Dynam. and Control Syst.*, **1** (3) (1995), 427–445.

[104] M. Ratner. Interactions between ergodic theory, Lie groups, and number theory. In *Proceedings of the International Congress of Mathematicians, Zurich, 1994*, ed. S. D. Chatterji (Basel: Birkhäuser, 1995), pp. 157–182.

[105] A. V. Safonov, A. N. Starkov, and A. M. Stepin. Dynamical systems with transitive symmetry group. Geometric and statistical properties. *Itogi Nauki i Tekhn. Sovrem. Probl. Mat. Fund. Napravleniya*, **66** (1991), 187–242.

[106] A. N. Starkov. A new progress in the theory of homogenous flows. *Uspekhi Mat. Nauk*, **52** (4) (1997), 87–192.

[107] G. Margulis. *Oppenheim conjecture* (Rover Edge–New York: World Scientific Publ., 1997).

[108] A. N. Starkov. *Dynamical Systems on Homogeneous Spaces* (Moscow: Fazis, 1999).

[109] K. Kuperberg. A smooth counterexample to the Seifert conjecture. *Ann. Math.*, **140** (3) (1994), 723–732.

[110] G. Kuperberg and K. Kuperberg. Generalized counterexamples to the Seifert conjecture. *Ann. Math.*, **144** (2) (1996), 239–268.

[111] I. Tamura. *Topology of Foliations* (Tokyo: Iwanami Shoten, 1976) [in Japanese]; (Moscow: Mir, 1979) [Russian translation].

[112] H. Hofer. Pseudoholomorphic curves in symplectization with applications to the Weinstein conjecture in dimension three. *Invent. Math.*, **114** (3) (1993), 515–563.

[113] S. R. Fenley. The structure of branching in Anosov flows of 3-manifolds. *Comment. Math. Helv.*, **73** (1997), 259–297.

[114] P. Pansu. The geodesic flow of negatively curved Riemannian manifolds. In *Seminar Bourbaki*, vol. 1990/91, exposes 730–744, Asterisque nos. 201–203 (1991) (Montrouge: Societe Mathematique de France, 1992), exp. no. 738, pp. 269–298.

[115] G. Knieper. The uniqueness of the measure of maximal entropy for geodesic flows on rank 1 manifolds. *Ann. Math.*, **148** (1998), 291–314.

[116] G. Besson, G. Courtois, and S. Gallot. Les variétés hyperboliques sont des minima locaux de l'entropie topologique. *Invent. Math.*, **117** (3) (1994), 403–445.

[117] L. Barreira, Ya. Pesin, and J. Schmeling. Dimension and product structure of hyperbolic measures. *Ann. Math.*, **149** (3) (1999), 755–783.

[118] M. L. Byalyi and L. V. Polterovich. Geodesic flows on the two-dimensional torus and "commensurability–incommensurability" phase transitions. *Funkts. Analiz i Ego Prilozh.*, **20** (4) (1986), 9–16.

[119] L. V. Polterovich. Geodesics on the two-dimensional torus with two rotation numbers. *Izv. AN SSSR. Ser. Mat.*, **52** (4) (1988), 774–787.

[120] V. Bangert. Mather sets for twist maps and geodesics on tori. In *Dynamics Reported*, eds. U. Kirchgraber and H. O. Walter, vol. 1 (Chichester–Stuttgart: John Wiley and B. G. Teubner, 1988), pp. 1–56.

[121] A. B. Katok. Some remarks on Birkhoff and Mather twist map theorems. *Ergod. Theory and Dyn. Systems*, **2** (2) (1982), 185–194.

[122] A. Fathi. Theoreme KAM faible et theorie de Mather sur les systemes lagrangiens. *C. R. Acad. Sci. Paris, Ser. I Math.*, **324** (9) (1997), 1043–1046.

[123] R.Mañé. Lagrangian flows: the dynamics of globally minimizing orbits. *Bol. Soc. Brasil. Mat. (N.S.)*, **28** (2) (1997), 141–153.

[124] G. Contreras, R. Iturriaga, G. P. Paternain, and M. Paternain. Lagrangian graphs, minimizing measures and Mane's critical values. *Geometric and Functional Analysis*, **8** (5) (1998), 788–809.

[125] A. D. Bruno. *Power Geometry in Algebraic and Differential Equations* (Moscow: Nauka–Fizmatlit, 1998) [in Russian].

[126] E. F. Mishchenko and N. Kh. Rozov. *Differential Equations with a Small Parameter and Relaxation Oscilaltions* (Moscow: Nauka, 1975) [in Russian].

[127] E. F. Mishchenko, Yu. S. Kolesov, A. Yu. Kolesov, and N. Kh. Rozov, *Periodic Motions and Bifurkation Processes in Singularly Perturbed Systems* (Moscow: Nauka–Fizmatlit, 1995) [in Russian].

[128] P. Cartier. Singular perturbations of ordinary differential equations and nonstandard analysis. *Uspekhi Mat. Nauk*, **39** (2) (1984), 57–76.

[129] A. K. Zvonkin and M. A. Shubin. Nonstandard analysis and singular perturbations of ordinary differential equations. *Uspekhi Mat. Nauk*, **39** (2) (1984), 77–127.

[130] W. Eckhaus. Relaxation oscillations including a standard chase on french ducks. In *Asymptotic Analysis II* (Berlin: Springer, 1983), pp. 449–494.

[131] *Dynamic Bifurcations*, ed. E. Benoit (Berlin: Springer, 1991).

[132] A. I. Neishtadt. On tightening of stability loss under dynamic bifurcations. I. *Differents. Uravneniya*, **23** (12) (1987), 2060–2067.

[133] A. I. Neishtadt. On tightening of stability loss under dynamic bifurcations. II. *Differents. Uravneniya*, **24** (2) (1988), 226–233.

[134] A. I. Neishtadt, C. Simo, and D. V. Treschev. On stability loss delay for a periodic trajectory. In Nonlinear Dynamical Systems and Chaos, eds. H. Broer et al. (Basel: Birkhäuser, 1996), pp. 253–278.

[135] V. I. Bakhtin. On Averaging in Multifrequency Systems. *Funkts. Analiz i Ego Prilozh.* **20** (2) (1986), 1–7.

[136] V. I. Arnold, V. V. Kozlov, and A. I. Neishtadt. Mathematical aspects of classical and celestial mechanics. *Itogi Nauki i Tekhn. Sovrem. Probl. Mat. Fund. Napravleniya*, **3** (1985).

[137] D. V. Treshchev. *Introduction to the Theory of Perturbations of Hamiltonian Systems* (Moscow: Fazis, 1998).

[138] C. Simo. Averaging under fast quasiperiodic forcing. In *Hamiltonian Mechanics, Integrability and Chaotic Behaviour*, ed. J. Seimenis (New York: Plenum Press, 1994), pp. 11–34.

[139] V. F. Lazutkin. Analytic integrals of a semistandard mapping and separatrix splitting. *Algebra i Analiz*, **1** (2) (1989), 116–131.

[140] V. F. Lazutkin. An analytic integral along the separatrix of the semistandard map: existence and an exponential estimate for the distance betwen the stable and unstable separatrices. *Algebra i Analiz*, **4** (4) (1992), 110–142.

[141] A. Delshams, V. Gelfreich, A. Jorba, and T. M. Seara. Exponentially small splitting of separatrices under fast quasiperiodic forcing. *Comm. Math. Phys.*, **189** (1) (1997), 35–71.

[142] J. A. Ellison, M. Kummer, and A. W. Saenz. Transcendentally small transversality in the rapidly forced pendulum. *J. Dyn. Diff. Equations*, **5** (1993), 241–277.

[143] J. Pöschel. Nekhoroshev estimates for quasi-convex Hamiltonian systems. *Math. Z.*, **213** (2) (1993), 187–216.

[144] P. Loshak. Canonical perturbation theory: An approach based on simultaenous approximations. *Uspekhi Mat. Nauk*, **47** (6) (1992), 59–140.

[145] P. Lochak and A. I. Neishtadt. Estimates of stability time for nearly integrable systems with a quasiconvex Hamiltonian. *Chaos*, **2** (4) (1992), 495–499.

[146] A. B. Katok. The entropy conjecture. In *Smooth Dynamical Systems*, ed. D. V. Anosov (Moscow: Mir, 1977), pp. 181–203 [in Russian].

184   *D. V. Anosov*

[147] Y. Yomdin. Volume growth and entropy. *Israel J. Math.*, **57** (3) (1987), 285–300.

[148] Y. Yomdin. C^k-resolution of semialgebrai mappings. Addendum to "Volume growth and entropy." *Israel J. Math.*, **57** (3) (1987), 301–317.

[149] M. Gromov. Entropy, homology and semialgebraic geometry. *Asterisque*, nos. 145–146 5 (1987), 225–240.

[150] M. Audin. *Spinning Tops* (Cambridge: Cambridge University, 1996).

[151] J. Moser. *Integrable Hamiltonian Systems and Spectral Theory* (Izhevsk: Udmurdskii Universitet, 1999) [in Russian].

[152] V. V. Kozlov, Integrability and nonintegrability in Hamiltonian dynamics. *Uspekhi Mat. Nauk*, **38** (1) (1983), 3–67.

[153] V. V. Kozlov, *Symmetry, Topology, and Resonances in Hamiltonian Mechanics* (Izhevsk: Izd. UdGU, 1995) [in Russian].

[154] I. A. Taimanov. Topologiya rimanovykh mnogoobrazii s integriruemymi geodezicheskimi potokami. *Trudy Mat. Inst. Steklov*, **205** (1994), 150–163.

[155] L. Butler. A new class of homogeneous manifolds with Liouville-integrable geodesic flows. Math. Preprint no. 1998–8, Queen's Univ. at Kingstone (1998).

[156] A. V. Bolsinov and I. A. Taimanov. Integrable geodesic flows on the suspension of toric automorphisms. Preprint Sfb 288 no. 426 (1999).

[157] C. Conley. *Isolated Invariant Sets and the Morse Index* (Providence, RI: Amer. Math. Soc., 1978).

[158] K. Mischaikow. Conley index theory. In *Dynamical systems (Montecatini Terme, 1994)* (Berlin: Springer, 1995), pp. 119–207.

[159] J. N. Mather and R. McGehee. Solutions of the collinear four-body problem which become unbounded in finite time. In *Dynamical Systems, Theory and Applications (Rencontres, Battelle Res. Inst., Seattle, Wash., 1974)* (Berlin: Springer, 1975), pp. 573–597.

[160] Z. Xia. The existence of noncollision singularities in newtonian systems. *Ann. Math.*, **135** (3) (1992), 411–468.

[161] M. Grayson, C. Pugh, and M. Shub. Stably ergodic diffeomorphisms. *Ann. Math.*, **140** (2) (1994), 295–329.

[162] C. Pugh and M. Shub. Stably ergodic dynamical systems and partial hyperbolicity. *J. of Complexity*, **13** (1) (1997), 125–179.

[163] C. Pugh and M. Shub. Stable ergodicity and partial hyperbolicity. In *Int. Conf. on Dynamical Systems, Montevideo, 1995, a Tribute to R. Mañé*, eds. F. Ledrappier et al. (Harlow: Longman, 1996), pp. 182–187.

[164] M. Shub and A. Wilkinson. Pathological foliations and removable zero exponents. *Invent. Math.*, **139** (3) (2000), 495–508.

[165] R. Adler, B. Kitchens, and M. Shub. Stably ergodic skew products. *Discr. and Contin. Dynam. Systems*, **2** (3) (1996), 349–350.

[166] W. Parry and M. Pollicott. Stability of mixing for toral extensions of hyperbolic systems. *Trudy Mat. Inst. Steklov*, **216** (1997), 354–363.

[167] A. B. Katok, Ya. G. Sinai, and A. M. Stepin. The theory of dynamical systems and general transformation groups with invariant measure, *Itogi Nauki i Tekhn. Matematicheskii Analiz* **13** (1975), 129–262.

[168] H. Furstenberg, Y. Katznelson, and D. Ornstein. The ergodic theoretical proof of Szemerédy's theorem. *Bull. Amer. Math. Soc.*, **7** (3) (1982), 527–552.

[169] H. Furstenberg. *Recurrence in Ergodic Theory and Combinatorial Number Theory* (Princeton, N.J.: Princeton University Press, 1981).

[170] R. J. Zimmer. Extensions of ergodic group actions. *Illinois J. of Math.*, **20** (3) (1976), 373–409.

[171] R. J. Zimmer. Ergodic actions with generalized discrete spectrum. *Illinois J. of Math.*, **20** (4) (1976), 555–588.

[172] H. Furstenberg. Poincaré recurrence and number theory. *Bull. Amer. Math. Soc.*, **5** (3) (1981), 211–234.

[173] G. R. Goodson. A survey of recent results in the spectral theory of ergodic dynamical systems. *J. of Dynam. and Control Syst.*, **5** (2) (1999), 173–226.

[174] O. N. Ageev. The spectrum multiplicity function and geometric representations of rearrangement, *Mat. Sb.*, **190** (1) (1999), 3–28.

[175] O. N. Ageev. On the spectrum multiplicity function for dynamical systems. *Mat. Zametki*, **65** (4) (1999), 619–621.

[176] I. P. Kornfel'd and Ya. G. Sinai. The entropy theory of dynamical systems. *Itogi Nauki i Tekhn. Sovrem. Probl. Mat. Fundam. Napravleniya*, **2** (1985), 44–70.

[177] S. Ferenczi. Systems of finite rank. *Colloq. Math.*, **73** (1) (1997), 35–65.

[178] J.-P. Thouvenot. Some properties and applications of joinings in ergodic theory. In *Proc. of the 1993 Alexandria Conference. Ergodic Theory and its Connections with Harmonic Analysis* (Cambridge: Cambridge University Press, 1995), pp. 207–235.

[179] V. V. Ryzhikov. Around simple dynamical systems. Induced joinings and multiple mixing. *J. of Dynam. and Control Syst.*, **3** (1) (1997), 111–127.

A. A. Razborov

Foundations of computational complexity theory

Lecture on April 23, 1998

This lecture is intended for those who are not acquainted with the theory of computational complexity. For this reason, I shall talk only about the foundations of this theory and its very first results. I shall try to deliver the main ideas which guide the researchers in this field of science.

The setting of my narrative are stories about one personage called M (for "mathematician"). I shall start with the following story.

1 Prehistory

Once upon a time, M sat at home trying to prove some (maybe important, or maybe not) theorem T. He tried to prove this theorem for a week, two weeks, a month, ..., but with no result. In the end, he gave up and asked quite the natural question:

Why, can this theorem be proved at all?

The question was addressed to nobody; most likely, it would fly away if another personage, L (for "logician"), would not pass by.

L heard the question, came into the room, and explained that such questions started to interest mathematicians some time in the beginning of the twentieth century. In a general setting, this question is contained in the famous *Hilbert's Program* devoted to the notion of mathematical proof.

This program, in particular, included the following three items.

Formalization of the notion of proof. Before asking whether or not an assertion can be proved, we must give a rigorous mathematical definition of provability.

Completeness. After the notion of proof is formalized, it is necessary to establish the *completeness* of the constructed *formal theory*. This means that any true proposition T must be provable in this formalization. In particular, for certain historical reasons, Hilbert himself was largely interested in the question about the provability of a properly formalized statement about the consistency of mathematics.

Decidability. The next goal of the program was to construct a computing device capable of determining whether a theorem T formulated in some formal language is provable (it was assumed that, according to the second item of the program, provability was equivalent to being true).

It is well known that the first goal of the program has been accomplished successfully. At present, the majority of mathematicians use the word "theorem" for a proposition that can be proved by means of the Zermelo–Fraenkel set theory (even if not all of them recognize this).

As to the remaining two items of the program, the situation happened to be much worse. The first shock, which rocked the whole mathematical community in the 1930s, was the result of Kurt Gödel that no sufficiently strong theory where the set of axioms is specified by an explicit list can be complete. More precisely, if such a theory is consistent and provides means for formalizing all arguments about positive integers, then there exists a proposition which cannot be neither proved nor disproved; moreover, an example of such a proposition is the statement that Hilbert was interested in, namely, that the theory under consideration is consistent.

Today, we deal with computations, so of greater interest to us is the 1936 theorem of Church. Church proved that there exists no algorithm capable of automatically verifying whether or not a given proposition is provable.

Thus, mathematician M learned from logician L that the question about the provability of a mathematical statement cannot be answered in full generality.

By that moment, however, he had already begun to doubt that theorem T was true, and he decided to look for a counterexample. M went to a computer room and wrote a program P which would successively check all input strings until a counterexample for T would be found. He launched the program and waited for its termination.

The program was running for an hour, two hours, a day, a week, ... Again, the mathematician worried and asked another question, namely,

Will P ever terminate?

The mathematician already knew whom to ask. He found the logician and asked him: Does there exist a method for learning whether a given program would terminate or it would be running eternally? Again, L gave quite a qualified answer. By a theorem proved by Turing in the 1930s, there exists no algorithm determining whether or not a given program will terminate, i.e., the termination problem is undecidable.

2 Point of departure

After that, the mathematician went home. He had a child, who studied geometry at school; as it happens sometimes, the kid asked dad to help him with the homework. People believe that mathematicians are able to calculate well and solve quadratic equations and various problems from elementary geometry (in reality, this is not always true).

Thus, the mathematician started to solve the problem suggested by his child, and realized that he had completely forgotten his geometry from elementary school. He, however, remembered that any geometric problem can be written using Cartesian coordinates in the language of real numbers.

A question occurred to him: Does there exist a universal algorithm for solving elementary geometry problems? After the conversation with the logician, he knew that if the positive integers are expressible in some theory, then this theory is undecidable. Intuition suggested that the theory dealing with real numbers, which are much more abundant, must be undecidable too. Still, to be on the safe side, he telephoned the logician to make sure that he was right. Strange as it seems, it turned out that he was not. The classical 1948 result of Tarski asserts the existence of an algorithm checking the provability of statements of elementary geometry.[1]

The mathematician rejoiced. He had nearly believed that logic was of no use; now, he saw that logic had very practical applications to real life.

Thus, M asked the kid to wait and went to a software shop. He found two compact disks with programs for solving geometric problems,

Tarski for Windows 95   and   Collins for Windows 95.

The first CD was $30 and the second $300. Naturally, M tried to find out the reason for such a difference in the price. He failed to get any explanations, so he bought the cheaper disk, Tarski for Windows 95.

M came home and inserted the CD in the computer. He wanted to test the program on a problem from the school textbook. However, the same thing happened again. The program was running for an hour, two hours, ... but it was not apparently going to output a solution. The mathematician halted the program and asked it to prove some elementary theorem, e.g., that the sum of angles in a triangle equals 180 degrees. Very little changed; the program continued thinking for an hour, two hours, a day, another day, ... Then M telephoned the logician and asked him (somewhat angrily) what was going on.

---

[1] And, moreover, of any propositions about real numbers formulated in terms of arithmetic operations, elementary logical connectives (negation, conjunction, disjunction, and implication), and quantifiers over the set of all real numbers: "for all (real numbers)" and "there exists (a real number) such that..."; the set of such propositions is called the Tarski algebra.

L answered that it was none of his business. Tarski proved the theorem about the existence of an algorithm, and surely, the algorithm had been correctly implemented in the program on the CD. What happens then bears no relation to mathematics or logic.

This is the point where computational complexity theory begins. We are interested not merely in the existence of algorithms for solving a problem but also in their efficiency.

Of course, the story told above is somewhat stylized, but it reflects the actual development of research that has led to the modern state of the art. In particular, the idea that some algorithms may be better than others and this is important was clearly understood in the 1960s. At that time, the first real computers (they had the scary name *electronic computing machines*) were designed, such as BESM (probably, many of you do not even know what it is). It became clear that a mathematical theory was necessary.

3 Foundations of the theory: basic notions

We put aside stories from M's life for a while; we shall keep returning to them later.

Let us try to give a few definitions.

The first observation made by M when he tried to understand this theory was that the overwhelming majority of algorithmic problems can be encoded as problems of evaluating some mapping

$$f\colon \{0,1\}^* \to \{0,1\}^*$$

from the set of finite binary words to itself. It is such mappings that we shall deal with.

The second question is: What devices are we going to employ for computing our mapping? There exists a huge diversity of algorithmic languages and computer architectures, but to what degree are the differences between them essential? It is the answer to this question that distinguishes computational complexity theory from other related fields. Thus, let us consider this question in more detail.

As an example we take the program most frequently used by mathematicians who are not engaged in programming. This is the TEX word processor. The choice of TEX is of no special importance; we could consider any other program instead.

Thus, we have a mapping f; in the case under consideration, this is a mapping

$$\boxed{\texttt{paper.tex}} \longrightarrow \boxed{\texttt{paper.dvi}}$$

transforming a file `paper.tex` into a file `paper.dvi` (we treat both files as long binary words). This transformation is implemented by an algorithm `texdvi.exe` translated for execution on a 286 processor (those who tried to run TEX on a 286 processor remember how it was).

At the moment, we are interested in the running time of an algorithm. It is this *complexity measure* (in the language of the science we deal with) that will be most often considered today. There are other characteristics, like memory (this complexity measure is second in importance), but we cannot discuss them in detail because of time limitations.

So far as our problem is concerned, the progress was made in two directions. First, as everybody knows, processors were improved:

$$\text{Intel 286, Intel 386, Intel 486, } \ldots$$

More and more powerful models were designed, diverse engineering contrivances accelerating their operations were invented, etc. Secondly, the algorithm itself was improved, and its new versions were released:

$$\texttt{texdvi1.exe, } \ldots \texttt{ texdvi10.exe, } \ldots$$

Subject to certain restrictions, the following formula is valid:

$$\left\langle \begin{array}{c} \text{total running} \\ \text{time} \end{array} \right\rangle = \left\langle \begin{array}{c} \text{number of operations} \\ \text{in the algorithm} \end{array} \right\rangle \times \left\langle \begin{array}{c} \text{time taken by} \\ \text{one operation} \end{array} \right\rangle.$$

The terms of this simple formula improve more or less independently. Roughly speaking, software is responsible for the first factor and hardware for the second.

What is going on from the mathematical point of view? Suppose that we have an algorithm involving t operations. When the performance of our processor improves, the time required for solving our problem is multiplied by a constant.

For this reason, there is the tradition in computational complexity theory to measure the running time of an algorithm up to a multiplicative constant ("up to $O(\cdot)$," as mathematicians say). This is very important. Such an approach allows us to disregard the choice of a particular computational model, the time required to perform one operation, the system of commands used to operate the computer, and so on. It is this approach that allows us to build a rather beautiful mathematical theory.

The theory of computational complexity somewhat differs from the all-embracing (by definition) Marxism–Leninism theory (apparently, not all of you know what it is, which is good): when a formula involves two terms and our theory is by no means responsible for the second one, this is declared explicitly. The improvement of processors is other people's business; we are concerned with improving algorithms up to a multiplicative constant.

We proceed to build a rigorous theory. One of the immediate advantages of the convention introduced above is that we are not very much concerned with the choice of a particular model. As a rule, changing a model (this time, I mean an abstract mathematical model) improves or slows down performance only by a constant factor, and we agreed to ignore such changes.

The standard definition of a computational model is as follows. An algorithm is executed on a machine with addressable memory whose cells are indexed by positive integers, each cell can store positive integers, the machine can perform arithmetic operations, and so on. The details are of little interest, because our $O(\cdot)$-convention allows us not to pay that much attention to them. If we replace "up to $O(\cdot)$" with "up to a polynomial," then all realistic computing devices whatsoever will become equivalent.

So, fix some computational model. Given a machine M, which computes a function f, and input data x (some binary word), we can define our basic function $T(M,x)$ equal to the number of operations (elementary steps) needed for the machine M to process the input word x.

4 The speed-up theorem

Now, let us study this function $T(M,x)$. The first thing that comes to mind is to choose the best algorithm for solving our algorithmic problem f and define the complexity of this problem as the complexity of this best algorithm.

It turns out that, unfortunately, such an intuitively obvious approach cannot be used because of the *Blum speed-up theorem*. Loosely speaking, this theorem asserts that the notion of "the best machine" for a given problem cannot be defined (at least for some problems).

The Blum theorem began a series of theorems underlying the modern theory of computational complexity. All these theorems were proved around 1970; for example, Blum proved his speed-up theorem in 1967. At approximately the same time, the concept of NP-completeness was developed; a description of this concept will conclude our introductory story about computational complexity theory.

In what follows, we shall need yet another important notion. The function $T(M,x)$ behaves very irregularly. Consider the example of the TEX processor. For the overwhelming majority of files, the program terminates at the very beginning because of the noncompliance with the input TEX format, and for some files termination may never occur because of infinite looping. In the general case, there is no way of studying this function, because it is too loose. We want to extract from this function another function of a positive integer argument, i.e., to obtain a function of positive integers to positive integers which would reflect the behavior of $T(M,x)$. There are several approaches to this task. We shall consider only the most popular one; the corresponding mapping of positive integers is called the *worst-case complexity*. It is defined by the formula

$$t_M(n) = \max_{|x| \leqslant n} T(M, x).$$

Among all words of bit length not exceeding n we choose the word for which the machine works worst of all (i.e., most slowly). The time taken to process such a word is called the worst-case complexity and denoted by $t_M(n)$. The machine surely terminates on any input word of length not exceeding n within time $t_M(n)$. Of course, for some words it may happen earlier.

Below, we give a simplified version of Blum's theorem; in fact, $\log t$ can be replaced by an arbitrary "reasonable" function tending to infinity.

Theorem (Blum, 1967). *There exists a computable[2] function f such that any machine M computing f can be sped up in the sense that there exists another machine M', which also computes f, such that*

$$t_{M'}(n) \leqslant \log t_M(n)$$

for almost all n.

The function mentioned in the statement of the Blum theorem is fairly exotic (the theorem implies that the worst-case time of its computation grows very rapidly for any machine). One of the "ideological" problems in computational complexity theory is to develop the theory to get rid of such pathological phenomena whenever possible. The proof of the Blum theorem involves a construction using the technique of diagonalization, and the resulting function bears no relation to real-life computations or to mathematics in general. But nevertheless, since we develop a mathematical theory, there is nothing we can do about it; we must admit that the chosen approach is not suitable and try others.

5   Complexity classes

Thus, we cannot hope to design the best machine computing a given function for every function. The alternative is the notion of *complexity classes*, which is one of the central notions of complexity theory.

There is a somewhat loose analogy with the definitions of integrals in the senses of Riemann and Lebesgue. If we cannot integrate in the sense of Riemann, we change the axis and start summation over a different axis. In the

---

[2] A computable function is a function such that there exists at least one algorithm evaluating it.

situation under consideration, we cannot say what the best machine for a function is; so, let us change the axis. We consider the set of all acceptable machines and call the class of all functions computable by such machines a *complexity class*. Perhaps, it is easier to give an example right away rather than launch into a long discussion:

$$\mathrm{DTIME}(t(n)) := \{f \mid \exists\, M : (M \text{ calculates } f) \,\&\, (t_M(n) = O(t(n)))\}.$$

This is one of the central definitions in complexity theory. The letter D denotes deterministic algorithms (other algorithms exist too) and TIME means precisely what you think. Given an arbitrary function $t(n)$ of a positive integer argument, we form the complexity class consisting of all functions f such that there exists a machine M computing f for which the time-signalizing function is bounded by the initial function $t(n)$ up to a multiplicative constant. The Blum theorem cited above is valid only for some special functions. But if we want to speed up computation by a factor of, say, 10, we can do this for any function by, e.g., increasing the number of available computer commands. This is why the O-estimate appears on the right-hand side of the definition.

Now, we shall define one of the most important complexity classes, namely,

$$\mathrm{P} = \bigcup_{k \geqslant 0} \mathrm{DTIME}(n^k).$$

The class P consists of the functions that can be computed by the machines under consideration and the time taken to compute them is polynomial in the input length. It is very convenient both practically and theoretically. In practice, it gives a rather good approximation for the class of functions that can be computed in reasonable time on real-life computers (some exceptions are discussed below). From the mathematical point of view, this class is extremely convenient because it is closed with respect to superposition. We shall see later on that it is this fact that makes a computability theory for the class P possible.

There are similar classes of languages recognizable in exponential time

$$\mathrm{EXPTIME} = \bigcup_{k \geqslant 0} \mathrm{DTIME}\left(2^{n^k}\right);$$

we can also define the double exponential time

$$\mathrm{DOUBLEEXPTIME} = \bigcup_{k \geqslant 0} \mathrm{DTIME}\left(2^{2^{n^k}}\right),$$

and so on.

6 The hierarchy theorem and complexity of elementary geometry

Now, let us return to mathematician M. It turns out that the running time of the Tarski algorithm, which M used when trying to solve problems of elementary geometry, belongs to none of the classes considered above. It is bounded from below as

$$t_{\text{Tarsky}}(n) \geqslant 2^{2^{\cdot^{\cdot^{\cdot^{2}}}}} \Big\} \ \epsilon n \ .$$

As you remember, there was yet another compact disk, Collins for Windows 95, with software for verifying propositions from the Tarski algebra.

Theorem (Collins). *The Tarski algebra belongs to the complexity class* DOUBLEEXPTIME.

Now, M was able to understand the difference between the CDs: the running time of Collins' algorithm is much, much less than that of Tarski's algorithm (although it may still be very large; the doubly exponential upper bound does not even guarantee that the time required to prove the theorem about the sum of angles in a triangle is less than the existence time of the Universe).

The question arises: Is it possible to further improve the decision algorithm for the Tarski algebra? A more general question is: Is it possible to improve any algorithm whatsoever? It might happen that, for example, any computable function belongs to the class P. Or, at least, any function from DOUBLEEXPTIME belongs to P.

In other words, the question is whether we have an object of study at all or the Blum speed-up theorem applies to *all* functions.

The second corner-stone of complexity theory is the *hierarchy theorem*.

The formulation of the hierarchy theorem given below, just as the formulation of the speed-up theorem given above, is far from being most general.

Theorem (Hartmanis, Stearns (1965)). P $\neq$ EXPTIME.

Thus, not all complexity classes coincide, so we do have an object of study.

I cannot deny myself the pleasure of giving an almost complete proof of this theorem. If you still remember mathematician M, in the second episode of his misadventures, he asked whether the program would terminate *ever*. Now, taught by bitter experience, he asked the following question:

Will this program terminate before New Year's Day?

It turns out that, if the time remaining to New Year's Day is precisely exponential, then this problem is what separates EXPTIME from P. Because there is a very simple algorithm for verifying whether the program will terminate before New Year's Day; namely, we should simply wait until New Year's Day, and the problem will be solved automatically. The hierarchy theorem asserts that this problem cannot be solved substantially faster.

Of course, the proof contains some technical details, but its essence is as described.

After that, mathematician M was ready to comprehend the following theorem.

Theorem (Fisher–Rabin (1974)). *The Tarski algebra does not belong to the class* P. *The running time of any decision algorithm for the Tarski algebra is at least* $2^{\epsilon n}$, *where ϵ is an absolute constant.*

Such a large *lower bound* explains why our mathematician failed to solve problems in the Tarski algebra in practice.

The most complicated and, apparently, most important domain of complexity theory is precisely that related to obtaining lower bounds. In the English-language literature, the part of complexity theory which deals with designing algorithms is called the theory of algorithms, and complexity theory *per se* refers to establishing lower bounds. Thus, complexity theory tries to prove that there exist no efficient algorithms.

Already the example of Tarski algebra exhibits two difficulties which people trying to prove lower bounds must overcome. Look, there was the Tarski theorem; all the efforts to improve it yielded no result. It was natural to suppose that the estimate was optimal. After that, an algorithm based on completely different ideas was designed. The algorithms are diverse: there are complicated algorithms, there are various algorithms. Still, we try to prove that more efficient algorithm can never be designed and at the same time analyze properties of algorithms from a fairly large class.

7 Reducibility and completeness

Let us proceed with our theory. The notion of complexity classes was introduced not to make the statements of theorems more economical (after all, the Fischer–Rabin theorem can be stated without mentioning any complexity classes – it is sufficient to leave only the second sentence in its formulation). The notion of complexity classes becomes important at the moment when the notion of *reducibility* emerges; this is the second central notion of the modern theory of computational complexity. There exist several versions of reducibility; I shall consider only the most important ones.

Karp reducibility

This reducibility is very simple. First, recall that we deal with computation of functions that map finite words to finite words. But, in many cases, it is much more convenient (and, as a rule, this involves no loss of generality) to consider so-called *languages*. A language L is a set of words $L \subseteq \{0,1\}^*$; it can also be interpreted as a mapping of the form $\phi \colon \{0,1\}^* \to \{0,1\}$ (then $L = \phi^{-1}(1)$). The passage to languages is not very restrictive: any function from words to words can be associated with a language, namely, with the set of pairs (x, i) such that the ith bit of $f(x)$ equals 1.

Definition. A language L_1 reduces to a language L_2 in the sense of Karp (this is denoted by $L_1 \preceq_p L_2$) if there exists a function f from P such that

$$\forall x : (x \in L_1 \equiv f(x) \in L_2) \,.$$

Reducibility means that, in order to recognize words from the language L_1, we can run a subprogram converting the initial word x into $f(x)$ and apply an algorithm recognizing the words from the language L_2 to $f(x)$. Under this definition of reducibility, the recognition program for L_2 is called in only once; allowing arbitrarily many calls to this subprogram, we obtain a different reducibility (in the sense of Turing). At the moment, the difference does not matter to us.

The reducibility relation is a preorder. It is reflexive and transitive. Its most fundamental property is that, if $L_2 \in$ P, then $L_1 \in$ P. It is important here that the class of polynomials is closed with respect to the superposition operation.

For example, EXP-reducibility will not work out. There would be no transitivity with respect to such a reducibility. The closure would be the finite towers of exponents, i.e., the class of *elementary recursive functions*.

Among the natural classes containing at least one reasonable function and closed with respect to superposition we can mention the class of quasipolynomials $2^{(\log n)^{O(1)}}$ and the class of quasilinear functions $n \log^{O(1)} n$. In recent years, quasilinear functions have become a focus of attention because of the belief that good algorithms are precisely those having quasilinear running time.

Anyway, the class P is central to the theory, so we shall consider polynomial reducibility.

It is easy to see that the class EXPTIME is closed with respect to this reducibility. Therefore, it is natural to ask whether there exist the most complex languages in this class, that is, such that any other language from this class reduces to them. If such languages exist, then we can solve any problem from EXPTIME by using an arbitrary recognition algorithm for such a complex language and the polynomial reducibility. The languages from a certain complexity class to which any language from this class can be reduced are called *complete* (with respect to the given class and the given type of reducibility). Omitting the requirement that the language itself must belong to the class under consideration, we obtain the definition of a *hard* language.

The Fischer–Rabin theorem is proved in precisely this way. The nonexistence of a polynomial algorithm for the Tarski algebra is not proved directly; instead, it is proved that the Tarski algebra is hard for the class EXPTIME. Thus, a polynomial algorithm for the Tarski algebra would provide a polynomial algorithm for all other problems from this class. But we know that there are problems in EXPTIME that cannot be solved in polynomial time (as that of termination of a program before New Year's Day).

Such an argument is typical of complexity theory. Problems are reduced to each other rather than solved directly. Naturally, the more problems from a given class reduce to the problem under consideration, the better is the situation.

The success of this area is judged by practice: numerous complete problems arise in various situations, and they are more natural than the termination problem.

8   Are all polynomial algorithms good?

Now, it is the right time to return to our poor mathematician M and talk about exceptions to the rule "the class P = the class of effectively computable functions." One day, for some purposes, M needed to solve a system of linear inequalities

$$\sum_{j=1}^{n} a_{ij}x_j \leqslant b_i \qquad (i = 1, \ldots, m).$$

In other words, he needed a linear programming package. Experts in computational complexity theory always study the literature before purchasing software, and M, taught by bitter experience, decided to adopt this rule too. Of course, he had heard about the simplex method used to solve linear programming problems almost everywhere, including the military department of Moscow University. But M read a paper proving that the simplex method is not polynomial. After that, M found a 1979 paper of Khachiyan where a polynomial algorithm for solving the linear programming problem was constructed. Thus, he went to the Mitino radio market and looked for something like **Khachiyan for Windows 95**. Surprisingly, he could not find anything like that. All the CDs for sale contained software based on the simplex method and its variations. It turns out that, although the algorithm for solving linear programming problems by the simplex method is exponential and Khachiyan's algorithm is polynomial, in practice, the former is faster than the latter. It is pretty hard to construct an example of the linear programming problem such that solving it by the simplex method takes a long time, while the running time of the polynomial algorithm of Khachiyan is approximately the same for all input data (the exponent is 6, which is pretty very bad).

This is the best-known exception to the rule that polynomial algorithms are good and exponential algorithms are bad. But in reality, this exception only confirms the rule: although nobody uses Khachiyan's algorithm to solve linear programming problems, this algorithm solves problems that cannot be solved by the simplex method in principle. For example, suppose we are given a convex body K and a direction. It is required to maximize the corresponding linear form on K. We know nothing about the convex body: it is specified with the use of a black box, or an *oracle*. This means that, when you choose a point p, the oracle says whether or not this point belongs to the body K and, if not, specifies a hyperplane separating p from K. The oracle's answers are correct to a certain accuracy ϵ. The solution must have the same accuracy. Clearly, the simplex method, which works by searching vertices, does not apply to this problem, because there are no vertices at all. Both the Khachiyan algorithm and the science emerged from it perfectly (that is, polynomially) copes with such problems. The role of the parameter n describing the input size is played by $d \cdot (\log \epsilon^{-1})$, where d is the dimension and ϵ is the precision of computation.

Such exceptions are very few. The second well-known example is primality testing.[3] As a rule, if you have an algorithm which theoretically works well, and this is an algorithm for a normal problem which arose from somewhere rather than constructed with hooligan purposes, then this algorithm works well in practice, too. In particular, the exponent k in the bound n^k for the running time of an algorithm is usually small, and when it is large, it can be reduced by various tricks. For the overwhelming majority of natural problems, the exponent does not exceed 3.

9   Nondeterministic computations

Apparently, those who came to this lecture, wish to learn something about the best-known open problem in this domain, P $\neq$ NP. I have already explained a little what P is. Now, let us proceed to NP.

For this purpose, we again return to M. While he was learning complexity theory, his son entered a university and started to study mathematical logic. As

---

[3] In 2002, a polynomial-time algorithm for this problem was designed, which removed it from the list of exceptions. (*A.R.* – added in proof)

you know, the course in mathematical logic begins with the celebrated propositional calculus. Suppose, you have a propositional formula $\Phi(p_1, \ldots, p_n)$; it is called a tautology if it is always true, no matter what $p_1, \ldots, p_n$ may be. One of the unpleasant exercises when learning propositional calculus consist in verifying whether or not a given propositional formula is a tautology. It was this question that M's son asked his dad this time. Naturally, M, being an expert in complexity theory, did not go to shop for software but tried to fit the problem into one of the known complexity classes. To use the standard notation, we shall consider the dual problem SAT of satisfiability: Is there at least one assignment to variables that makes the formula true?

Superimposing this problem on the whole picture of complexity classes, M saw at once that SAT $\in$ EXPTIME. An algorithm for solving the problem SAT in exponential time is evident: there are 2^n possible assignments to the variables, and the time required to evaluate the formula for any given assignment is polynomial. The next step is to classify the problem: Does it belong to P, or is it complete in EXPTIME? This question, which arose in the early 1970s years, is still open. It is hard to explain (at least, I would not undertake this task) why the experts who were trying to solve this problem for almost 30 years failed to construct a polynomial algorithm. But the fact that the proof of the completeness of this problem in EXPTIME is elusive does have an explanation. Looking at real exponential algorithms, such as the decision algorithm for the analogue of the Tarski algebra over the complex field, and comparing them with the above childish argument, we see the difference at once with the naked eye: the algorithm suggested above for solving SAT uses exponential time very little, only for the *exhaustive search* of the exponential number of possibilities, whereas every possibility is processed in a polynomial time.

An explosion in computational complexity theory was initiated by defining the class NP of languages that can be recognized by such search algorithms. More scientific name for search algorithms is nondeterministic algorithms. Thus, NP is the class of languages that can be recognized within nondeterministic polynomial time.

Now, let us give a definition of this class. This definition, as well as the definition of the class P, contains the word "machine"; more precisely, it involves a nondeterministic Turing machine (NTM), although people dealing with real-life machines may feel displeased by the use of the word "machine" in this context. By a nondeterministic machine, we mean a machine which operates as a usual machine with the only difference that it can put a question mark in some cell, after which its operation forks into two branches, 0 and 1:

(the branches correspond to writing 0 and 1 in the cell). Then the machine continues to work. At some moment, the operation may branch again. A computation tree arises. Along each branch of the computation tree, the NTM operates as a usual computer, but the result obtained by such a machine depends on the results obtained along all the branches. Let us determine this result in the case of verifying whether a word x belongs to a language L. Each computational branch gives one of the two possible answers "yes" and "no." An NTM recognizes the language L when a word x belongs to L if and only if the answer "yes" is obtained on *at least one* computational branch.

Let us illustrate the power of NTMs on the example of the satisfiability of propositional formulas. Very frequently, when a student (polynomial deterministic Turing machine) fails to solve a problem, the teacher plays the role of a nondeterministic machine and communicates the correct answer (prepared in advance), which can be easily verified. In other words, an NTM "is eager" to prove the assertion $x \in L$, and at the moments of branching, it acquires an unlimited intellectual ability and makes the best choice. If there exists a computation branch for which the answer is "yes," then $x \in L$. Otherwise, no branching leads to a positive result (a word does not belong to the language only when the NTM has no possibility whatsoever to prove the converse).

Do not ask me about the physics of the process, that is, how branching goes on, where such a machine is, whether it is possible to look at it... Such machines do not exist in reality. One of the most significant recent achievements in complexity theory was the development of a quantum computation model. Quantum computers are one of the candidates for the role of such a machine in the real world. At least, physicists have no essential objections to the existence of quantum computers and experts in complexity theory have no imperative objections to the possibility of simulating nondeterministic machines by quantum computers. Moreover, abstract quantum computers are already capable of solving some of the most important search problems, such as factorization of numbers.[4]

I emphasize again that a nondeterministic machine is a purely theoretical notion. The convenience offered by this notion, however, is truly fantastic.

---

[4] At the last International Congress (Berlin, August 1998), American mathematician P. Shor was awarded the Nevanlinna Prize for this research.

That was the very first fictitious model in computational complexity theory. At present, there are a lot of such models, which are much more complicated, such as interactive proofs and so on. These models do not arise by themselves but are constructed for the purposes of defining complexity classes and classifying natural problems.

It is intuitively clear that the notion of an NTM is perfectly adjusted to modeling search algorithms. In fact, we have already shown how to do this. If we have an algorithm which tries to search through some set of possibilities, then our machine can guess which of these possibilities is good and, then, simulate the second part (a polynomial verification of a particular choice). Proving the converse assertion is somewhat more complicated, but still, it is fairly simple. An NTM generates a calculation tree, and we must perform an exhaustive search over all possible branches.

The modern theory of computational complexity was initiated by works of Cook, Karp, and Levin (who obtained his results independently) in the early 1970s (1970–1972). They proved the following series of theorems.

1. The *satisfiability* problem (SAT) is complete for the class NP (this is a theorem of Cook). Thus, the problem which mathematician M tried to solve (whether a polynomial algorithm for solving the SAT problem can be designed) is equivalent to the question whether the classes P and NP coincide (P $\neq$ NP). If there is no polynomial algorithm, then these classes do not coincide, since the problem SAT separates them; if there is such an algorithm, then any problem from the class NP can be solved efficiently. The satisfiability problem is responsible for the class NP. Roughly speaking, NP is nothing but the problems that reduce to the satisfiability of propositional formulas.

2. Naturally, such a result may meet some rejection, because the satisfiability problem is not that important, and it is hardly expedient to construct a whole theory of solving it. But the next step was a 1971 paper of Karp, where 21 complete problems for the class NP were formulated. All these problems are equivalent.

At present, the list of NP-complete problems, which arise in literally all the domains of mathematics, contains thousands of problems. Search problems arise wherever algorithms do. This is no surprise, because programmer's work largely consists in choosing a better possibility. It is much more surprising that, very often (in fact, with some exceptions, almost always), particular search problems are comparatively easy to classify as polynomial-time computable or NP-complete.

This is the way the theory of search problems was developed. It has acquired great importance, because search algorithms arise almost anywhere. In the opinion of American topologist Smale, the P $\neq$ NP problem will be one of the most important questions of mathematical science in the forthcoming century.

The story about what makes, so to speak, the core of computational complexity theory is nearing an end. It should not leave the impression that the entire complexity theory only works on the relation P ≠ NP. The scheme of research described above, which unites problems into complexity classes and studies them by means of reducibility, proved surprisingly efficient and fruitful in many diverse situations.

Let us briefly mention several most important possibilities (in addition to quantum calculations already mentioned).

There is algebraic complexity. For example, if we are interested in computation of some polynomial, it is determined by the number of additions and multiplications involved in the computation. The details of bitwise computation and of the implementation of arithmetic operations are ignored.

There is also geometric complexity related to the Voronoi diagrams and similar things, but I shall not speak about it.

Finally, there is Boolean complexity, which differs from the complexity considered today in that it deals with functions $f: \{0,1\}^n \to \{0,1\}$ defined on words of fixed length.

In each of these domains, there are plenty of very interesting problems which fit into the framework of this general ideology.

Acknowledgments

I am very grateful to M. N. Vyalyi and M. V. Alekhnovich for qualified help with making the lecture readable and Professor J. Krajíček for valuable comments.

S. P. Novikov

The Schrödinger equation and symplectic geometry

Lecture on June 25, 1998

This lecture is concerned with fairly elementary things. I shall acquaint you with some useful notions of mathematical physics. Before proceeding to what I am going to talk about (this is some exotic operators on graphs), let me remind you of the Schrödinger equation. This is the equation

$$-\psi'' + u(x)\psi = \lambda\psi;$$

the value $u(x)$ is called a potential. Sometimes, this equation is considered formally, but if the Schrödinger equation arises from quantum mechanics, then it is usually assumed that $\psi \in L_2(\mathbb{R})$, i.e.,

$$\int_{\mathbb{R}} |\psi(x)|^2 dx < \infty.$$

The space $L_2(\mathbb{R})$ is Hilbert. The spectral problem is considered for vectors from this space.

Unlike in the traditional matrix situations where the problem on eigenvalues arises, in the case under consideration, the notion of continuous spectrum arises.

We assume that the function $\psi(x)$ sufficiently rapidly tends to zero as $x \to \infty$. To obviate the necessity of substantiating the convergence, we shall even assume that the function $\psi(x)$ is compactly supported, i.e., it identically vanishes outside a bounded domain.

1 Quantum scattering

We are interested in the notion of quantum scattering. When the potential is compactly supported, it is natural to assume that there exist solutions $\psi_\pm$ such that $\psi_\pm(x) \to e^{\pm ikx}$ as $x \to -\infty$ (here $k^2 = \lambda$). This is one basis of the solution space. The second basis is obtained by considering solutions $\phi_\pm$ such that $\phi_\pm(x) \to e^{\pm ikx}$ as $x \to +\infty$.

The solution space of a second-order differential equation with an arbitrary fixed λ is two-dimensional; therefore,

$$a\psi_+ + b\psi_- = \phi_+, \qquad c\psi_+ + d\psi_- = \phi_-. \tag{1}$$

The matrix $T = \begin{pmatrix} a & b \\ c & d \end{pmatrix}$ arises. In some purely mathematical books, this matrix is incorrectly called the scattering matrix. The correct term is the *monodromy matrix*. Monodromy matrices are encountered in various fields of mathematics, e.g., in the complex theory of differential equations, very frequently. A monodromy matrix is the matrix of transfer from $-\infty$ to $+\infty$ along the x-axis, i.e., the matrix of transfer from left to right. This matrix has a whole series of very interesting properties. In particular, consider the question about scattering. Consider a $\lambda \in \mathbb{R}$. We know that self-conjugate matrices have real spectrum. The same is true of operators if they are self-conjugate in a certain reasonable sense. Moreover, it is easy to see that, in the case under consideration, a spectrum of interest to us occurs only for positive λ. For $k^2 = \lambda$, this means that $k \in \mathbb{R}$. We refer to the domain $\lambda \geqslant 0$ on the real line as the *scattering zone*.

Now, I shall say what the scattering matrix is. For $k \in \mathbb{R}$, we have $\overline{\psi}_+ = \psi_-$ and $\overline{\phi}_+ = \phi_-$, and the matrix T has the form $T = \begin{pmatrix} a & b \\ \overline{b} & \overline{a} \end{pmatrix}$; moreover, $\det T = |a|^2 - |b|^2 = 1$ for certain reasons, which we are especially interested in. Such matrices form the group $SU(1,1)$. These are the special unitary matrix, which preserve indefinite Hermitian inner product with one positive square and one negative square. I recommend proving that $SU(1,1) \cong SL_2(\mathbb{R})$ as a useful algebraic exercise. The point is that we have obtained a complex matrix for a purely real equation: we require that the function $u(x)$ be real (usually, u has the meaning of an electric potential). If we took $\cos kx$ and $\sin kx$ instead of $e^{\pm ikx}$ at $\pm\infty$ when selecting a basis, then the monodromy matrix would be a real matrix from the group $SL_2(\mathbb{R})$.

We have changed the basis ψ_+, ψ_- for the basis ϕ_+, ϕ_-. It can be proved that, when the basis ϕ_-, ψ_- is changed for the basis ϕ_+, ψ_+, the transition matrix is unitary (although not every unitary matrix can be obtained in this way). This matrix is denoted by $S(\lambda)$ and called the *scattering matrix*.

This is elementary material, which is commonly taught to second-year physics students. Mathematics students do not always study it at all. But the course in ordinary differential equations for students in theoretical physics follows precisely this scheme.

The transition from the matrix $T(\lambda) \in SU(1,1)$ to the matrix $S(\lambda) \in U(2)$ is called the *Cayley transformation*.

There is yet another useful exercise; I borrowed it from the good textbook of V. I. Arnold entitled *Additional Chapters of the Theory of Ordinary Differential Equations*. Consider a pair of vectors e_1 and e_2 and the matrix $T \in SL_2(\mathbb{R})$ mapping these vectors to vectors e_1' and e_2'. In four-dimensional space $\mathbb{R}^4$, we introduce the basis e_1, e_2, e'_1, e'_2 and define a skew-symmetric inner product in such a way that $\langle e_1, e_2 \rangle = -\langle e'_1, e'_2 \rangle = 1$ and the remaining products of basis elements vanish (certainly, $\langle e_2, e_1 \rangle = -\langle e'_2, e'_1 \rangle = -1$). The space $\mathbb{R}^4$ with this skew-symmetric inner product is called the *space of symplectic geometry*, or the *space of symplectic linear algebra*. Note that the graph of the mapping T on $\mathbb{R}^4$ is a two-dimensional subspace of this space.

Exercise. The graph of the mapping T is a Lagrangian subspace, i.e., if η is an arbitrary vector and $\xi = T\eta$, then $\langle \eta, \xi \rangle = 0$.

Now, I can tell about analogues of scattering theory which arise in graph theory. A *graph* Γ is a one-dimensional simplicial complex; it has only edges and vertices. We require that the graph have no double edges, i.e., that the intersection of any two edges either be empty or consist of one vertex. In addition, we require that the graph have no end-vertices. This means that, if there is an edge going to a vertex, then there must be another edge going from this vertex. For a vertex P, we denote the number of edges going from this vertex by n_P. We assume that $n_P < \infty$; the graphs themselves may sometimes be infinite.

The simplest example of an infinite graph is the discretized line $\mathbb{R}$ with marked integer points.

2 Schrödinger operators on graphs

There are two Schrödinger operators on graphs. One of them acts on functions of vertices, and the other acts on functions of edges.

Definition. The Schrödinger operator acts on the space of functions whose variables are vertices of a graph by the rule

$$(L\psi)_P = \sum_{P'} b_{P,P'} \psi_{P'},$$

where $b_{P,P'} = b_{P',P} \in \mathbb{R}$ and $b_{P,P'} \neq 0$ only if $P = P'$ or $P \cup P' = \partial R$ is the boundary of an edge R.

The number $b_{P,P'}$ is called a *potential*. Only neighboring vertices can interact. There are thousands of papers on probability theory and combinatorics which study the Schrödinger operator (it is also called the second-order operator) acting on functions of graph vertices.

The Schrödinger operator acting on functions of edges is defined similarly as

$$(L\psi)_R = \sum_{R'} d_{R,R'} \psi_{R'}.$$

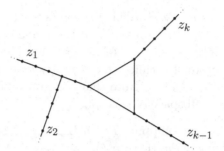

Figure 1. A graph with k tails

The condition is also similar, namely, $d_{R,R'} \neq 0$ only for edges being nearest neighbors of each other. By nearest neighbors, we mean either coinciding edges or edges having a common vertex.

The edge operators do not reduce to vertex operators.

The simplest example of the Schrödinger operator on a simplicial complex (acting on simplices of arbitrary dimensions) is the so-called Laplace–Beltrami operators; they have been extensively used by topologists, starting with the famous paper of Singer and his coauthors, where the so-called Ray–Reidemeister–Singer torsion was studied. Thus, the multidimensional situation has been considered.

Consider the simplest situation of the discretized line. The Schrödinger and Sturm–Liouville equations were discretized in computational mathematics since the advent of this science; naturally, the discrete Schrödinger has been extensively studied. In the case of the discretized line, the edge and vertex Schrödinger operators coincide. To every edge R_n and vertex P_n we can assign the number n. The Schrödinger equation on the discretized line is then written as

$$c_{n-1}\psi_{n-1} + c_{n+1}\psi_{n+1} + v_n\psi_n = \lambda_n\psi_n.$$

In the theory of solitons, this equation was extensively used to study the so-called discrete systems. In classical computational mathematics, it was assumed that $c_n = 1$ for all n. This constraint is inconvenient. Afterwards, it turned out that, in the theory of solitons and in quantum physics, it is convenient to consider the general class of Schrödinger operators.

On the discretized line, as well as in the continuous case, each λ determines a two-dimensional solution space. A monodromy operator naturally arises. If the coefficients approach certain constants at $\pm\infty$, then a scattering matrix with the same properties as above arises. This case differs little from that of the continuous line.

Consider a graph Γ with k tails $z_1, \ldots, z_k$ (Fig. 1). Each tail is a discretized half-line. We require that $c_n = 1$ and $v_n = 0$ outside a bounded domain, i.e., the Schrödinger equation has the form

$$\psi_{n-1} + \psi_{n+1} = \lambda \psi_n. \tag{2}$$

What the solutions to equation (2) look like? The answer is very simple: we can take $\psi_{n,\pm} = a_\pm^n$, where

$$a_\pm = \frac{1}{2} \left(\lambda \pm \sqrt{\lambda^2 - 4} \right).$$

The scattering zone (the zone of continuous spectrum) is $|a_\pm| = 1$ (or, equivalently, $-2 \leqslant \lambda \leqslant 2$). It is for this zone that we want to construct scattering.

In each tail z_j, we choose solutions $\psi_{+,j}$ and $\psi_{-,j}$. It may happen that these solutions cannot be extended over the entire graph. The solutions $\psi_\pm$ are of exponential type. We can also introduce solutions c_j and s_j being analogues of cosine and sine. They are obtained as follows:

$$c_j + a_\pm s_j = \psi_{\pm,j}.$$

This is a convenient real basis.

Let us introduce the space of asymptotic states

$$\mathbb{R}^{2k} = \bigoplus_{j=1}^k \mathbb{R}_j^2,$$

where $\mathbb{R}_j^2 = \{c_j, s_j\}$ is the space of solutions on the jth tail.

Consider the value (we call it *Wronskian*) defined by

$$W_{\vec{R}}(\phi, \psi) = b_{P,P'}(\phi_{P'}\psi_P - \psi_{P'}\phi_P)$$

for the vertex operator and by

$$W_{\vec{R}}(\phi, \psi) = \sum_{R' \cap \vec{R} = P} d_{R,R'}(\phi_{R'}\psi_R - \psi_{R'}\phi_R)$$

for the edge operator. Here $\vec{R} = PP'$ is an oriented edge.

What is the continuous analogue of this value? Given the Schrödinger operator $-\psi'' + u(x)\psi = \lambda \psi$, the Wronski determinant $W(\phi, \psi)$ for a pair of solutions (ϕ, ψ) has the form

$$W(\phi, \psi) = \phi'\psi - \phi\psi'.$$

In the continuous case, the main property of the Wronskian is that $W(\phi, \psi) = \text{const}$. This theorem has a direct analogue for Schrödinger operators on graphs.

Theorem 1. (a) $W_{\vec{R}}(\phi, \psi)$ *is a well-defined function whose arguments are a pair of solutions and an oriented edge. This function is skew-symmetric: it changes sign under the permutation of the solutions and under the change of the orientation of the edge;*
  (b) $\partial W = 0$, *i.e., W is a cycle.*

The physical interpretation of W being a cycle is Kirchhoff's first law. If $\Phi = \phi + i\psi$ is interpreted as a field, then $-2iW(\phi, \psi) = j_\Phi$ is a current, and the equality $\partial W = 0$ expresses the first law of Kirchhoff.

The cycle W is open; this means that, if the graph has an infinite tail, then this cycle has no finite support. The assertion that W is a cycle corresponds to the constancy of the Wronskian in the continuous case. Indeed, for the line, any open cycle is merely a line with some coefficient.

We interpret W as a skew-symmetric inner product on pairs of solutions. This inner product takes vector values. Thus, we treat W as an element of the one-dimensional homology group of the graph $H_1(\Gamma, \mathbb{R})$.

This may be useful only if there are more than one solutions at the given λ. Indeed, any skew-symmetric bilinear form on one-dimensional space identically vanishes. Very frequently, the solution space has dimension larger than 1 provided that the graph has symmetry. The second case which I want to explain is that of a graph without symmetry but with k infinite tails. In this case, the solution space is at least k-dimensional.

My point of departure in studying these questions was as follows. The classics in number theory, starting with Selberg, and modern geometers, including Sarnak, considered the Laplace–Beltrami operator in various domains on the Lobachevskii plane. I mean domains related to discrete groups, for which the fundamental domains have finite areas. The corresponding domains in the Lobachevskii plane have finitely many ends. As early as the 1950s, I. M. Gelfand called attention to the results of Selberg and mentioned that they were worth translating into the language of scattering theory. This task was accomplished by the members of Faddeev's Leningrad school, who exploited the idea of Gel'fand, but only in the case of no more than two tails. M. Gromov developed and advertised a similar general, thought fairly evident, idea. If you look at the hyperbolic geometry from infinity, it appears as a purely one-dimensional formation. The nearer the vertices of a triangle to the boundary, the smaller its area and the more closely it resembles a graph. This suggests that it is natural to model spectral theory, especially for the continuous spectra of the Laplace–Beltrami operators, with the use of the theory of graphs with tails.

Suppose given a graph Γ with k tails $z_1, \ldots, z_k$. I have introduced the space of asymptotic states $\mathbb{R}^{2k} = \bigoplus_{j=1}^{k} \mathbb{R}_j^2$. This is the space of solutions to the

Schrödinger equation on the tails under the assumption that the coefficients of the operator on the tails tend to the standard values which I have written out above. Namely,

$$\psi_{\pm,j,n} = c_{j,n} + a_{\pm} s_{j,n}.$$

Here j is the number of the tail and n is the number of the vertex.

The Wronskian can be used to turn the space of asymptotic states into a symplectic space as follows. We require that $\langle c_j, s_{j'} \rangle = \delta_{j,j'}$ and introduce the corresponding skew-symmetric inner product. It coincides with the Wronskian for two solutions on each tail.

Now, let us make the following observation. Suppose that ϕ and ψ are a pair of solutions to the equation $L\psi = \lambda\psi$. To the (real) solution ψ we assign its asymptotic value being an element of the space $\mathbb{R}^{2k}$:

$$\psi \mapsto \psi^{\mathrm{as}} \in \mathbb{R}^{2k}.$$

This correspondence is defined as follows. Every solution to the Schrödinger equation on a graph is equal to something on each tail. We take the direct sum of these states. I am considering real solutions, but it is always possible to pass to complex solutions by means of complexification.

Theorem 2. *If $L\psi = \lambda\psi$ and $L\phi = \lambda\phi$, then $\langle \phi^{as}, \psi^{as} \rangle = 0$.*

This theorem means that, if two sets of asymptotic values specified arbitrarily on each tail can be extended to solutions to the Schrödinger equation on the entire graph, then their inner product always vanishes. In reality, this property completely determines all the unitarity property of the scattering process. In other words, as symplectic geometers say, the set of all asymptotic values of solutions existing in reality forms a Lagrangian subspace of dimension k in a space of dimension $2k$.

In essence, this fact is purely topological. Suppose given a graph Γ with k tails $z_1, z_2, z_3, \ldots, z_k$. Consider a pair of solutions $L\psi = \lambda\psi$ and $L\phi = \lambda\phi$. It does not matter whether the Schrödinger operators under consideration are edge or vertex, because they do not differ on the tails. Consider the Wronskian W. Its boundary vanishes.

Exercise. The Wronskian W being a cycle implies that W can be represented in the form

$$W = \sum_{i=1}^{k} \alpha_i z_i + \text{something finite};$$

i.e., if W contains an edge of a tail with some coefficient, then all other edges of this tail have the same coefficient in W.

Note also that one tail can never be extended to a cycle on the entire graph. Only differences of tails can; yet, we can extend them in different ways. The differences of the extensions are some finite cycles in the graph.

Exercise. $W = \sum\limits_{j=2}^{k} \beta_j(z_1 - z_j) + \text{something finite.}$

Comparing the last two expressions, we see that $\sum \alpha_j = 0$. Indeed, $\alpha_1 = \sum_{j=2}^{k} \beta_j$ and $\alpha_j = -\beta_j$ for $j > 1$. The equality $\sum \alpha_j = 0$ is equivalent to $\langle \phi^{as}, \psi^{as} \rangle = 0$. Thus, the fact that the asymptotic solutions form a Lagrangian plane in a neighborhood of infinity is fundamental. Next, it is easy to show that this plane is determined by k equations; therefore, its dimension is at least k. On the other hand, a Lagrangian plane cannot have dimension larger than k.

Therefore, the operator assigning asymptotic values to solutions determines a k-dimensional subspace for any λ. It may have a kernel, which consists of the eigenfunctions identically vanishing on all tails. They have interesting analogues in the spectrum related geometry, the study of which was initiated by Selberg. For generic graphs, they can be removed by a small perturbation. But for graphs with group symmetry, they cannot be removed.

The scattering matrix is constructed as follows. In the Lagrangian plane, we choose a basis of vectors of the form

$$\psi_j = \psi_{+,j} + \sum_q s_{jq}(\lambda)\psi_{-,j}.$$

The scattering matrix is $(s_{jq}(\lambda))$ (by definition).

If the basis chosen in the Lagrangian space is purely real, then the scattering matrix S is unitary and symmetric. This can be explained as follows. Take a unitary matrix A and write the matrix S in the form $S = AA^t$. Let us see how many such matrices exist. If we replace A by AO, where O is a matrix from the real orthogonal group, then AA^t will be replaced by $AOO^tA^t = AA^t$. Therefore, $S \in U_k/O_k$. As is known, the set of all Lagrangian planes is isomorphic to precisely U_k/O_k.

For the first time, the discretized line with complete set of coefficients arose in integrating the so-called Todd chain; such a Schrödinger operator proved to have much better properties than its special cases considered in computational mathematics.

Miles Reid

Rings and algebraic varieties

Lecture on February 19, 1999

Let k be an algebraically closed field (e.g., $k = \mathbb{C}$). We shall consider only rings of the form $R = k[x_1, \ldots, x_n]/I$, where I is a prime ideal. In other words, we shall consider finitely generated algebras without zero divisors over an algebraically closed field.

A ring R can be assigned its *affine scheme* $\operatorname{Spec} R$, which is the set of all prime ideals of R. For a reasonable ring, the maximal ideals are close to the geometric notion of point. Prime ideals are close to the geometric notion of irreducible subvariety, and in the cases we are interested in are completely determined by the maximal ideals. (More precisely, every prime ideal is an intersection of maximal ideals.)

We regard the affine scheme as an algebraic variety V in affine space $\mathbb{A}^n$ (if $k = \mathbb{C}$, then $\mathbb{A}^n = \mathbb{C}^n$), rather than as a set. An algebraic variety V is the set of all points in the space $\mathbb{A}^n$ at which all polynomials from the ideal I vanish. An identification of $\operatorname{Spec} R$ with V corresponds to a choice of coordinates $x_1, \ldots, x_n$.

Conversely, to an algebraic variety V, there corresponds the ring $k[V]$ of polynomial functions on V; this ring coincides with R.

Any mathematician has heard of algebraic varieties as zero sets of some polynomials. There are two approaches:

(1) given equations, it is required to determine a set of points;

(2) given a geometric object, it is required to specify it as an algebraic variety (i.e., immerse it in affine space and specify by equations).

Example. Suppose that, on affine space $\mathbb{C}^2$, a finite group $G \subset \operatorname{GL}(2, \mathbb{C})$ of linear transformations acts. We want to endow the quotient space $\mathbb{C}^2/G$ with the structure of an algebraic variety.

For example, what is the ring of functions? A likely candidate is as follows. Take the ring of invariant functions $k[x, y]^G$ in the polynomial ring $k[x, y]$. This ring has the form $k[x, y]^G = k[u_1, \ldots, u_n]/I$. Indeed, we choose an invariant polynomial u_1, then u_2, and so on, until there remain no new (independent) polynomials. The ideal I corresponds to the relations between the chosen invariant polynomials.

Miles Reid, Professor at Warwick University (Great Britain).

If the group G is generated by a mapping $x, y \mapsto -x, -y$, then all invariant polynomials are expressed in terms of $x^2 = u$, $xy = v$, and $y^2 = w$. These polynomials are related by $uw = v^2$. As a result, we obtain a mapping $\mathbb{C}^2 \to Q \subset \mathbb{C}^3$, where Q is the quadric defined by the equation $uw = v^2$. This mapping identifies all points on each orbits and no other points.

This example can be generalized. Let $\epsilon = \exp(2\pi i / r)$. Consider the group generated by the mapping $x, y \mapsto \epsilon x, \epsilon^{-1} y$. The invariant polynomials are $x^r = u$, $xy = v$, and $y^r = w$. They are related by $uw = v^r$. We obtain a mapping $\mathbb{C}^2 \to X \subset \mathbb{C}^3$, where X is defined by the equation $uw = v^r$. This equation determines a singularity of type A_{r-1}.

We can also consider the group generated by the mapping $x, y \mapsto \epsilon x, \epsilon^3 y$, where $\epsilon^7 = 1$. In this case, the invariant polynomials are $x^7 = u_1$, $x^4 y = u_2$, $xy^2 = u_3$, and $y^7 = v$. The relations between them can be written in the form

$$\mathrm{rk} \begin{pmatrix} u_1 & u_2 & u_3^2 \\ u_2 & u_3 & v \end{pmatrix} \leqslant 1.$$

Exercise. Prove that the polynomials specified above generate the whole ring of invariant polynomials, and the relations generate the whole ideal of relations.

Now, we proceed to define a projective variety. In this case, R is a graded ring, i.e., $R = \bigoplus_{n \geqslant 0} R_n$, where $R_0 = k$ and it is assumed that $x_n y_m \subset R_{n+m}$ for homogeneous polynomials. We also assume that the ring is finitely generated and has no zero divisors. Thus, $R = [x_1, \ldots, x_n] / I$, where the generator x_i has weight a_i (not necessarily equal to 1) and I is a prime homogeneous (with respect to the weights) ideal.

The graded ring R is assigned $\mathrm{Proj}\, R$, which is the set of all homogeneous prime ideals $P \subset R$ not containing the trivial prime ideal $m_0 = \bigoplus_{n > 0} R_n$. In the situation under consideration, an ideal that is maximal with respect to this condition correspond to a geometric of $\mathrm{Proj}\, R$, and all the homogeneous prime ideals are determined in terms of these.

The projective variety $X = \mathrm{Proj}\, R$ is a union of affine varieties. Namely, $X = \bigcup X_f$, where the affine variety X_f ($0 \notin X_f \subset R_n$) has the form $X_f = (R[\frac{1}{f}])_0 = \{\frac{g}{f^k}\}$ (the subscript 0 indicates that only elements of degree 0 are taken); here $\deg g = nk$.

Choosing coordinates, we embed X in the weighted projective space $\mathbb{P}(a_0, a_1, \ldots, a_n)$ defined as follows. Let us introduce the equivalence relation

$$(x_0, x_1, \ldots, x_n) \sim (\lambda^{a_0} x_0, \lambda^{a_1} x_1, \ldots, x_n \lambda^{a_n})$$

on the set $\mathbb{C}^{n+1} \setminus \{0\}$ and consider the quotient space modulo this equivalence. If all the a_i are equal to 1, then we obtain the usual projective space.

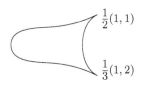

$$\frac{1}{2}(1,1)$$

$$\frac{1}{3}(1,2)$$

Figure 1

Example. The weighted projective space $\mathbb{P}(1,2,3)$ is the projective plane $\mathbb{P}^2$ with two horns (Fig. 1). These horns are the quotient singularities $\frac{1}{2}(1,1)$ and $\frac{1}{3}(1,2)$. (They are the very same quotient singularities as those we have just encountered.)

The equivalence relation in the space $\mathbb{C}^3$ has the form $(x,y,z) \sim (\lambda x, \lambda^2 y, \lambda^3 z)$ in the case under consideration. There arise exceptional orbits. For example, consider the restriction to the y axis, i.e., the orbit of the point $(0,1,0)$. The action of the element $\lambda = -1$ leaves the point $(0,1,0)$ fixed. This means that the orbit has a nontrivial stabilizer. But when the stabilizer increases, quotient singularities occur.

Why is the quotient singularity equal to one half? On the orbit under consideration, the only nonzero function equals to y (up to proportionality). Therefore, the denominator contains only y:

$$\left(R\Big[\frac{1}{f}\Big]\right)_0 = \left(k[x,y,z]\frac{1}{y}\right)_0 = \Big\{\frac{g}{y^k}\Big\}.$$

Here y has degree 2; hence the degree of the polynomial g must be even rather than is arbitrary.

In other words, if we set $\eta = \sqrt{y}$, then the local coordinates in a neighborhood of the point $(0,1,0)$ are x/η and z/η^3. These are coordinates on a cyclic covering of the variety rather than on the variety itself.

The situation in a neighborhood of the point $(0,0,1)$ is considered similarly.

The weighted projective space $\mathbb{P}(1,2,3)$ can be embedded in the usual projective space of larger dimension. Namely, we can embed it in $\mathbb{P}^6$. For this purpose, we take the ring $k[x,y,z]^{[6]}$ of polynomials whose degrees are multiples of 6. The basis polynomials are

$$x^6 \qquad x^4y \qquad x^2y^2 \qquad y^3$$

$$x^3z \qquad xyz$$

$$x^2$$

The nine relations between these monomials specify a del Pezzo surface $S_6 \subset \mathbb{P}^6$.

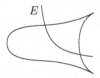

Figure 2

Applications

To the ring $R = k[X]$ (of all polynomial functions on X) we assign the affine algebraic variety $X = \operatorname{Spec} R$. To obtain a projective variety, we must fix an ample divisor; then $X = \operatorname{Proj} R$. But this is a complicated matter. We refer the interested reader to the second chapter of Hartshorne's book.

We shall consider only one simple case in which E is an elliptic curve with a given point $P \in E$. The elliptic curve can be understood as either a nonsingular plane cubic curve or a compact Riemannian surface of genus 1. The problem is to immerse it in projective space and find an equation specifying it.

Let $\mathcal{L}(nP)$ be the spaces from the Riemann–Roch theorem, i.e., the spaces of global meromorphic functions having poles only at the point P of multiplicity not higher than n. According to the Riemann–Roch theorem,

$$\dim \mathcal{L}(nP) = \begin{cases} n & \text{for } n > 0, \\ 1 & \text{for } n = 0. \end{cases}$$

Consider the graded ring $R(E, P) = \bigoplus_{n \geqslant 0} \mathcal{L}(nP)$. (We can do this for an arbitrary algebraic variety with a given ample divisor.)

The spaces R_0 and R_1 correspond to the same functions. Indeed, if an elliptic function is allowed to have only one simple (nonmultiple) pole, then such an elliptic function is constant. Therefore, R_0 and R_1 contain only constant functions. But we consider the direct sum; hence the same function corresponds to different elements of the ring $R(E, P)$, depending on whether we consider this function as an element of R_0 or of R_1. For convenience, we set $R_0 = k$ and $R_1 = k \cdot x$ (here x corresponds to the function identically equal to 1 treated as an element of R_1).

Theorem. $R(E, P) = k[x, y, z]/(f_6)$, where $\dim x = 1$, $\dim y = 2$, and $\dim z = 3$. The ideal (f_6) is generated by a homogeneous relation of the form $z^2 + \cdots = y^3 + \ldots$. (Any such relation can be reduced to the form $z^2 = y^3 + a_4(x)y + b_6(x)$.)

According to this theorem, E is embedded in $\mathbb{P}(1, 2, 3)$ as a hyperplane section. The presence of the term z^2 shows that, under this embedding, the elliptic curve is disjoint from the singular points of the weighted projective space $\mathbb{P}(1, 2, 3)$ (Fig. 2).

Proof of the theorem. The space $\mathcal{L}(0)$ is generated by the element 1.

The space $\mathcal{L}(P)$ is generated by the element x.

The space $\mathcal{L}(2P)$ is generated by the elements x^2 and y.

The space $\mathcal{L}(3P)$ is generated by the elements x^3, xy, and z.

Up to $\mathcal{L}(5P)$, the number of monomials equals the dimension of the space. But $\dim \mathcal{L}(6P) = 6$, whereas there are the seven monomials displayed above, and therefore there is a relation between them.

The monomials are linearly independent because the mapping $E \to \mathbb{P}^6$ determined by x^2 and y is not constant (i.e., x^2 and y are algebraically independent). $\qquad\square$

The process can be algorithmized. Consider a graded ring R for which $\dim R_n = l(n)$, where $l(n) = n$ for $n > 0$ and $l(0) = 1$. To this graded ring we can assign the Poincaré series $L(t) = \sum l(n) t^n$. For a finitely generated ring, the Poincaré series is a rational function. The form of this rational function depends on the dimensions of the generators and on the number of relations and their dimensions. For example, in the case under consideration,

$$L(l) = \frac{1 - t^6}{(1 - t)(1 - t^2)(1 - t^3)}. \tag{1}$$

This means that there are three generators of dimensions 1, 2, and 3 and one relation of dimension 6.

Equality (1) can be proved, e.g., as follows. By definition, $L(t) = 1 + t + 2t^2 + 3t^3 + \dots$. Therefore,

$$(1 - t)L(t) = 1 + t^2 + t^3 + t^4 + \dots,$$

whence

$$(1 - t^2)(1 - t)L(t) = 1 + t^3 = \frac{1 - t^6}{1 - t^3}.$$

Experts in commutative algebra know that this is a very general construction. If M is a graded module and the ring has an element v not being a zero divisor, then multiplication by v induces the exact sequence

$$0 \to M \xrightarrow{v} M/v.$$

Here M/v is a module over a smaller ring, namely, over the ring obtained by substituting $v = 0$. The multiplication of the series considered above corresponds precisely to this reduction.

All the issues in this lecture are dealt with in much greater detail in two introductory chapters of my projected book; see

http://www.maths.warwick.ac.uk/~miles/surf

A. B. Katok

Billiard table as a playground for a mathematician

Lecture on March 10, 1999

The title of this lecture can be understood in two ways. Literally, in a somewhat facetious sense: mathematicians are playing by launching billiard balls on tables of various forms and observing (and also trying to predict) what happens. In a more serious sense, the expression "playground" should be understood as "testing area": various questions, conjectures, methods of solution, etc. in the theory of dynamical systems are "tested" on various types of billiard problems. I hope to demonstrate convincingly that at least the second interpretation deserves serious attention.

The literature concerning billiards is rather large, including scientific papers as well as monographs, textbooks, and popular literature. Short brochures by G. A. Galperin and A. N. Zemlyakov [4] and by G. A. Galperin and N. I. Chernov [5] are written in a rather accessible manner, and touch a broad circle of questions. An introduction to problems related with billiards for a more advanced reader is contained in Chapter 6 of the book [9]. The next level is represented by a very well written book of S. Tabachnikov [14], whose publication in Russian is unfortunately delayed. The book by the author and B. Hasselblatt [8] contains a rather detailed modern exposition of the theory of convex billiards and twisting maps. A serious but rather accessible exposition of modern state of the theory of parabolic billiards is contained in a survey paper by H. Masur and S. Tabachnikov which will be published (in English) in spring 2002 [11]. The collection of papers [12] contains rich material on hyperbolic billiards and related questions. More special references will be given below during the exposition.

1 Elliptic, parabolic, and hyperbolic phenomena in dynamics

The problem of motion of a billiard ball is stated in a very simple way. One has a closed curve $\Gamma \subset \mathbb{R}^2$. Inside the domain $\mathcal{B}$ bounded by the curve, one has a uniformly moving point which covers segments of straight lines, and when the point meets the curve, it is reflected according to the rule "the angle of incidence equals the angle of reflection." The problem is to understand the nature of this motion at a large time.

Here we have a dynamical system which is in general not everywhere defined. For example, if in a domain with a piecewise smooth boundary the point hits an angle, then it is unclear how to continue the trajectory. There are also more delicate effects: for some initial conditions it is possible that during a finite time an infinite number of hits of the boundary occurs and the motion cannot be continued. But these effects are pathological; one can say that we have a dynamical system.

The solution of the problem of motion of a ball depends on the domain. One of the reasons of interest to this problem is that the formal description of the motion is very simple and only the essential part is to be investigated. The second, more serious reason has been already mentioned. It is related to the fact that if one attempts to somehow classify problems of theory of dynamical systems, then, in a somewhat rough manner, they can be divided into elliptic, parabolic, and hyperbolic ones (see Fig. 1). Thus, a billiard table is a testing area on which one can test methods, conjectures, questions, arising in various fields of theory of dynamical systems.

There is nothing new in using these words for expressing some trichotomy. The corresponding classification in theory of partial differential equations is well known. But for dynamical systems, such a classification seemingly has not been carried out systematically.

In the case of billiards, elliptic effects arise, for example, for an ellipse. This coincidence is not completely accidental, but it cannot be extended to billiards inside a parabola or a hyperbola. A more general situation in which elliptic effects occur, is as follows: the curve is smooth (of a sufficiently large class of smoothness), convex, and its curvature nowhere vanishes. The study of the billiard problem inside such domains gives a good example for demonstration of problems and results related to elliptic behavior of dynamical systems.

In a parabolic situation, the domain is a usual polygon. For simplicity one can even take a right-angled triangle whose angles differ from $30°$ and $45°$. A right-angled triangle with the angle $\pi/8$ already gives an example of a dynamical system with parabolic behavior.

The hyperbolic situation is well represented by three examples (see Fig. 1): a square with a small disk removed, a "stadium," and a cardioida.

The idea about at least a dichotomy which exists in the theory of dynamical systems has been widely accepted in the last years. One of the most remarkable books on the theory of dynamical systems written in the second half of the twentieth century is the book by Yorgen Moser "Stable and random motion in dynamical systems." "Stable" means elliptic effects, "random" means hyperbolic effects. Parabolic effects are not discussed in Moser's book.

To give some idea on the nature of the trichotomy arising here, let me explain

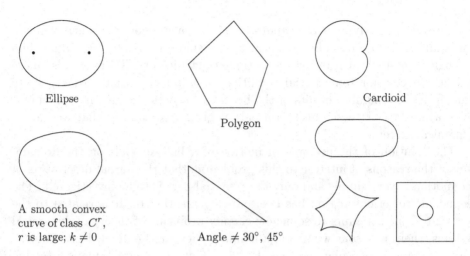

Ellipse

Polygon

Cardioid

A smooth convex
curve of class C^r,
r is large; $k \neq 0$

Angle $\neq 30°$, 45°

Figure 1. Elliptic, parabolic, and hyperbolic billiards

the origin of these terms. For linear mappings the corresponding trichotomy is well known. For a linear mapping $L: \mathbb{R}^{2n} \to \mathbb{R}^{2n}$ three main kinds of behavior are possible:

(1) Stable behavior. It arises when all the eigenvalues λ_i have absolute value 1 and there are no nontrivial Jordan blocks: $\mathrm{Sp} L \subset S^1$. In this situation all the orbits come back and are stable. This is elliptic behavior.

(2) Again $|\lambda_i| = 1$, but there are nontrivial Jordan blocks. A Jordan block has an eigenvector, therefore there are stable orbits. But in this situation, typical is the polynomial growth of distance between orbits. This is parabolic behavior.

(3) Hyperbolic behavior: $\mathrm{Sp} L \cap S^1 = \varnothing$. In this situation the distance between any two orbits exponentially grows either in the positive or in the negative direction.

Combinations of these three paradigms are also possible. For example, a rather important situation is what is called partially hyperbolic behavior, when the spectrum contains a hyperbolic component and something else. This is a very important paradigm in dynamics.

It would be very naive to attempt to construct a concept of nonlinear differential dynamics based only on these three models. What is the subject of nonlinear differential dynamics? It is the analysis of asymptotic behavior of smooth systems for which one has the notion of local infinitesimal behavior of the system and, on the other hand, due to compactness of the phase space, there is the phenomenon of returning of orbits arbitrarily closely to their initial position. Roughly speaking, dividing nonlinear dynamical systems into elliptic, parabolic, and hyperbolic ones corresponds to the situations when the linear behavior, which is more or less approximated by these three types, is combined with the nontrivial nature of returning.

This approach ignores a very essential part of problems of theory of dynamical systems, for example, such things as analysis of Morse–Smale systems or effects related with bifurcations. These are situations when returning is simple, and interesting phenomena are related, for example, to how the phase space is divided into basins of attraction to several attracting points or limit cycles. All of this is ignored. We are now speaking only on the part of dynamics which is related to recurrent behavior. Nonrecurrent behavior is more or less ignored by us now.

To interpret phenomena interesting to us correctly, one must understand what is a linearization of a dynamical system. Let $f: M \to M$ be a given map acting on the phase space. We assume that the phase space is a smooth object, hence one can speak about the action on tangent vectors. For any point $x \in M$ one has the linear map $Df_x: T_x M \to T_{f(x)} M$. Such a map is interesting by itself only in the case of a fixed point. But in the general case in dynamics one can consider the iterations Df_x^n. Introducing a Riemannian metric, one can speak about the asymptotic speed of growth of length of vectors. A Riemannian metric can be introduced not uniquely, but on a compact manifold any two metrics differ no more than by a multiplicative constant, therefore the speed of growth of vectors is defined correctly.

The elliptic behavior arises when in the linearized system either there is no growth of length of vectors at all, or it is slower than a linear one (a sublinear growth).

A Jordan block of minimal size 2 already corresponds to a linear growth. The parabolic behavior is a subexponential growth (usually a polynomial one).

The hyperbolic paradigm is the most well understood one. It corresponds to the situation when the system splits and in some directions one has exponential growth, and in the other directions one has exponential decay. When the time is reversed, these directions are exchanged with one another.

Sometimes mixed situations occur. For example, one can take the direct product of two systems of different type. But as a metastatement, one can say that the hyperbolic paradigm dominates: if there is a nontrivial hyperbolic effect and something else, then usually the behavior of the system can be understood based on its hyperbolic part. This is not true literally. For example, this is not true for the direct product of dynamical systems. But for a typical dynamical system the hyperbolic behavior dominates everything else.

One can also make the following interesting remark. When the dimension of the phase space is small, mixed behavior is impossible. For example, when the dimension of the phase space equals 2, the partially hyperbolic behavior is impossible, because for the hyperbolic behavior one needs at least one extending and one compressing direction. By the same reason in small dimensions the elliptic or parabolic behavior occurs more often.

The most pure example of elliptic behavior is the situation when one has a smooth isometry. In this case there is no growth. For smooth isometries dynamics can be understood rather easily. If on a compact phase space there is a smooth isometry, then the phase space splits into invariant tori, and on each torus a parallel translation arises (or a rotation, if one uses multiplicative notation). In particular, if the manifold itself is not a torus, then such motion is not transitive.

Of course, it is a particular case of what is well known in Hamiltonian mechanics, namely, it corresponds to completely integrable Hamiltonian systems. This is a good example of interaction of paradigms, because, if one looks at a completely integrable system naively, then it should be attributed to the parabolic paradigm. Indeed, the linear part of a completely integrable system is parabolic, because in the direction transversal to the invariant tori one has a twisting. On the other hand, the space splits into invariant tori, and on each torus analysis is carried out with elliptic methods.

This situation is typical. This is the reason why the elliptic paradigm is important. It is a rather rare case that the global behavior on the whole phase space is characterized by absence of growth. But rather frequently one has some elements inside the phase space, where the behavior can be described by means of the elliptic paradigm.

The hyperbolic situation is the most well studied. In a sense, it is the only universal paradigm of complex behavior in dynamics. It can be well understood with the help of Markov chains and simple stochastic models. From the viewpoint of applications of dynamics, if the hyperbolic behavior is established, then one can apply a rather powerful machinery which makes it possible to study the behavior of nonlinear systems. All this arises due to the interaction of a certain behavior of the linearized system with more or less a priori existing returning. In linear systems hyperbolicity is followed just by running away of the system to infinity. But if there is no space to run away, if one necessarily has to return, then the above mentioned and well understood types of complex behavior arise.

In contrast to elliptic and hyperbolic behavior, parabolic behavior is, firstly, unstable, and, secondly, it is characterized by absence of standard models. In the elliptic situation one has a universal model, namely, rotation on the torus (or some of its avatars), and in the hyperbolic situation one has the Markov model which describes everything. In the parabolic situation, seemingly, one even cannot say that there is a set of models to which everything is more or less reduced. Nevertheless, there are rather typical phenomena which occur in concrete classes of systems. One of these phenomena consists in that frequently the effect of moderate stretch can be replaced by the effect of cutting. For example, if one has a system which locally looks like an isometry but has discontinuities, then such a system is concerned with the parabolic paradigm.

A well known example is exchange of segments. We cut a segment into pairs and exchange them according to a permutation given *a priori*. Locally this system looks like an elliptic one, but there is the effect of cutting. It is rather easy to realize that this system should be considered as a parabolic one: during the iterations the number of segments grows in a linear way. This linearity is not the result of twisting, but it is the result of cutting. But the effect is approximately the same.

Thus, parabolic behavior is frequently related to the presence of moderate singularities in systems. So it is not occasional that a polygon was drawn on the picture illustrating parabolic behavior.

2 Billiards in smooth convex domains

George K. Birkhoff was the first to consider billiards systematically as models for problems of classical mechanics. Birkhoff considered billiards only in smooth convex domains. Of course, he did not think about billiards in polygons, and all the more in nonconvex domains.

First of all, one can perform the reduction to the billiard mapping. The initial dynamical system for a billiard is a system with continuous time. But the trajectory inside the billiard table can be easily reconstructed if one knows what happens at the moments of reflection. Therefore it suffices to consider the so-called billiard mapping. The phase space of the billiard mapping looks as follows. The vector v outcoming after reflection is characterized by the cyclic coordinate $\phi \in S^1$ which fixes the position of the point on the curve Γ, and the angle $\theta \in [0, 2\pi)$ between the tangent vector and the vector v (Fig. 2).

The phase space of the billiard mapping is a cylinder. After the reflection we obtain a new point ϕ_1 and a new outcoming vector, which corresponds to the angle θ_1. The map $T(\phi_0, \theta_0) = (\phi_1, \theta_1)$ is what is called the billiard mapping. It maps the open cylinder into itself; by continuity it can be extended to the closed cylinder. The points with $\theta = 0$ are fixed (we assume that the curve Γ does not contain straight segments).

Exercise 1. Show that the presence of straight segments in the boundary implies discontinuity of the billiard mapping.

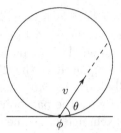

Figure 2. Vector coordinates after reflection

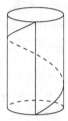

Figure 3. Twisting

Exercise 2. Find conditions under which the billiard mapping is differentiable (one or infinitely many times) on the boundary of the cylinder.

The billiard mapping possesses two important qualitative properties.

1) Conservation of area. The element of area

$$dA = \sin\theta d\theta d\phi = d\alpha d\phi, \qquad \text{where} \quad \alpha = \cos\theta,$$

is conserved. (To introduce coordinates in which area is conserved, one should take $\cos\theta = \alpha$ instead of θ.)

2) Twisting. Let us fix the coordinate ϕ_0 and change the coordinate θ. Then the coordinate ϕ of the image will change monotonously until it passes all the circle and comes back (Fig. 3). The image of a vertical line is twisted.

These two properties allow one to realize that one has elliptic behavior. The problems arising in connection with elliptic behavior are divided into two parts:

1) caustics,
2) Birkhoff orbits and Aubry–Maser sets.

Let me begin with the second part. We want to find periodic orbits of the billiard system. Periodic orbits can be various. They differ not only by the period, but also by some combinatorics. For example, two orbits of period 5 in Fig. 4 have different combinatorics. In the first case there is one rotation, and in the second case there are two. These orbits are regular: the order of

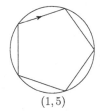 

$(1, 5)$ $\qquad\qquad\qquad\qquad$ $(2, 5)$

Figure 4. Orbits with equal periods and different combinatorics

points on the orbit is conserved; it is the same as in the rotation. It is these (regular) orbits that are called Birkhoff orbits. This name is related to the fact that he has proved a remarkable and relatively simple theorem on existence of regular orbits. Seemingly, this theorem was the starting point of application of variational methods in dynamics.

Theorem 1 (Birkhoff). *For any two coprime numbers p and q there exist at least two periodic orbits of the type (p, q).*

Sketch of the proof. The proof uses only convexity and smoothness. Consider various inscribed polygons with the required combinatorial properties. Let us call such polygons by states. The states form a finite-dimensional space. On the space of states there is the functional of length. If we allow the vertices of polygons to coincide, then we obtain a compact space. Hence the length functional has maxima.

Any extremal point of the length functional is a billiard orbit (if this point is not on the boundary). This is a local statement. It is easy to check that the derivative of the length vanishes if and only if the angles are equal. The linear part of variation of the functional depends just on the difference of the angles.

It is easy to prove that the maximum cannot be achieved on the boundary, i.e., the vertices of a polygon cannot coincide.

Thus, the longest polygon is a required periodic orbit. But this is still only the easiest part of the theorem. One has to find another periodic orbit. This can be done in the following way. Cyclic renumeration of the vertices of the orbit we have found gives q maxima. Let us deform one of these maxima into another one. If we go from one maximum to another one, then we have to go down. Let us try to lose a smallest possible height. In this case we need to pass a saddle (Fig. 5), because if at the lowest point we were not on a saddle, then we could change the trajectory a little and decrease the lack of height. A saddle is also a critical point, i.e., the required periodic orbit.

If we do not lose height at all, then in this case one has a whole family of periodic orbits. $\qquad\qquad\qquad\qquad\qquad\qquad\qquad\qquad\qquad\qquad\qquad\qquad$ $\square$

Figure 5. A saddle

This proof shows concisely how one can change a difficulty by another one. The difficult point of this argument is in how to keep aside from the boundary. This can be easily achieved if we just do not consider the boundary, and consider all states. Evidently, the function in question is bounded: any edge is no longer than the diameter of the curve. One can omit the condition of ordering of points and then prove that the global maximum is necessarily approached on a correctly ordered orbit. If, for instance, we consider globally maximal orbits which make two rotations during a full round, then they do it in a correct order. And instead we can prove that one can avoid the boundary inside an ordered family.

The importance of the Birkhoff theorem is in that we immediately find infinitely many periodic orbits.

Now an interesting story begins on how Birkhoff missed an important discovery.

Birkhoff presents his variational argument, and then he says that in exactly the same manner one can purely topologically prove the so-called last geometric theorem of Poincaré: "If the bases of a cylinder rotate in different directions with area being conserved, then such a diffeomorphism has at least two fixed points." Moreover, if the angles of rotation on the upper and lower bases are different, then for any rational angle of rotation one can find a corresponding periodic orbit, even without the condition of twisting.

Birkhoff was extremely proud to prove the last geometric theorem of Poincaré. But he missed a very remarkable conclusion of his own elementary proof. This conclusion is the following. Let us look what happens in passing to the limit $p_n/q_n \to \alpha$, where α is an irrational number. Usually in dynamics such tricks would not work, because the asymptotic behavior is unstable with respect to initial data, and one cannot pass to the limit. But here, just because we are dealing with the elliptic situation, a simple but surprising phenomenon arises. If we consider a Birkhoff orbit on the cylinder, then it consists of a finite number of points. If the number q_n is large, then the number of points will be also large and they will be strongly condensed. It is rather easy to prove that these points always lie on a Lipschitz graph (i.e., on the graph of a function satisfying the Lipschitz condition). The Lipschitz constant here is fixed, it does not depend on the length of an orbit. The set of Lipschitz functions with a given Lipschitz constant is compact, hence one can pass to a limit. In a

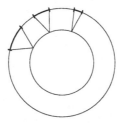

Figure 6. The Denjoy counterexample

somewhat different way, one can say the following: let us take finite orbits and consider their limit in the Hausdorff topology. In the Hausdorff topology closed subsets of a compact set form a compact set, hence the limit exists, which is not surprising. But the limit is an invariant set which is a subset of a Lipschitz graph, because in the Hausdorff topology subsets of Lipschitz graphs form a closed subset.

We still don't know what is the geometry of the obtained graph, but we know what is its dynamics. Its dynamics is the same as the one of the rotation on the angle α, nothing else can occur. Indeed, the order in which the points are transposed under rotation on the angle α, is uniquely determined by the orders in which the points are transposed under rotations on the angles p_n/q_n approximating the angle α. Hence on any finite segment in the limit combinatorics will be the same as needed, because on any finite segment combinatorics is stabilized and is the same as under rotation on the angle α.

Thus, a closed invariant set on a circle arises (since topologically the limit Lipschitz graph is a circle). This set has a dynamics which preserves the order and exactly reconstructs rotation on the angle α. From the times of Poincaré it is known when this is possible: either the invariant set is the whole circle, the orbits are dense and the transform is conjugate to a rotation of the circle (this is, of course, elliptic behavior, at least in the topological sense), or the circle contains an invariant Cantor set which arises in the so-called Denjoy counterexample (see, for example, Chapters 11 and 12 in [8]). The Denjoy counterexample looks as follows. Let us take a point on the circle and blow it up into an interval. Then its image and preimage should be also blown up into intervals (Fig. 6), etc. To have convergence, one needs these intervals to be smaller and smaller. This is rather easy to make topologically. As a result, one gets a transform of the circle which contains an invariant Cantor set and which is half-conjugated to a rotation (there exists a continuous mapping which makes it a rotation, but these intervals shrink into points).

For transforms of a circle such a behavior is exotic, because by Denjoy's theorem this is impossible in the class C^2, it is possible only in the class C^1.

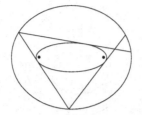

Figure 7. Confocal ellipses

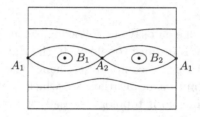

Figure 8. Trajectories in the phase space

But for twisting maps this is rather normal behavior (of course, if it did not happen that we get a full circle). Thus, an interesting alternative arises. When one has accumulation of Birkhoff orbits on an invariant set (this is exactly the Aubry–Maser set), this invariant set is either a Cantor set (possibly with some additions) or the whole circle. The case of the whole circle is called a caustic. One of these two cases holds always and for any rotation number. The corresponding Cantor set is unique, i.e., if one removes from the invariant set the wandering part corresponding to separate wandering points, then the remaining Cantor set is unique. But this does not obstruct the existence of other Cantor sets which have the same rotation number and on which the order is preserved, but this order, although compatible with the cyclic order, would not be compatible with the order on this set. The Cantor set constructed as the limit of Birkhoff orbits of maximal length is special; it is called minimal. It is the set of minimal energy.

The next question is the following: does it happen that one obtains the whole circle? The answer to this question is illustrated by Fig. 7. Let us take a big ellipse as a billiard table and consider the orbit which is tangent to an inner confocal ellipse. It turns out that this orbit will be tangent to this ellipse further. The same is true for confocal hyperbolas. However, a hyperbola consists of two branches, but if an orbit is tangent to one of the branches of a hyperbola, then it will be further tangent to this hyperbola, and the branches tangent to the orbit will alternate, one after another.

This is a picture in the configuration space. And what will be in the phase space? The picture in the phase space is also well known, it looks very much like the picture for a pendulum (Fig. 8; in this figure the cylinder is unwound). Namely, there are two orbits of period 2 corresponding to the large and small diameters of the ellipse. An "eight" trajectory corresponds to orbits passing through the foci of the ellipse (if an orbit passes through a focus, then it will further pass through the foci, one after another). The trajectories situated outside this eight correspond to orbits tangent to ellipses. And the trajectories inside the eight correspond to the orbits tangent to hyperbolas.

Which of these orbits correspond to Birkhoff and Aubry–Maser orbits, and which do not correspond to them? In other words, which of these orbits can be obtained by Birkhoff's and Aubry–Maser's constructions and which cannot be obtained? The orbits with rational rotation numbers tangent to ellipses are obtained by Birkhoff's construction, and the rest of the orbits tangent to ellipses are obtained by Aubry–Maser's construction. And hyperbolas cannot be obtained by such constructions. Indeed, for the rotation number 1/2 one has one minimal orbit and one minimax orbit.

It is interesting to understand what happens in passing to the inverse limit, i.e., when we pass from irrational to rational numbers. In the situation under consideration the answer is rather simple. We obtain an invariant circle, but it is not fully covered by Birkhoff orbits. It consists of Birkhoff orbits and asymptotical curves. This is a rather general phenomenon, with the exception that not always one obtains a full circle.

We have considered billiard tables of a rather special form. The following rather famous question has not yet got a definite answer: "What can be other billiard tables for which at least a neighborhood of the upper and lower base of the cylinder is fibered into invariant curves?" In other words, when the system is completely integrable? It is assumed that this can happen only for an ellipse.

Much more fundamental is the following question: when at least some curves are conserved? It is rather remarkable that necessary and sufficient conditions of the existence of at least one invariant curve are rather simple. Of course, we mean an invariant curve passing around the cylinder. Only such a curve can arise as the limit of Birkhoff orbits. It is not difficult to prove that if one has an invariant curve on which the order is preserved and the rotation number equals α, then such a curve is unique if α is irrational, and it is the limit of Birkhoff orbits.

If we are interested in the question: what arises as the limit of Birkhoff orbits, whether it is a curve or a Cantor set, then it is natural to ask when it is a curve. Let us assume that the curve which bounds the table is sufficiently smooth, for example, of the class C^∞ (it suffices to require that the curve be of the class C^6). In this case, a theorem proved by Vladimir Fedorovich

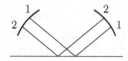

Figure 9. Reordering of points

Lazutkin (1941–2000) [10] states that an invariant circumference exists when the curvature of the boundary vanishes nowhere. Actually in this situation there are infinitely many invariant circumferences.

Lazutkin's proof is an adaptation for this case of the celebrated Kolmogorov theorem on perturbations of Hamiltonian systems. Formally the Kolmogorov theorem does not cover this situation, because here we deal with behavior of a degenerating system. One must appropriately change the coordinates to apply anything. The nonzero curvature is needed just to make this change of coordinates possible.

If the curvature vanishes, then there are no invariant curves. This much more simple fact has been proved by John Maser. In fact, one can prove a stronger statement. Namely, if the boundary contains a flat point, then no Aubry–Maser set can pass through this point. And on an invariant circle one must have not only points corresponding to Birkhoff orbits but also points corresponding to Aubry–Maser orbits. (See, for example, Section 13.5 in [8].)

This argument is rather simple. A reflection with respect to a line changes the order of points (Fig. 9). If first is the point 1 and second is the point 2, then after a reflection first is the point 2 and second is the point 1. In Fig. 9 the lines are parallel, but the same effect holds if the lines are different. Thus, after a reflection with respect to a line the order of points must change. Infinitesimally the same happens during reflection with respect to a curve at a point with the zero curvature.

Here some interesting geometric effects arise. Consider the inverse problem: how to construct a billiard table for which caustics exist? To this end, one can use a construction which is well known for the case of an ellipse. One can take an ellipse and throw on it a lace whose length is greater than the length of the ellipse. Then one should stretch this lace and draw a curve (Fig. 10). As a result one obtains a confocal ellipse. For the larger ellipse the smaller one will be a caustic.

The same construction works for an arbitrary curve. If one takes an arbitrary curve and a lace longer than the curve, and then stretches the lace and draws a new curve, then for the new curve the initial curve will be a caustic.

Sometimes a nonsmooth inner curve yields a smooth billiard table. For example, if for the inner curve one chooses an astroida, then as a result one

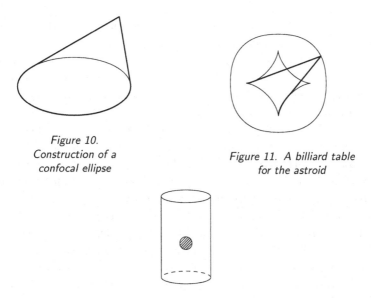

Figure 10.
Construction of a
confocal ellipse

Figure 11. A billiard table
for the astroid

Figure 12. The cut out circle

obtains a smooth table for which the astroida is a caustic (Fig. 11).

It is rather clear why elliptic billiards can be considered as a testing area. Firstly, they give examples of twisting maps. Billiards give some geometric intuition which can be developed and then used for arbitrary twisting maps, and twisting maps cover many interesting cases besides billiards. And, secondly, billiards give an example of a standard difficulty in dynamics. How can one take into account the Lagrangian structure? The picture on a cylinder, where one has the coordinate and the impulse, is a Hamiltonian picture, it is a picture in the phase space. And a billiard is a Lagrangian structure. The Lagrangian structure is not invariant, it is related with division into coordinates and impulses.

For instance, let us consider the following question. Can a billiard have an open set of periodic orbits? For a Hamiltonian twisting map an example is constructed very easily. One should cut out a small circle from the cylinder, make a rational rotation, and then glue the circle back (Fig. 12). There are no Hamiltonian obstructions. However, there are some reasons to expect that for billiards nothing similar is possible. And this is not an idle question, because, for example, estimates of remainder terms in the Weyl asymptotics for eigenfunctions of the Laplace operator depend on the assumption that in a billiard the set of periodic orbits has the zero measure. This is proved only for orbits of period 3.

We will finish the discussion of elliptic effects by describing a natural "bridge" to the parabolic case.

Consider a convex polygon P possessing the property that the group generated by reflections with respect to its sides generates a "covering" of the plane. In other words, the images of P under the action of elements of this group cover the plane and if two such images intersect each other, then they coincide. There are just a few such polygons: rectangles, right triangles, right-angled triangles with the angles 45° and 30°. The group generated by reflections with respect to sides of such a polygon contains a normal subgroup of finite index consisting of parallel translations. In the four cases the indices equal respectively $4, 6, 8$, and 12. Taking representatives of congruence classes of the subgroup of translations and acting by them on the initial polygon, we obtain a fundamental domain for the subgroup of translations, and this domain is a torus. Let us make a partial unfolding of the billiard flow by means of the chosen fundamental domain, i.e., instead of reflecting the trajectory let us reflect the polygon. Some pairs of parallel sides will be then identified by translations, and the billiard flow will be thus represented as the free motion of a particle on a (flat) torus: each tangent vector moves in its direction with the velocity equal to one. This is a completely integrable system: the initial angle is a first integral, the phase space is fibered into invariant tori, and on each torus a flow of isometries acts. Each such flow is a standard elliptic system.

3 Parabolic behavior: billiards in polygons

A simplest parabolic billiard table is a right-angled triangle with the angle $\pi/8$. When a trajectory meets the boundary, let us reflect the triangle instead of reflecting the trajectory. In this concrete case everything stops rather soon. If one takes 16 copies of the triangle and makes from them an octagon (Fig. 13), then the motion turns into the parallel flow on this octagon, the opposite sides being identified. The obtained object is a Riemann surface (in this case of genus 2) with a quadratic differential. When the vertices of the octagon are glued together, one obtains the angle 6π. To resolve this singularity one should take the cubic root. Then one can obtain a Riemann surface with a field of directions. The field of directions has one singular point which is a saddle with 6 separatrices. This field of directions can be realized by means of a quadratic differential.

This flow has a first integral, it is the angle (in the octagon the direction of movement is preserved). This first integral has singularities.

Exercise 3. Analyze, in a similar manner, billiards in a right hexagon and a "gnomon."

Such a construction holds in all cases when the angles of the triangle are

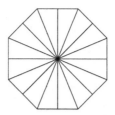

Figure 13. The simplest elliptic billiard

commensurable with π. In this case one can construct a Riemann surface with a quadratic differential from a finite number of copies of the billiard table. This flow has a first integral, and what is obtained can be studied using rather powerful methods from Teichmuller theory. As a result one achieves a rather good understanding of what is going on. Here one meets typical parabolic effects. For example, on all invariant manifolds (in this case, for the fixed value of the angle), the system is topologically transitive; and also on almost all invariant manifolds the system is strictly ergodic, i.e., the invariant measure is unique. And in the exceptional cases, when the invariant measure is non-unique, the number of invariant measures does not exceed the genus of the surface. These are typical parabolic effects; the invariant measure is not always unique, but usually the number of nontrivial invariant measures is finite.

Thus, a billiard system in a polygon with the angles commensurable with π, namely $\pi p_i/q_i$, where p_i and q_i are coprime integers, generates a one-parameter family of flows on some surface whose genus is determined by the geometry of the polygon and arithmetic properties of the numbers p_i/q_i. One should not fall into the illusion that the structure of these flows is rather simple. For example, the genus of the surface (and hence, in typical cases, the number of fixed points of the flow) is proportional to the least common multiple of the denominators q_i.

Nevertheless, these one-parameter families possess more complicated versions of some properties of the family of linear flows on a torus (which, as it was explained above, correspond to billiards in rectangles and some simple triangles). As I have already mentioned, for almost all values of the first integral the flow has the unique invariant measure (point supported measures corresponding to equilibrium states are not taken into account). But, in contrast to the case of flows on a torus, the set of exceptional values of the parameter is noncountable. Recall that on a torus one has a simple dichotomy between the slope angles whose tangents are rational, when all the orbits are closed, and the angles whose tangents are irrational, when the invariant measure is unique and hence any orbit is uniformly distributed with respect to the Lebesgue measure.

In the case of families of flows generated by quadratic differentials on surfaces of genus greater than one (in particular, for the families of flows arising from billiards in rational polygons), the situation is more complex. Still there is a countable number of "rational" values of the parameter for which all the trajectories are closed. Note that, in contrast to the case of the torus, there are several different homotopic types of closed orbits. The number of such types can be estimated using the simple argument that the orbits from different families do not intersect each other and hence their number does not exceed the genus of the surface. Besides that, there exists a set of values of the parameter which has zero measure but positive Hausdorff dimension, for which the flow is quasiminimal (i.e., any semitrajectory which does not tend to a fixed point is dense), but there exist more than one non-atomic invariant measure.

A more deep consideration shows that this difference is a reflection of the dichotomy between *Diophantine* irrational numbers or vectors, for which the speed of rational approximation is not very high, and *Liouville* numbers or vectors, for which "anomalous good approximation" arises. In the case of linear flows on a torus, for Diophantine slope angles, time averages for sufficiently smooth functions converge very rapidly. Moreover, Diophantine flows are rather stable: time changes and even small nonlinear perturbations preserving the rotation number of such flows can be "straightened." For Liouville slope angles, time averages can behave rather irregularly: from time to time they can be very close to the integral or rather far from it, so that the speed of convergence over some sequences of moments of time is very high, and over other sequences it is rather low. Respectively, even smooth changes of time can essentially change large time dynamics: for example, eigenfunctions, even measurable, can disappear, and the flow becomes weakly mixing.

For flows arising from quadratic differentials and for billiards in rational polygons, the values of parameters for which there is more than one invariant measure, correspond to the slope angles with irrational Liouville tangents. Therefore it is not surprising that similar but more bright phenomena arise: instead of slow convergence of averages to the integral by the Lebesgue measure, there is no convergence at all. On the other hand, for a set of values of the parameter of full measure corresponding to slope angles with Diophantine tangents, one has similar though much more complex stability phenomena. They were discovered and studied during the last five years by the young mathematician Giovanni Forni; his papers constitute one of the most bright modern achievements in the theory of dynamical systems. The central observation due to Forni is that although the invariant measure is unique, there are also invariant distributions (generalized functions), i.e., invariant continuous linear functionals defined on smaller spaces of functions than all continuous functions. For functions of a given class of smoothness, the space of invariant distributions is finite-dimensional, but the dimension tends to infinity with the growth of smoothness class. Combination of strict ergodicity (uniqueness of an invariant measure) with the existence of an infinite set of independent invariant *distributions* is rather typical for dynamical systems with parabolic behavior. The simplest example, in which full investigation can be carried out with the help of elementary Fourier analysis, is the affine mapping of the two-dimensional torus

$$(x, y) \mapsto (x + \alpha, x + y) \pmod 1,$$

where α is an irrational number. A more interesting example which is studied by means of the theory of infinite-dimensional unitary representations of the group $SL(2, \mathbb{R})$, is the oricyclic flow on a surface of constant negative curvature.

Returning to flows on surfaces, note that according to Forni's results, invariant distributions determine the speed of convergence of time averages. Roughly speaking, one has some typical power speed; if the first group of invariant distributions vanishes, then this speed increases, and this happens several times, until one obtains the maximal possible speed of decreasing of averages which is inverse proportional to time. Vanishing of a sufficient number of invariant distributions also guarantees that the flow obtained by a change of time can be straightened.

Even in the case of polygons with rational angles the description of the billiard is not completely reduced to considering separately the flows on invariant manifolds. For example, let us consider the question on the growth of the number of periodic trajectories of length no greater than T as a function of T. Of course, periodic orbits arise in families which consist of "parallel" orbits of equal length. Hence one should count the number $P(T)$ of such families. In the case of a billiard in a rectangle (which, as we mentioned several times, is reduced to the geodesic flow, i.e., the free motion of a particle on a flat torus), this problem amounts, after a suitable renormalization, to counting the number of points with integer coordinates in the circle of radius T with center at the origin. Therefore,

$$\lim_{T \to \infty} \frac{P(T)}{\pi T^2} = 1.$$

For general rational billiards, the growth of $P(T)$ is also quadratic, i.e.,

$$0 < \liminf_{T \to \infty} \frac{P(T)}{T^2} \leqslant \limsup_{T \to \infty} \frac{P(T)}{T^2} < \infty.$$

Besides that, it is known that periodic orbits are dense in the phase space. The question on existence of the limit $\frac{P(T)}{T^2}$ as $T \to \infty$ for an arbitrary rational rectangle still remains open. The positive answer is obtained, on the one hand, for some special polygons which amount to quadratic differentials on surfaces with a large number of symmetries (Veech surfaces), and on the other hand, for generic quadratic differentials. It is rather likely that there exist polygons with pathological behavior of the function $P(T)$. Note that our first nontrivial example of a billiard in a right-angled triangle with the angle $\pi/8$ and the hypotenuse 1 yields a Veech surface, and for it we can find $\lim_{T\to\infty} \frac{P(T)}{T^2}$.

For billiards in polygons in which not all angles are commensurable with π, surprisingly little is known. Such billiards are good examples of parabolic systems of sufficiently general kind. One has to say that the methods of analysis available now are insufficient for serious investigation of such systems. Indeed, successful study of parabolic systems is related with two special situations:

(1) flows on surfaces discussed above, where the dimension of the phase space is very small (in complement to the dimension corresponding to orbits one has only one transversal direction), and

(2) flows on homogeneous spaces, where one has large local symmetry.

Two main open questions concerning arbitrary billiards, are the description of global complexity of behavior of trajectories and the asymptotic behavior of typical trajectories with respect to the Lebesgue measure. Let us begin with the second question. Here a lot is known, and at the same time very little. If one fixes the type of the billiard table (for instance, convex polygons with a given number of edges), then the angles are the natural parameters in the space of such billiards. Billiards with angles commensurable with π, for which, as it was explained above, a lot is known, form a dense set in this space. Starting with ergodicity of rational billiards on most invariant submanifolds and taking into account the fact that for large denominators each such manifold almost uniformly covers the phase space, one can show by rather standard categorical arguments that for a dense G_δ in the space of parameters the billiard is ergodic in the whole phase space. However, this topologically ample set of billiards is rather thin from the metric point of view: not only its Lebesgue measure but also its Hausdorff dimension in the space of parameters equals zero. This set reminds one of the set of numbers admitting a rational approximation with extremely high speed, like a triple exponent. It is assumed that for typical Diophantine values of the vector of angles the billiard is ergodic. Up to now no serious approaches to this problem are known. Also more subtle statistical properties, such as mixing, are not known for any irrational billiards including the Liouville situation described above for which ergodicity is proved. The structure of singular invariant measures for irrational billiards is also not known.

Of course, a particular case of this last question is description of periodic trajectories, since each such trajectory generates a singular ergodic invariant measure. On the one hand, it is unknown whether for an arbitrary polygon

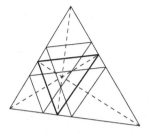

Figure 14. Orbits of periods 3 and 6 in an acute triangle

there exists at least one periodic billiard trajectory. As it was mentioned above, for rational polygons there are infinitely many such trajectories and they are dense in the phase space. However, one has not succeeded in passing to a limit for irrational polygons, even of a special kind. The problem here is that as the denominator increases, invariant manifolds become surfaces of a very high genus and periodic orbits have very complicated homotopical type, and hence are very long. However, there are some special situations when periodic orbits with simple combinatorics arise which are conserved during small perturbations of angles. A classical example is the orbit of period 3 formed by the bases of altitudes in an arbitrary acute-angled triangle. Of course, this orbit admits a variational description. But, in contrast to Birkhoff orbits in convex billiards, the triangle formed by the bases of altitudes has the minimal perimeter among all inscribed triangles. And the maximal and minimax triangles degenerate into the double maximal altitude. The orbit of period 3 thus constructed is surrounded by a family of parallel orbits of period 6 (see Fig. 14). Note that these are the only periodic orbits whose existence is known for all acute-angled triangles. The question on the density or at least the existence of an infinite number of parallel families of periodic orbits remains open.

For an arbitrary right-angled triangle the existence of periodic orbits has been proved just a few years ago. Unfortunately, these orbits are somewhat disappointing. These are trajectories which are reflected orthogonally from one of the sides and after a finite number of reflections return to the same side also in the orthogonal direction. Evidently, such an orbit is reflected then back and repeats its way in the backward direction. This is an example of orbits with stable combinatorics. It turns out that for almost any initial position the orbit orthogonal to a side returns back to this side in the orthogonal direction and therefore is periodic. This is a rather easy consequence of conservation of measure and of the fact that possible directions of an orbit form the unique trajectory of the infinite dihedral group generated by reflections with respect to the two nonperpendicular sides of the triangle. This argument can be generalized to some polygons "close" to rational ones, i.e., those for which the values of angles modulo π lie in a one-dimensional space over rational numbers. For an arbitrary obtuse-angled triangle this argument cannot be applied, and the existence of even one periodic orbit is unknown.

The existence of periodic orbits is closely related to the question on the global complexity of the behavior of trajectories. The growth of the number of distinguishable trajectories with the time can be estimated in different ways. The most natural way is related to coding. To each trajectory one assigns a sequence of symbols in correspondence with the reflections with respect to sides of the polygon, so that each side is denoted by its own symbol. Of course, in this way one naturally encodes billiard maps, i.e., returning maps of the billiard flows to the boundary. In order to obtain full information on the flow, one should also indicate the time between two consecutive reflections. The growth of complexity for the billiard mapping (respectively, flow) is given by the function $S(N)$ (respectively, $\mathcal{S}(T)$) equal to the number of different codes of length n (respectively, to the number of different codes arising for segments of trajectories of length T). Obviously, each family of parallel periodic orbits generates an infinite periodic code, and it is almost also obvious that, vice versa, each infinite periodic code corresponds to a family of parallel periodic orbits. These orbits can be closed either after one period or after two periods (as orbits of period 6 parallel to the Faniano triangle in an acute-angled triangle).

In the case of polygons with rational relative to π angles, both functions admit a quadratic estimate:

$$0 < \liminf_{N \to \infty} \frac{S(N)}{N^2} \leqslant \limsup_{N \to \infty} \frac{S(N)}{N^2} < \infty$$

and

$$0 < \liminf_{T \to \infty} \frac{\mathcal{S}(T)}{T^2} \leqslant \limsup_{T \to \infty} \frac{\mathcal{S}(T)}{T^2} < \infty.$$

Note that in this case a positive part of all admissible codes is realized by periodic trajectories.

An alternative way of describing complexity is to compute the number of ways in which codes can change. Obviously, the code changes when the trajectory hits an angle. It is also obvious that there is only a finite number of segments of trajectories of bounded length which hit angles both in the positive and negative directions. By rather evident reasons, such singular trajectories are called *generalized diagonals* of a polygon. Define $D(N)$ (respectively, $\mathcal{D}(T)$) as the number of generalized diagonals with $\leqslant N$ edges (respectively, as the number of generalized diagonals of length $\leqslant T$). As above, these quantities admit a quadratic estimate for rational polygons.

It is natural to assume that for arbitrary polygons the growth of trajectories should not be much more rapid than for rational ones, since the local geometric structure of the billiard flow is the same in both cases. However, the only known fact in this direction consists in much more weak subexponential estimates:

$$\lim_{N\to\infty} \frac{\log S(N)}{N} = \lim_{N\to\infty} \frac{\log D(N)}{N} = \lim_{T\to\infty} \frac{\log \mathcal{S}(T)}{T} = \lim_{T\to\infty} \frac{\log \mathcal{D}(T)}{T} = 0.$$

4 Hyperbolic behavior: billiards of Sinai, Bunimovich, Wojtkowski and other authors

As we have already mentioned, hyperbolic behavior is rather common and allows one to establish the basic elements of stochastic or "chaotic" behavior. The domination of hyperbolic behavior is natural by analogy with linear systems. Indeed, a randomly chosen matrix most likely has no eigenvalues whose absolute value equals one. Even if one *a priori* restricts oneself to matrices with the determinant equal to one, this is still true for matrices of size 3×3 or more. Although this analogy cannot be literally transferred to nonlinear systems, it at least shows the importance of the hyperbolic paradigm.

Historically the first examples of hyperbolic behavior of billiards were found by Ya. G. Sinai [13]. The simplest examples of a Sinai type billiard are, firstly, a square with a circle cut out, and, secondly, a convex polygon whose sides are replaced by arcs convex inwards (see Fig. 1). From the point of view of rigorous mathematical analysis, the second example turns out to be somewhat more easy than the first one. Hyperbolic behavior in Sinai billiards is related to the phenomenon of scattering of light well known from geometric optics: a parallel or divergent flow of light becomes more divergent after reflection with respect to a convex mirror. Not too complicated computations show that if the reflection is sufficiently regular, then the angle measure of a sheaf grows exponentially. This yields hyperbolicity of the linearized system.

In the analysis of scattering billiards two technical difficulties arise.

Firstly, one must achieve sufficient regularity of reflections with respect to convex inwards parts of the boundary. It is clear why to this extent the second example is better than the first one: in the second example the time between two consecutive reflections is bounded. And in the first example there are periodic trajectories parallel to the sides of the square which do not meet the obstacle at all. Of course, such trajectories form a set of zero measure, but trajectories which form very little angles with them meet the obstacle only in a very large time. This phenomenon is called infinite horizon; respectively, boundedness of the time between two reflections corresponds to finite horizon.

Infinite horizon implies non-uniformity of hyperbolic estimates over the phase space. Although this yields essential technical complications in proofs of ergodicity, mixing and other stochastic properties, this also confirms the role of billiards as an important testing area for various methods and tools of analysis of dynamics. Indeed, non-uniform hyperbolicity is much more common than uniform one. For example, global uniform hyperbolic behavior for classical conservative systems imposes restrictions on the topology of the phase space. But non-uniform hyperbolicity is compatible with any topology. This fact, although predicted rather long ago, has been established in full generality only recently by D. Dolgopyat and Ya. Pesin [6].

The second difficulty in the analysis of dispersing billiards is the presence of singularities (discontinuities and unboundedness of derivatives) in a system. Here is the difference between these systems and billiards in smooth convex domains, considered above, where the billiard mapping is smooth. Singularities arise at tangent points of trajectories with convex inwards parts of the boundary. Of course, they also arise when a trajectory hits an angle. Singularities of the second type arise also in parabolic billiards, and in the case of scattering billiards they yield not significant complications. Such singularities yield discontinuities of the first kind for functions representing dynamics: a surface of discontinuity arises, and functions are smooth on both sides of the surface. Thus, the differential along a billiard trajectory which does not hit a discontinuity point behaves rather regularly. For trajectories tangent to the boundary from inside, the derivatives near these trajectories are unbounded, so that the discontinuities are more serious. Note that elastic collisions and more complex effects of this kind naturally arise in many important problems of classical mechanics, for example, in the problem of n bodies. The influence of such phenomena on the large time behavior of trajectories is one of the central problems in mechanics. Here also billiards, and especially their multidimensional analogs, play the role of an important testing area.

Scattering billiards are rather essential for the mathematical background of models of statistical physics. This is an important and interesting subject, which however we will not touch here. From the view point of geometry, scattering billiards possess some defects, for example, inavoidable singularities on the boundary. However, if one considers billiards not on plain domains but on domains on a flat torus, this defect can be avoided. For example, the billiard on a torus with a circle removed is a classical example of a Sinai billiard. Nevertheless, it is interesting to know how hyperbolic behavior can arise in other ways than scattering by convex inwards parts of the boundary. The first answer to this question is given by a rather celebrated example of a "stadium," i.e., two semicircumferences connected by segments of common tangent lines (see

Figure 15

Fig. 15). This is an example of the so-called Bunimovich billiards [2], where hyperbolic behavior arises as a result of consecutive focusing of sheaves of orbits. From the point of view of configuration space this picture dramatically differs from the case of scattering billiards; however, in the phase space, where both coordinates and velocities are taken into account, uniform exponential growth arises.

Bunimovich billiards were discovered in an interesting way. In the beginning of the 70's L. A. Bunimovich, who was then a graduate student of Sinai, was working on extending the class of billiards with exponential running away and stochastic behavior of orbits. He discovered that if one adds little "pockets" to a scattering billiard, then the billiard on the thus obtained table in which convex parts are followed by concave ones, possesses exponential running away of trajectories. Actually Bunimovich discovered a new important mechanism of hyperbolicity. However, he himself firstly considered his work as just a little generalization of the results on scattering billiards. During Bunimovich's talk on a seminar at MIAS[1] directed by D. V. Anosov and the author, the natural question on the mechanism of hyperbolicity arose, and in particular on whether the presence of any scattering components is necessary. I draw the speaker's attention to the fact that his arguments seemingly did not imply this necessity, and proposed a stadium as a model for verification of this conjecture. The rest of Bunimovich's geometric conditions were satisfied, at least if full circles did not intersect each other (see Fig. 15). After thinking a little Bunimovich said that his arguments should hold in this case, and in the next version of his paper he stated conditions which did not require the presence of scattering components. Moreover, it turned out that the initial geometric conditions can be weakened, so that, for example, in the case of a stadium, the distance between the circles can be arbitrarily small.

Among Bunimovich billiards there are a lot of other interesting and rather simple forms, but they all have the common property that the boundary can contain, except scattering parts, only segments of straight lines and arcs of circumferences. The natural question, how essential is this condition, has been studied by specialists for about ten years. A technical difficulty is the following. Hyperbolicity is established with the help of a system of cones in tangent spaces

---

[1] Mathematical Institute of the Academy of Sciences (Moscow).

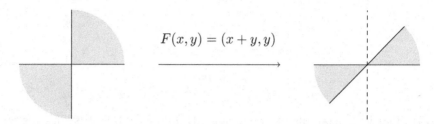

$$F(x,y) = (x+y,y)$$

Figure 16. The action of the parabolic transformation on the cone

to points of the phase space, which are transformed to themselves under the action of dynamics. For simplicity and geometric visualization, it is better to think about the billiard mapping rather than the flow. In this case the phase space is two-dimensional, and the cones in question are the interior parts of two opposite angles formed by a pair of lines intersecting at the origin. The system of cones which is invariant, both in scattering billiards and Bunimovich billiards is the same. Geometrically, these cones are defined as the sets of infinitesimal dispersing pieces of trajectories. For hyperbolicity it is necessary that the cone together with its boundary be mapped strictly inside the corresponding cone in the image. Of course, this holds in the case of scattering billiards already after one reflection. And in the case of flat and circumference mirrors, the cone goes into itself, but one of its sides is left invariant. This is a typically parabolic effect, since in this way unipotent matrices act. Let us take, for instance, the matrix $\begin{pmatrix} 1 & 1 \\ 0 & 1 \end{pmatrix}$. The cone in question is defined by the condition $x_1 x_2 > 0$, i.e., it is the union of the first and the third quadrant on the plane. Its image is the cone $|x_1| > |x_2|$, $x_1 x_2 > 0$ (see Fig. 16). After further iterations the image becomes more and more thin, but it is still "glued" to the horizontal axis. In order to get hyperbolicity, Bunimovich uses a geometric condition which yields strict invariance of cones after the reflection with respect to *different* circumference parts of the boundary (as in the case of a stadium). Since the trajectory which is reflected from the circumference part under a very small angle, continues to do this many times, it seemed that the explicit form of the iteration after reflection with respect to circumference parts (integrability of the billiard in a circle) played an essential role. This way I explained for myself rigid Bunimovich's conditions.

However, it turned out that one can overcome this difficulty. Billiards with convex parts of the boundary can be hyperbolic by many reasons. As soon as the method based on the use of systems of invariant cones was invented, the problem of finding new classes of hyperbolic billiards became easy. Note that Bunimovich used another technique which is formally equivalent to the system of invariant cones, but is much less visual. The pioneers of the use of the method of invariant cones in dynamics were V. M. Alekseyev (1932–1980) and Yorgen Moser (1928–1999). An essential step was introducing this method in non-uniformly hyperbolic situation. The author used this method for constructing examples of smooth systems with stochastic behavior on various manifolds. However, the most essential progress here is due to Maczei Wojtkowski. And again billiards turned out to be an ideal testing area. Having understood the key role of systems of cones, Wojtkowski realized that the problem could be solved in the inverse order, namely, to find classes of billiard tables corresponding to a given system of cones. The preprint of his key paper on this subject [3] was called "Principles for the design of billiards with nonvanishing Lyapunov exponents." As a square or torus with a removed circle is a quintessence of the phenomenon discovered by Sinai, and the stadium symbolizes Bunimovich billiards, in the same way a typical example of Wojtkowski billiards is given by cardioida (see Fig. 1). The importance of Wojtkowski's result in the theory of billiards is in that he discovered classes of hyperbolic examples which are open in C^2 topology, and thus [11] this property does not depend on small variations of mirrors.

As I have already mentioned, constructing new classes of hyperbolic billiards became possible with the use of the method of invariant cones. As an example of flexibility of this method, let us mention the following result due to Victor Donnay [7]: any sufficiently small piece of a convex curve is a part of the boundary of a piecewise smooth convex hyperbolic billiard. Note also that the use of the method of invariant cones allowed one to obtain many remarkable examples of classical dynamical systems with non-uniform hyperbolic behavior.

Important unsolved problems are related with existence of hyperbolic billiards with smooth (at least twice differentiable) boundary. Note that the boundary of the stadium is differentiable, but the curvature (and hence the second derivative) is discontinuous. Even twice differentiable examples with nonconvex or non-simply-connected boundary are unknown.

What does hyperbolicity give? It allows one to show that in many cases a deterministic dynamical system behaves in many respects as a sequence of independent random quantities. In a sense, this statement is true literally: under some (often rather easily checked) conditions, in complement to (even non-uniform) hyperbolicity, the phase space of a system conserving finite volume can be divided into a finite number of sets $A_1, \ldots, A_n$ of positive measure so that, firstly, each point of the phase space is encoded by the sequence of hits to these sets at positive and negative moments of time, and, secondly, these sets are completely independent with respect to the dynamics F, i.e.,

$$\text{vol}\left(\bigcap_{k=0}^{n} F^k(A_{i_k})\right) = \prod_{i=0}^{n} \text{vol} A_{i_k}.$$

Although these sets are of exotic nature, this property, which is naturally called the Bernoulli property, implies many important properties: convergence of time averages to the space average (ergodicity), decrease of correlation (mixing), asymptotic independence of the future from the past (the K-property, or the Kolmogorov property).

References

[1] G. D. Birkhoff. *Dynamical Systems* (New York: Amer. Math. Soc., 1927).

[2] L. A. Bunimovich. On the ergodic properties of nowhere dispersing billiards. *Comm. Math. Phys.* **65** (3) (1979), 295–312.

[3] W. Wojtkowski. Invariant families of cones and Lyapunov exponents. *Ergodic Theory Dynam. Systems*, **5** (1985), 145–161.

[4] G. A. Galperin and A. N. Zemlyakov. *Mathematical Billiards* (Moscow: Nauka, 1990) [in Russian].

[5] G. A. Galperin and N. I. Chernov. *Billiards and Chaos* (Moscow: Znanie, 1991) [in Russian].

[6] D. Dolgopyat and Ya. Pesin. Every compact manifold carries a completely hyperbolic diffeomorphism. To appear in *Ergodic Theory Dynam. Systems*.

[7] V. J. Donnay. Using integrability to produce chaos: Billiards with positive entropy. *Comm. Math. Phys.* **141** (2) (1991), 225–257.

[8] A. B. Katok and B. Hasselblatt. *Introduction to Modern Theory of Dynamical Systems* (Cambridge: Cambridge Univ. Press, 1995).

[9] I. P. Kornfeld, Ya. G. Sinai, and S. V. Fomin. *Ergodic Theory* (Moscow: Nauka, 1980) [in Russian].

[10] V. F. Lazutkin. Existing of caustics for the billiard problem in a convex domain. *Izv. Akad. Nauk SSSR, Ser. Mat.*, **37** (1) (1973), 186–216.

[11] H. Masur and S. Tabachnikov. Rational billiards and flat structures. To appear in *Handbook in Dynamical Systems 1A* (Amsterdam: Elsevier).

[12] *Hard Ball Systems and the Lorentz Gas*, ed. D. Szász (Berlin: Springer-Verlag, 2000).

[13] Ya. G. Sinai. Dynamical systems with elastic reflections. Ergodic properties of dispersing billiards. *Usp. Mat. Nauk*, **25** (2) (1970), 141–192.

[14] S. Tabachnikov. Billiards. *Panor. Synth.*, **1** (1995).

A. N. Rudakov

The Fibonacci numbers and simplicity of $2^{127} - 1$

Lecture on April 3, 1999

1 Introduction

For a long time, from 1877 to 1951, the number $M = 2^{127} - 1$ had been a champion, being the largest known prime. The primality of $2^{127} - 1$ was established by Édouard Lucas. He invented a remarkable method for proving primality. For $M = 2^{127} - 1$, the computation took about 100 hours, but it involved no divisions by smaller primes. I am going to describe the mathematical part of Lucas' algorithm and discuss some elegant results of finite arithmetic. We shall not perform the computation proper.

I personally consider this theme a very good illustration of the general idea that constructing a good algorithm needs a good theory. Though, there are expositions of Lucas' result which employ a significantly smaller amount of theory than my exposition in this lecture (see [1, 2]).

A detailed historical study of the work of Lucas and of other works on finding primes is contained in [3].

2 The Fibonacci numbers and the main theorem

As is known, the sequence of Fibonacci numbers is obtained as follows: we take $u_1 = 1$ and $u_2 = 1$ and define each next number by the formula $u_{n+1} = u_n + u_{n-1}$. The Fibonacci numbers

$$1, 1, 2, 3, 5, 8, 13, 21, 34, 55, \ldots$$

have many remarkable properties. For example, the first two numbers are odd, the next number is even, then two odd numbers go again, etc. A simple way to see this is to consider Fibonacci numbers modulo 2. Knowing u_{n-1} and u_n modulo 2 and taking into account that u_{n+1} is congruent to their sum modulo 2, we can write the sequence

$$u_3 \equiv 1 + 1 \equiv 0 \pmod 2,$$
$$u_4 \equiv 0 + 1 \equiv 1 \pmod 2,$$
$$u_5 \equiv 1 + 0 \equiv 1 \pmod 2,$$

$$\ldots\ldots\ldots,$$

or $1, 1, 0, 1, 1, 0, 1, 1, 0, \ldots$. Thus, every third number is even and the numbers going before and after it are odd.

It can be shown that every fifth number is divisible by 5. For this purpose, it suffices to consider the Fibonacci numbers modulo 5. This yields the sequence

$$1, 1, 2, 3, 0, 3, 3, 6, 9 \equiv -1, 0, -1, -1, -2, -3, 0, -3, -3, \ldots .$$

After the 20th term, the sequence repeats, and every fifth position is occupied by zero.

Problem 1. *Show that every fourth Fibonacci number is divisible by* 3.

Problem 2. *Show that, if m divides u_k, then m divides u_{2k}, u_{3k}, u_{4k}, $\ldots$.*

There exists a formula for the Fibonacci numbers. It is very well known, but let me remind you of the argument. Let us forget the "initial data" $u_1 = 1$ and $u_2 = 1$ and consider only the transition equation

$$x_{n+1} = x_n + x_{n-1}. \tag{1}$$

Certainly, there are many sequences satisfying this equation. One of them, which is sometimes called *Lucas numbers*, is

$$v_1 = 1, \ v_2 = 3, \ v_3 = 4, \ \ldots, \ v_{n+1} = v_n + v_{n-1}.$$

There also exist other such sequences, and if $\{a_n\}$ and $\{b_n\}$ are two of them, we can construct one more by taking their linear combination with some coefficients, e.g., $c_n = 2a_n + 3b_n$. We took the coefficients 2 and 3, but they can be arbitrary. In particular, if

$$\alpha = \frac{1 + \sqrt{5}}{2} \quad \text{and} \quad \beta = \frac{1 - \sqrt{5}}{2},$$

i.e., if α and β are the roots of the equation $x^2 = x + 1$, then the sequences $a_n = \alpha^n$ and $b_n = \beta^n$ satisfy the transition equation (1), and therefore, all their linear combinations have this property. Since $\alpha + \beta = 1$ and $\alpha^2 + \beta^2 = 3$, the sum of these sequences gives the Lucas numbers:

$$\alpha^n + \beta^n = v_n. \tag{2}$$

To obtain the Fibonacci numbers, we must select the coefficients more carefully. As a result, we obtain

$$u_n = \frac{\alpha^n - \beta^n}{\alpha - \beta}. \tag{3}$$

In particular, this implies $u_{2n} = u_n \cdot v_n$.

Now, I would like to state the main theorem, which is essentially the Lucas theorem (1886), although Lucas formulated it differently. A modern exposition of historical details is contained in [3].

Theorem 1. *Suppose that q is a prime of the form $4k + 3$ and $M = 2^q - 1$. Then M is a prime if and only if $v_{\frac{M+1}{2}} \equiv 0 \pmod{M}$.*

This result is a base of the algorithm for establishing the primality of the number $2^{127} - 1$; we must only add a "fast" method for computing $v_{\frac{M+1}{2}}$. We shall discuss it later on.

3   Complex numbers in finite arithmetics

Let us slightly modify the terminology: instead of "a is comparable with b modulo m," or "$a \equiv b \pmod{m}$," we shall say "a equals b in arithmetic modulo m," or "$a =_{(m)} b$." Formally, this changes nothing, we simply use slightly different words; but this provokes us to imagine that there are some numbers of "arithmetic modulo m," which are denoted by the same symbols as the integers but differ from them otherwise. For example, 6 and -1 are two denotations for the same number of "arithmetic modulo 7."

This approach leads us to the question as to whether the realm of numbers can be extended by adding, e.g., "complex numbers." After all, complex numbers are pairs of reals, and pairs of numbers can be considered in finite arithmetics too. Let us consider "complex numbers modulo 7." We define such a complex number z as a pair $z = (a, b)$, where a and b are "numbers modulo 7." Addition and multiplication are defined in the usual way, addition by

$$(a_1, b_1) + (a_2, b_2) = (a_1 + a_2, b_1 + b_2)$$

and multiplication by

$$(a_1, b_1) \cdot (a_2, b_2) = (a_1 a_2 - b_1 b_2, a_1 b_2 + a_2 b_1).$$

It is fairly easy to show that this is a good definition, that is, there are zero and identity elements, associativity, commutativity... We can also calculate inverse elements: if $z = (a, b)$, then

$$z^{-1} = \left(\frac{a}{a^2 + b^2}, \frac{-b}{a^2 + b^2} \right).$$

However, it may happen that $a^2 + b^2 =_{(7)} 0$, and the inverse element is not defined.

This can be verified directly, because 7 is a very small number. All elements modulo 7 are easy to write out; these are

$$0, 1, 2, 3, 4, 5, 6.$$

The squares are 0, 1, 4, and 2. That is all. The sums of two squares are

$$0, 1, 4, 2; \quad 1, 2, 5, 3; \quad 4, 5, 3, 6; \quad 2, 3, 6, 4.$$

We see that 0 occurs only once as $0 = 0^2 + 0^2$; all the remaining sums are nonzero. Therefore, each nonzero complex number has an inverse. We have constructed a good arithmetic with all the four operations; such an object is otherwise called a field, or, to be more precise, a field which is the quadratic extension of the prime field with seven elements.

A propos, 7 cannot be replaced by 5, because $1^2 + 2^2 =_{(5)} 0$ in arithmetic modulo 5. We need to think what numbers in arithmetic modulo p are to be considered "negative." Recall that, to construct the usual complex field, we take -1, i.e., a negative number for which a square root does not exist, and formally "add" this square root, that is, write $z = a + bi$, where $i^2 = -1$. Then, the rules for extending the operations emerge by themselves, from multiplying out:

$$(a_1 + b_1 i) + (a_2 + b_2 i) = (a_1 + a_2) + (b_1 + b_2)i,$$

$$(a_1 + b_1 i) \cdot (a_2 + b_2 i) = (a_1 a_2 + b_1 b_2(-1)) + (a_1 b_2 + a_2 b_1)i.$$

We could take another negative number, -2 say, and represent complex numbers in the form $z = a + bj$, where $j^2 = -2$. The result is the same, in particular, operations also emerge by themselves, from multiplying out:

$$(a_1 + b_1 j) + (a_2 + b_2 j) = (a_1 + a_2) + (b_1 + b_2)j,$$

$$(a_1 + b_1 j) \cdot (a_2 + b_2 j) = (a_1 a_2 + b_1 b_2(-2)) + (a_1 b_2 + a_2 b_1)j.$$

The only difference occurs at the place where j^2 appears. The formula for the inverse element is

$$(a + bj)^{-1} = \frac{a}{a^2 + 2b^2} + \frac{-b}{a^2 + 2b^2} j;$$

since $a^2 + 2b^2 \neq 0$ whenever $(a, b) \neq (0, 0)$, there are no problems.

Problem 3. *Verify that a quadratic extension of the prime field with 5 elements can be constructed by considering the numbers $a + bj$, where $j^2 = -2$. All the four arithmetic operations are well-defined.*

Let me introduce the following definition: we shall say that an element a is negative in arithmetic modulo p, where p is a prime, if the equation $x^2 =_{(p)} a$ has no solutions; otherwise, we say that it is positive (provided that $a \neq_{(p)} 0$). For example, the numbers 1 and 4 modulo 5 are positive, while 2 and 3 are negative. Since $-1 =_{(5)} 4$, the number -1 is also positive; strange as it seems, this is so. But modulo 7, the numbers 1, 2, and 4 are positive and -1, -2, and -4 (or 6, 5, and 3) are negative. What might naturally be called the sign of an element is historically referred to as the Legendre symbol $\left(\frac{a}{p}\right)$. By definition,

$$\left(\frac{a}{p}\right) = \begin{cases} +1 & \text{if } a \text{ is positive modulo } p, \\ -1 & \text{if } a \text{ is negative modulo } p, \\ 0 & \text{if } a =_{(p)} 0. \end{cases}$$

It can be verified that, for an odd prime p, precisely half $\left(\text{i.e., } \frac{p-1}{2}\right)$ of nonzero numbers modulo p are positive and precisely half are negative, and that the product of two negative numbers is always positive.

Problem 4. *Prove that, if* $\left(\frac{a}{p}\right) = -1$ *and* $\left(\frac{b}{p}\right) = -1$*, then* $\left(\frac{ab}{p}\right) = +1$.

Problem 5. *Verify that, if t is negative modulo p, then the numbers $a + bj$, where $j^2 = t$, form a quadratic extension of the prime field with p elements (where all the four arithmetic operations are well-defined).*

The main application of what is said above to our considerations is as follows. Let p be a prime. Whenever 5 is positive or negative modulo p, the numbers $\alpha = \frac{1+\sqrt{5}}{2}$ and $\beta = \frac{1-\sqrt{5}}{2}$ are defined, in the latter case, as complex numbers modulo p (i.e., as elements of the quadratic extension), and formulas (2) and (3) for the Lucas and Fibonacci numbers still make sense modulo p.

4 Complex conjugation for numbers modulo p

An essential structural component of the usual complex numbers is the operation of complex conjugation: if $z = a + bi$, then $\bar{z} = a - bi$. We know that the conjugate of a sum is the sum of conjugates, and the same is valid for products:

$$\overline{(z_1 + z_2)} = \bar{z}_1 + \bar{z}_2, \quad \overline{(z_1 \cdot z_2)} = \bar{z}_1 \cdot \bar{z}_2.$$

We easily conclude that, if α is a complex root of an equation

$$x^2 + ax + b = 0$$

with real coefficients, then $\bar{\alpha}$ is also a root of this equation.

We can define conjugation in the quadratic extension of the field with p elements by

$$\overline{(a + b \cdot j)} := a - b \cdot j.$$

Clearly, the conjugate of a sum is the sum of conjugates, and the conjugate of a product is the product of conjugates. In addition, the following remarkable formula holds.

Let p be a prime; suppose that t is negative modulo p, i.e., $\left(\frac{t}{p}\right) = -1$. We construct complex numbers as the numbers of the form $a + b \cdot j$, where a and b are numbers modulo p and $j^2 = t$.

Proposition 1. *Under these conditions, if $z = a + bj$ and $\bar{z} = a - bj$, then*

$$\bar{z} = z^p. \tag{4}$$

In particular, $z^{p+1} = z\bar{z} = a^2 - tb^2$; thus, the $(p+1)$th power of a "complex" number is necessarily a "real" number.

Let us prove (4). Note that

$$(x + y)^p =_{(p)} x^p + y^p$$

for our numbers. This is so because the coefficients of Newton's binomial, $\binom{p}{i} = \frac{p!}{i!(p-i)!}$, are integers divisible by p for $0 < i < p$. Thus, we can write

$$(a + b \cdot j)^p =_{(p)} a^p + b^p \cdot j^p.$$

Using Fermat's little theorem, we conclude that $a^p =_{(p)} a$ and $b^p =_{(p)} b$. It remains to calculate j^p. Obviously,

$$j^p = j^{p-1} \cdot j = t^{(p-1)/2} \cdot j.$$

We must show that $t^{(p-1)/2} =_{(p)} -1$ for a negative element t. Note that the number $(p-1)/2$ is integer, and if s is a positive element, then $s = a^2$ and

$$s^{(p-1)/2} =_{(p)} a^{p-1} =_{(p)} 1;$$

the latter equality follows from Fermat's theorem. Thus, the positive elements provide $(p-1)/2$ roots of the polynomial equation

$$x^{(p-1)/2} = 1$$

in the field of "elements modulo p." By Bézout's theorem, the number of roots of a polynomial equation cannot exceed its degree; therefore, for a negative element t, we have

$$t^{(p-1)/2} \neq_{(p)} 1.$$

At the same time, $t^{p-1} =_{(p)} 1$, and the relation

$$t^{p-1} - 1 =_{(p)} (t^{(p-1)/2} - 1)(t^{(p-1)/2} + 1)$$

leaves the only possibility $t^{(p-1)/2} =_{(p)} -1$. This completes the proof of the formula.

Corollary. *Suppose that p is a prime and 5 is negative modulo p. Then, for $\alpha = \frac{1+\sqrt{5}}{2}$ and $\beta = \frac{1-\sqrt{5}}{2}$, the following relations hold:*
  (i) $\alpha^p =_{(p)} \beta$ and $\beta^p =_{(p)} \alpha$;
  (ii) $\alpha^{p+1} =_{(p)} \beta^{p+1} =_{(p)} \alpha \cdot \beta =_{(p)} -1$.

Applying this result to the Fibonacci and Lucas numbers, we obtain

$$u_{p+1} = \frac{\alpha^{p+1} - \beta^{p+1}}{\alpha - \beta} \equiv 0 \pmod{p};$$

$$v_p = \alpha^p + \beta^p \equiv \alpha + \beta \equiv 1 \pmod{p}.$$

To use these congruences, we must be able to determine whether 5 is positive or negative modulo a given p. Let us try to learn to do this.

5   The square root of 5 modulo p

The assertion which I want to state easily follows from more general and fairly deep results about the Legendre symbol; together, they constitute the quadratic reciprocity law. We need only a special case of this general law, which was discovered by Euler and Legendre and proved in full generality by Gauss and which is a pearl of "elementary" number theory.

Proposition 2.

$$\left(\frac{5}{p}\right) = \begin{cases} +1 & \text{if } p \equiv \pm 1 \pmod{5}, \\ -1 & \text{if } p \equiv \pm 2 \pmod{5}. \end{cases}$$

First, we state two general lemmas.

Lemma 1 (Legendre).

$$a^{(p-1)/2} \equiv \left(\frac{a}{p}\right) \pmod{p}.$$

Actually, this means that the $((p-1)/2)$th power of a modulo p is equal to $+1$ if a is positive and to -1 is a is negative. We have discussed this in the preceding section.

Note that any nonzero number modulo p is equal to one of the numbers $1, 2, \ldots, (p-1)/2$ up to sign. Let us denote the set of these numbers by $\mathcal{P}$:

$$\mathcal{P} = \{1, 2, \ldots, (p-1)/2\};$$

then, for any nonzero x modulo p, either $x \in \mathcal{P}$ or $-x \in \mathcal{P}$. Take a p and an $a \neq_{(p)} 0$.

Lemma 2 (Gauss). *Suppose that, for* $k = 1, 2, \ldots, (p-1)/2$, *the numbers* ϵ_k *are equal to* $+1$ *or* -1 *and* $a \cdot k \cdot \epsilon_k \in \mathcal{P}$ *modulo* p. *Then*

$$\left(\frac{a}{p}\right) = \prod_{k=1}^{(p-1)/2} \epsilon_k.$$

Indeed, note that, if numbers k' and k'' from $\mathcal{P}$ are different, then the products $a \cdot k' \cdot \epsilon_{k'}$ and $a \cdot k'' \cdot \epsilon_{k''}$ are also different. They might coincide only if we had $a \cdot k' =_{(p)} a \cdot k''$ or $a \cdot k' =_{(p)} -a \cdot k''$, but both these relations are impossible. Therefore, as k runs over the set $\mathcal{P}$, the products $a \cdot k \cdot \epsilon_k$ also run over this set. Let K be the product of all elements from $\mathcal{P}$. We have

$$K = \prod_{k=1}^{(p-1)/2} a \cdot k \cdot \epsilon_k =_{(p)} a^{(p-1)/2} \cdot K \cdot \prod_{k=1}^{(p-1)/2} \epsilon_k.$$

Cancelling K, we obtain $1 =_{(p)} a^{(p-1)/2} \cdot \prod_{k=1}^{(p-1)/2} \epsilon_k$; this and Legendre's lemma prove the lemma of Gauss.

Remark 1. Our proof may be said to generalize one of the well-known proofs of the Fermat little theorem.

Now, we can proceed to prove the proposition. We have $a = 5$. For an odd p,

$$p \equiv \pm 1 \pmod 5 \iff p = 10n + 1 \quad \text{or} \quad p = 10n + 9,$$
$$p \equiv \pm 2 \pmod 5 \iff p = 10n + 3 \quad \text{or} \quad p = 10n + 7.$$

Let us apply Gauss' lemma to $p = 10n + 1$. Here $(p-1)/2 = 5n$, and we must consider $k = 1, 2, \ldots, 5n$.

For $k = 1, 2, \ldots, n$, we have

$$5k = 5, 10, \ldots, 5n \quad \text{and} \quad \epsilon_k = +1.$$

For $k = n+1, \ldots, 2n$, we have

$$5k = 5n + 5, \ldots, 10n \quad \text{and} \quad \epsilon_k = -1.$$

For $k = 2n+1, \ldots, 3n$, we have

$$5k = (10n + 1) + 4, \ldots, (10n + 1) + 5(n-1) + 4 \quad \text{and} \quad \epsilon_k = +1.$$

Similarly, for $k = 3n+1, \ldots, 4n$, we have $\epsilon_k = -1$, and for $k = 4n+1, \ldots, 5n$, we have $\epsilon_k = +1$ again.

Thus, -1 occurs $2n$ times, and $\prod \epsilon_k = +1$. This means that if $p = 10n + 1$, then $\left(\frac{5}{p}\right) = +1$.

If $p = 10n + 3$, the argument is quite similar. In this case, $(p-1)/2 = 5n+1$. For $k = 1, \ldots, n$, we have $\epsilon_k = +1$.

For $k = n + 1, \ldots, 2n$, we have $\epsilon_k = -1$.

For $k = 2n + 1, \ldots, 3n$, we have $\epsilon_k = +1$.

For $k = 3n + 1, \ldots, 4n$, we have $\epsilon_k = -1$.

If $k = 4n + 1$, then $5k = 20n + 5 = (10n + 3) + (10n + 2)$, which gives $\epsilon_k = -1$.

For $k = 4n + 2, \ldots, 5n + 1$, we have $\epsilon_k = +1$.

As a result, we obtain $2n + 1$ negative ones, which is larger by one than the number of ones. Therefore, $\left(\frac{5}{p}\right) = -1$ in this case. We leave the two remaining cases to the reader and regard Proposition 2 as being proved.

6   Proof of the main theorem

We return to the proof of the main theorem stated at the end of Section 2. Let $N = 2^{q-1} = \frac{M+1}{2}$, i.e., $M + 1 = 2N$.

First, note that we can calculate $M \pmod 5$. We know that $2^4 \equiv 1 \pmod 5$ and that

$$M = 2^q - 1 = 2^{4k+3} - 1 \equiv 2^3 - 1 \equiv 2 \pmod 5.$$

Let us memorize this: under the conditions of the theorem, $M \equiv 2 \pmod 5$.

Now, suppose that M is a prime. Then $\left(\frac{5}{M}\right) = -1$, and we can apply the lemma and the corollary from Section 4. In particular,

$$\alpha^{M+1} \equiv \beta^{M+1} \equiv -1 \pmod M,$$

and hence $v_{M+1} \equiv -2 \pmod M$.

Note that

$$(v_N)^2 = (\alpha^N + \beta^N)^2 = \alpha^{2N} + \beta^{2N} + 2(\alpha\beta)^N = v_{2N} + 2 \cdot (-1)^N. \tag{5}$$

We know that N is even; therefore,

$$(v_N)^2 = v_{2N} + 2 \equiv -2 + 2 \equiv 0 \pmod M.$$

This proves the theorem in this case.

Conversely, suppose, we know that $v_N \equiv 0 \pmod M$. It is required to prove that M is a prime. We can assert at once that not all prime divisors p of M have the form $p \equiv \pm 1 \pmod 5$ (because $M \equiv 2 \pmod 5$); there exists a prime divisor p for which $p \equiv \pm 2 \pmod 5$, i.e., $\left(\frac{5}{p}\right) = -1$. Therefore, the number 5 is negative modulo p, and we can apply the results of Section 4. In particular,

$\alpha^{p+1} =_{(p)} \beta^{p+1} =_{(p)} -1$. Since p divides M and $v_N \equiv 0 \pmod{M}$ (and hence $v_N \equiv 0 \pmod{p}$), we have

$$v_N = \alpha^N + \beta^N =_{(p)} 0.$$

Let $\epsilon = \alpha/\beta$. It follows from the above considerations that

$$\epsilon^N =_{(p)} -1, \tag{6}$$

but, according to the corollary, $\epsilon^{p+1} =_{(p)} 1$.

Note that (6) implies $\epsilon^{2N} =_{(p)} +1$.

Lemma 3. *Suppose that $\epsilon^a = 1$, $\epsilon^b = 1$, and division with a remainder yields $a = b \cdot c + r$. Then $\epsilon^r = 1$.*

Indeed, $1 = \epsilon^a = (\epsilon^b)^c \cdot \epsilon^r = 1 \cdot \epsilon^r = \epsilon^r$.

Let d be the minimum positive number for which $\epsilon^d =_{(p)} 1$. Then, by the lemma, d divides $2N$ (and therefore, $d = 2^s$) and d divides $p+1$. If $s < q$, and d divides $N = 2^{q-1}$, whence $\epsilon^N = (\epsilon^d)^{N/d} =_{(p)} 1$, which contradicts (6). Thus, $s = q$, and $d = 2^q = p+1$ divided $2N = M+1$. But $p \leqslant M$; therefore, $p = M$, and we conclude that M is prime. This completes the proof of the theorem.

Thus, the primality of the number $M = 2^q - 1$ depends on the value of $v_{2^{q-1}}$ modulo M.

Remarkably, formula (5) can be applied to calculate v_{2^i}. Put $r_i = v_{2^i}$. Then $r_0 = v_1 = 1$. Formula (5) gives $r_1 = r_0^2 + 2 = 3$ (here N is odd). At $i \geqslant 1$, applying (5) with an even N, we obtain

$$r_{i+1} = r_i^2 - 2, \quad r_1 = 3,$$

and the main result takes the following form.

Theorem 2. *If q is a prime of the form $4k + 3$, then $M = 2^q - 1$ is a prime if and only if $r_{q-1} \equiv 0 \pmod{M}$.*

7   Organization of computation. Examples

It is convenient to compute r_i in the binary system, i.e., with the use of binary representations. We have $r_1 = 11$.

For r_2, we obtain

$$
\begin{array}{r}
1\ 1 \\
\times \quad 1\ 1 \\
\hline
1\ 1 \\
+ \quad 1\ 1 \\
\hline
\cdots \cdots \\
- \quad 1\ 0 \\
\hline
r_2 = \quad 1\ 1\ 1
\end{array}
$$

Thus, $r_2 = 111$, which is 7 in the decimal representation.

For r_3, we have

$$
\begin{array}{r}
1\ 1\ 1 \\
\times\ 1\ 1\ 1 \\
\hline
1\ 1\ 1 \\
1\ 1\ 1 \\
1\ 1\ 1 \\
\hline
-\qquad\qquad 1\ 0 \\
\hline
r_3 = 1\ 0\ 1\ 1\ 1\ 1 \\
\end{array}
$$

Thus, $r_3 = 101111$, which is the decimal 47.

This already gives a special case of the theorem: for $q = 3$, $M = 7$ is prime and $r_2 = 7 \equiv 0 \pmod{7}$.

The next q is $q = 7$, for which $M = 2^7 - 1 = 127$. Certainly, we could verify the primality of the number 127 by mere divisions, but let us see how the algorithm works.

We must calculate r_4, r_5, and $r_6 \pmod{127}$. A pleasant feature of the algorithm is that we can reduce by 127 even in the process of computation; such a reduction corresponds to the "shift" of the binary representation by seven digits:

$$2^7 \equiv 1 \pmod{2^7 - 1}, \quad \text{therefore}, \quad 2^{7+k} \equiv 2^k \pmod{2^7 - 1}.$$

Thus, r_4 modulo 127 can be computed as follows:

$$
\begin{array}{r}
:\ \ 1\ 0\ 1\ 1\ 1\ 1 \\
:\ 1\ 0\ 1\ 1\ 1\ 1 \\
+\qquad 1\ :\ 0\ 1\ 1\ 1\ 1 \\
1\ 0\ :\ 1\ 1\ 1\ 1 \\
1\ 0\ 1\ 1\ :\ 1\ 1 \\
\hline
-\qquad\qquad\qquad 1\ 0 \\
\end{array}
$$

After rearrangement, we obtain

$$
\begin{array}{r}
:\ \ 1\ 0\ 1\ 1\ 1\ 1 \\
:\ 1\ 0\ 1\ 1\ 1\ 1\ 0 \\
:\ 0\ 1\ 1\ 1\ 1\ 0\ 1 \\
:\ 1\ 1\ 1\ 1\ 0\ 1\ 0 \\
:\ 1\ 1\ 0\ 1\ 0\ 1\ 1 \\
\hline
-\qquad\qquad\qquad 1\ 0 \\
\end{array}
$$

Now, we must compute "cyclically," carrying each one that occurs on the left of the seven digits to the right. We obtain

$$
\begin{array}{r}
1\ 0\ 1\ 1\ 0\ 1 \\
1\ 0\ 1\ 1\ 1\ 1\ 0 \\
+\ 0\ 1\ 1\ 1\ 1\ 0\ 1 \\
1\ 1\ 1\ 1\ 0\ 1\ 0 \\
1\ 1\ 0\ 1\ 0\ 1\ 1 \\
\hline
0\ 1\ 1\ 0\ 0\ 0\ 0
\end{array}
$$

Thus, $r_4 \equiv 0110000$ (mod 127).

Now, for r_5, we have

$$
\begin{array}{r}
0\ 0\ 0\ 0\ 1\ 1\ 0 \\
+\ 0\ 0\ 0\ 1\ 1\ 0\ 0 \\
\hline
-\ 0\ 0\ 0\ 0\ 0\ 1\ 0 \\
\hline
0\ 0\ 1\ 0\ 0\ 0\ 0
\end{array}
$$

This means that $r_5 \equiv 2^4$ (mod 127). Next, $r_6 \equiv 2^8 - 2 \equiv 2 - 2 \equiv 0$ (mod 127). Therefore, 127 is a prime.

The primality of the number $M = 2^{127} - 1$ was verified similarly, but the binary numbers subjected to cyclic additions had length 127. According to Williams [3], Lucas made a checkerboard and arranged numbers on its rows by placing pieces in the positions of ones and leaving the squares of zeros empty. The cyclic additions can be implemented as a game with a few simple rules. Lucas computed r_{127} modulo $2^{127} - 1$ by playing this game for about 100 hours.

References

[1] J. M. Bruce. A really trivial proof of the Lucas–Lehmer test. *Amer. Math. Monthly*, **100** (1993), 370–371.

[2] M. I. Rosen. A proof of the Lucas–Lehmer test. *Amer. Math. Monthly*, **95** (1988), 855–856.

[3] H. C. Williams. *Édouard Lucas and Primality Testing* (New York: Wiley, 1998).

Stephen Smale

On problems of computational complexity

Lecture on May 20, 1999

We shall discuss one problem which has elementary form and yet illustrates the main difficulties of computational complexity theory. For a polynomial $f \in \mathbb{Z}[t]$, we define a number $\tau(f)$ as follows. Consider the sequence $(1, t, u_1, \ldots, u_m = f)$ where each successive term is obtained from some two preceding terms, i.e., $u_k = u_i \circ u_j$ for some $i, j < k$; the symbol $\circ$ denotes one of the three arithmetic operations (addition, subtraction, and multiplication). The invariant $\tau(f)$ is equal to the least possible m.

There is the Shub–Smale conjecture that the *number of different integer roots of a polynomial f does not exceed $\tau(f)^c$, where c is some absolute constant.*

Example. The sequence $1, t, t^2, t^{2^2}, \ldots, t^{2^k}, t^{2^k} - 1$ witnesses that $\tau(t^{2^k} - 1) \leqslant k + 1$. But the polynomial $t^{2^k} - 1$ has 2^k different roots. Therefore, for different complex roots, the conjecture is false.

A similar example can be constructed with the use of the Chebyshev polynomials. The Chebyshev polynomials are calculated by a simple recursive formula. They also give an example of polynomials of high degree with small τ. All roots of the Chebyshev polynomials are real and pairwise different. Thus, for different real roots, the conjecture is false too.

Theorem 1 (Shub–Smale). *The Shub–Smale conjecture implies* $\mathrm{P} \neq \mathrm{NP}/_{\mathbb{C}}$.

Now, I must explain what $\mathrm{P} \neq \mathrm{NP}/_{\mathbb{C}}$ means.

First, note that in algebra, nontrivial problems usually begin with Diophantine equations corresponding to algebraic curves, i.e., with two variables. We have problems even in the case of one variable.

If we forget about $\mathbb{C}$, we obtain the $\mathrm{P} \neq \mathrm{NP}$ problem, which is one of the key problems in computer science. Moreover, this problem, together with the Poincaré conjecture and the conjecture about the zeros of Riemann's zeta-function, is one of the most important problems of mathematics; this is a present from computer science.

Consider polynomials $f_1(z_1, \ldots, z_n), \ldots, f_k(z_1, \ldots, z_n)$ over $\mathbb{C}$. Do they have a common zero? This is a problem of recognizing a property: the conditions are the polynomials $f_1, \ldots, f_k$ (to be more precise, the several complex

Stephen Smale, Professor at the University of California (USA).

numbers being the coefficients of these polynomials), and the result should be one of the two answers, *yes* (there is a common zero) or *no* (there is no common zero).

Hilbert's Nullstellensatz (zero point theorem) gives the following answer: there is no common zero if and only if there exist polynomials $g_1, \ldots, g_k$ such that $\sum g_i f_i = 1$.

Hilbert's Nullstellensatz is a criterion rather than a method. It provides no algorithm. But approximately ten years ago, Brownawell[1] showed that, in Hilbert's Nullstellensatz, it can be assumed that

$$\deg g_i \leqslant \max(3, \max \deg f_i)^n,$$

and this result is unimprovable.

The theorem of Brownawell gives an algorithm, for it reduces the problem to solving a system of linear equations for the coefficients of the polynomials g_i.

Now, consider the question about the speed of this algorithm: How many arithmetic operations is required to answer the question? We shall refer to the number of the coefficients in the polynomials f_i as the *size* of the input data and to the number of arithmetic operations as the *running time* of the algorithm. We say that a given algorithm is *polynomial-time* if

$$\text{time} \leqslant (\text{size})^C, \tag{1}$$

where C is a constant.

Polynomial-time algorithms are precisely the algorithms which are sensible to implement on a computer. If, say, time exponentially depends on size, then, as the size of the input data increases, time quickly grows beyond all reasonable limits. The algorithm of Brownawell is an exponential-time algorithm. An exponential upper bound for the running time of this algorithm can easily be derived from, e.g., the Gaussian elimination method for solving systems of linear equations.

The conjecture is as follows: the problem $\mathrm{HN}/\mathbb{C}$ (of whether a system of polynomial equations over $\mathbb{C}$ has a common zero) is hard, i.e., there exists no polynomial-time algorithm for solving this problem.

By algorithms we mean algorithms over $\mathbb{C}$ rather than Turing machines. Namely, an algorithm is an oriented graph with one vertex to which no edge goes (the *input*). The graph may have cycles. It determines the operation of a computational machine as follows. The machine is fed by a sequence of complex numbers $(\ldots, 0, z_1, \ldots, z_n, 0, \ldots)$ infinite in both directions; among these

[1] W. Brownawell. Bounds for the degrees in the Nullstellensatz. *Ann. Math.*, **126** (3) (1987), 577–591.

numbers only $z_1, \ldots, z_n$ can be different from zero. There are no constraints on the value of n; thus, such a computational machine can process arbitrarily long sequences of numbers. The graph has vertices of three types:

Outputs. No edges go away from these vertices. When such a vertex is reached, processing terminates.

Computing nodes. One edge goes to a computing node and one edge goes out from it. At a computational node, an arithmetic operation on some terms of the sequence is performed, and one term of the sequence is replaced by the result. In addition, all terms of the sequence can be multiplied by the same number or shifted.

Branching nodes. One edge goes to a branching node and two edges with marks "yes" and "no" go from it. At a branching node, it is determined whether $z_i = 0$. If $z_i = 0$, then we go along the edge with mark "yes," and if $z_i \neq 0$, then we go along the edge with mark "no." (If the computations are over $\mathbb{R}$, then inequalities of the type $x_i > 0$ or $x_i \geqslant 0$ can be verified.)

The algorithm outputs a sequence of numbers. The algorithm $\mathrm{HN}/_\mathbb{C}$ of interest to us outputs only one nonzero element, which can take precisely two values corresponding to the answers "yes" and "no."

This definition of an algorithm was given by L. Blum, M. Shub, and S. Smale in the late 1980s. It is strange that nobody had thought out this very natural definition before. A more detailed exposition of the theory of such algorithms is contained in the book L. Blum, F. Cucker, M. Shub, and S. Smale. *Complexity and Real Computation* (Springer Verlag, 1997).

The algorithm described above is naturally associated with an input–output function. This function is defined on a certain set of input data (for instance, the computer cannot perform division by zero, so the algorithm terminates when being fed by certain input data).

The number n is referred to as the size of the input data and the length of the path from an input to an output is the running time for this input (the path may vary with the input). The polynomial-time algorithms satisfy inequality (1) with some constant C for all inputs. The class of such algorithms is denoted by $\mathrm{P}/_\mathbb{C}$.

After this definition is given, the question about the existence of a polynomial-time algorithm for the $\mathrm{HN}/_\mathbb{C}$ problem acquires a rigorous mathematical meaning. Note that the statement that there exists no such algorithm is equivalent to the statement $\mathrm{P} \neq \mathrm{NP}/_\mathbb{C}$ (I shall not give the definition of $\mathrm{NP}/_\mathbb{C}$ in this lecture).

Instead of the field $\mathbb{C}$, we can take any field K and define a computational machine over an arbitrary field. For example, the field $K = \mathbb{Z}_2$ corresponds to the definition of an algorithm conventional in logic and computer science.

We could also consider the problem about common zeros of polynomials over $\mathbb{Z}_2$. The conjecture that there is no polynomial-time algorithm for solving this problem is equivalent to the $P \neq NP$ conjecture in its classical setting.

For a number $m \in \mathbb{Z}$, we can define an invariant $\tau(m)$ by analogy with the invariant τ for polynomials. Namely, we take sequences $(1, m_1, \ldots, m_k = m)$ similar to those considered above and define $\tau(m)$ as the least possible k. Using the Stirling formula, we can prove that $\tau(m!) \leqslant (\ln m)^C$. There is the conjecture that a lower bound of the form $(\ln m)^{C'} \leqslant \tau(m!)$ holds too; this problem is related to prime decomposition.

At first sight, these two problems (about the invariant τ for polynomials and for numbers) are not related to each other.

Let us return to the problem $P \neq NP/_K$. We shall not define $NP/_K$; instead, we shall talk about an equivalent problem – Hilbert's Nullstellen-problem $HN/_K \notin P/_K$ over an arbitrary field K. (Hilbert's Nullstellensatz is not valid if the field is not algebraically closed, but the problem about common zeros of a system of polynomials makes sense for any field; I mean this problem here.)

In the case of a non-algebraically closed field, the following assertion is valid.

Theorem 2. *If a field K is not algebraically closed and* $\operatorname{char} K = 0$, *then* $P \neq NP/_K$.

For the field $\mathbb{Z}_2$, which is non-algebraically closed but has nonzero characteristic, the question remains open.

Let us return to algebraically closed fields. For an algebraically closed field K (with $\operatorname{char} K = 0$), the problem $P \neq NP/_K$ is equivalent to the problem $P \neq NP/_{\mathbb{C}}$. Thus, the problem reduces to considering one field, say $\mathbb{C}$ or $\overline{\mathbb{Q}}$ (this is the notation for the algebraic closure of the field $\mathbb{Q}$). This is one of the main results of the book mentioned above. Its proof uses the notion of the height of an algebraic number.

It is very likely that $P/_K = P/_{\mathbb{F}_2}$ for any finite field K. But for fields of finite characteristic, there are more questions than answers.

Consider the question about the equivalence of the problems over $\mathbb{C}$ and over $\overline{\mathbb{Q}}$. One of the main difficulties in the passage from complex to algebraic numbers is involved in getting rid of complex constants which may be used in computations. In general, they might strongly simplify the computations, but it is proved that they do not.

Classical computer science is concerned with the problem $P \neq NP/_{\mathbb{Z}_2}$. In the book mentioned above, the question about the relation between $P \neq NP/_{\mathbb{Z}_2}$

and $P \neq NP/_{\mathbb{C}}$ was not considered. In the preface to the book, Dick Karp conjectured that these problems are in no way related. But after the book had been written, Smale made the following observation.

Recall that the polynomial-time algorithms are so interesting because they can be implemented efficiently on a computer. But at present, one more important class of algorithms is used, the so-called BPP-algorithms. These algorithms are allowed to "toss a coin" and perform calculations depending on the result. It is required that the correct answer be obtained in a "qualified majority" of cases. Repeating computations many times, we can obtain a result which is correct with a very high probability. For example, if a correct result is obtained with probability 3/4, then after 50 repetitions the probability of error will amount to one divided by the number of atoms in the Universe.

From the mathematical point of view, the BPP condition imposes weaker constraints than the P condition, but in practice, BPP-algorithms are as good as P-algorithms.

Theorem 3 (Smale). *If* $BPP \not\supseteq NP$, *then* $P \neq NP/_{\mathbb{C}}$.

From the point of view of modern computer science, $BPP \not\supseteq NP$ resembles $P \neq NP$ very much.

Pierre Cartier

Values of the ζ-function

Lecture on May 21, 1999

The lecture has four parts:

    1. Values of the ζ-function;

    2. Polylogarithmic functions;

    3. Generalization of polylogarithmic functions and multiple values of the ζ-function (MZVs);

    4. Conjectures on the nature of some numbers.

1 Values of the ζ-function

The beginning is very classical. I start with calculating the sum of the series $\sum_{n=1}^{\infty} \frac{1}{n^2}$. In 1739, Euler proved that the sum of this series equals $\frac{\pi^2}{6}$. First he calculated the partial sums and then guessed the answer. It should be mentioned that this series converges very slowly. Thus, to evaluate the sum of this series, Euler had to develop special numerical methods.

Euler invented a kind of "proof," which I shall repeat in a few words. Let $P(x) = c_N x^N + c_{N-1} x^{N-1} + \cdots + c_0$ be a polynomial of degree N with roots $\xi_1, \ldots, \xi_N$. Then $\xi_1 \ldots \xi_N = (-1)^N \frac{c_0}{c_N}$ and $\sum_{i=1}^{N} \xi_1 \ldots \hat{\xi}_i \ldots \xi_N = (-1)^{N-1} \frac{c_1}{c_N}$. These two equalities imply $\sum_{i=1}^{N} \frac{1}{\xi_i} = -\frac{c_1}{c_0}$. Another way to prove this is to consider a polynomial with roots $\frac{1}{\xi_i}$. We could also prove the formula $\sum_{i<j} \frac{1}{\xi_i \xi_j} = \frac{c_2}{c_0}$ in a similar manner. Thus, if $c_1 = 0$, then $\sum_{i=1}^{N} \frac{1}{\xi_i^2} = -2\frac{c_2}{c_0}$. Euler, certainly, knew all these formulas.

Now, suppose that we have a "polynomial" with roots $1, 2, 3, \ldots$. Then we can use it to calculate $\sum_{n=1}^{\infty} \frac{1}{n^2}$. That this "polynomial" has infinitely many roots did not confuse Euler. Consider the function $\frac{\sin \pi x}{\pi x} = s(x)$. The roots of the "polynomial" $s(x)$ are the roots of the equation $\sin \pi x = 0$, where $x \neq 0$. The

---

Pierre Cartier, Institut de Mathématiques de Paris–Jussieu, CCNRS, Paris, France.

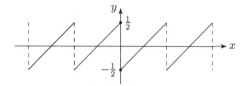

Figure 1. Saw-tooth function

positive roots are precisely what we need. But taking account of the negative roots, we obtain the doubled sum of the series $\sum\limits_{n=1}^{\infty} \frac{1}{n^2}$.

Expanding the sine in a series, we see that

$$s(x) = 1 - \frac{\pi^2}{6} x^2 + \dots .$$

Thus, $c_0 = 1$, $c_1 = 0$, and $c_2 = -\frac{\pi^2}{6}$. Therefore, $2 \sum\limits_{n=1}^{\infty} \frac{1}{n^2} = -2\frac{c_2}{c_0}$, and hence $\sum\limits_{n=1}^{\infty} \frac{1}{n^2} = -\frac{c_2}{c_0} = \frac{\pi^2}{6}$.

Certainly, this proof is not satisfactory. For instance, applying such an argument to the function e^{x^2}, we come to an absurd conclusion. Only Weierstrass (1860) and Hadamard (1895) had eventually cleared up this point by considering factorization of entire functions of a complex variable. For entire functions satisfying certain growth conditions at infinity, Euler's argument leads to a correct result.

There is another proof, which is easier to express in terms of Fourier series. The Fourier series were successfully used in the eighteenth century, although they were rigorously substantiated only a century later. Consider a saw-tooth function $\phi(x)$ (Fig. 1). It is periodic with period 1, and $\phi(x) = x - \frac{1}{2}$ for $x \in]0, 1[$. Clearly, $\int_0^1 \phi(x)\, dx = 0$. Introducing the Fourier coefficients $c_n = \int_0^1 \phi(x)e^{2\pi i n x}\, dx = \frac{1}{2\pi i n}$ (for $n \neq 0$), we obtain

$$\phi(x) = \sum_n c_n e^{-2\pi i n x} = \sum_{n \neq 0} \frac{e^{2\pi i n x}}{2\pi i n}.$$

The Parseval theorem shows that

$$\int_0^1 |\phi(x)|^2\, dx = \sum_{n \neq 0} |c_n|^2 = \frac{1}{2\pi^2} \sum_{n=1}^{\infty} \frac{1}{n^2}.$$

On the other hand,

$$\int_0^1 |\phi(x)|^2\, dx = \int_0^1 \left| x - \frac{1}{2} \right|^2 dx = \frac{1}{12}.$$

This is indeed a proof.

In textbooks, yet other proofs can be found. For instance, the Cauchy residue formula can be used. Namely, consider the meromorphic function $\frac{1}{x^2(e^{2\pi i x}-1)}$. This function has a pole of order 3 at the point $x=0$ and poles of order 1 at the points $x=n\neq 0$. Applying the Cauchy residue formula to this function, we obtain the same result.

Euler did more: he calculated the sums $1+2+3+\ldots$ and $1^2+2^2+3^2+\ldots$. I shall not repeat these calculations. They refer to what I call "mathemagic," when calculations are performed without a proper substantiation and some results are obtained. After that, many years pass before they are substantiated. In modern times, such is the situation with the Feynman path integrals. People evaluate them, but these evaluations are not corroborated. The calculations of Feynman resemble those of Euler. They have not been substantiated so far, but they will some time.

Now, let me introduce the classical Riemann zeta function $\zeta(s)=\sum\limits_{n=1}^{\infty}\frac{1}{n^s}$. It is assumed that $s\in\mathbb{C}$ and $\mathrm{Re}\,s>1$; in this case, the series absolutely converges. We have already found that $\zeta(2)=\frac{\pi^2}{6}$. Both methods of calculation (through polynomials and by using the Fourier series) apply to calculating $\zeta(4)$ and give the same result $\zeta(4)=\frac{\pi^4}{90}$. The substantiation of the calculation with the use of the Fourier series is even simpler, because, in this case, the Fourier series converges absolutely. Similarly, we can prove that if k is a positive integer, then $\zeta(2k)=\pi^{2k}r$, where r is a rational. But the Fourier series method gives nothing in the case of $\zeta(3),\zeta(5),\ldots$. The only thing known about these numbers is that $\zeta(3)\notin\mathbb{Q}$ (it is not even known whether the number $\zeta(3)$ is transcendental). The irrationality of $\zeta(3)$ was proved by R. Apéry in 1978. Then Don Zagier and H. Cohen simplified and clarified the proof of Apéry, and in August 1978 Cohen presented the proof of Apéry at the International Congress of Mathematicians in Helsinki.

An integral representation for the function $\zeta(s)$

Consider the gamma function $\Gamma(s)=\int_0^\infty e^{-x}x^{s-1}\,dx$; here $\mathrm{Re}\,s>0$. The change $x=n\xi$ yields

$$n^{-s}=\frac{1}{\Gamma(s)}\int_0^\infty e^{-n\xi}\xi^{s-1}\,d\xi,\quad n=1,2,\ldots.$$

Considering the sum $\sum\limits_{n=1}^{\infty} n^{-s}$, we reduce the right-hand side to an expression containing a geometric progression, which we can sum. As a result, we obtain

$$\zeta(s) = \sum_{n=1}^{\infty} n^{-s} = \frac{1}{\Gamma(s)} \int_0^{\infty} \frac{x^{s-1}}{e^x - 1} \, dx. \tag{1}$$

There is a method of Hadamard, which is well presented and developed in the book *Generalized Functions* by I. M. Gel'fand and G. E. Shilov. The method is as follows. Consider the integral $\Phi(s) = \int_0^{\infty} F(x) x^{s-1} \, dx$, where F is a function of class C^{∞} which has a Taylor expansion even at zero and rapidly decreases as $x \to \infty$ together with its all derivatives.

Example 1. $F(x) = e^{-x}$.

The function $\Phi = \Phi_F$ can be extended over $\mathbb{C}$ as a meromorphic function with simple poles at the points $0, -1, -2, \ldots$. This can easily be proved by integrating by parts. Indeed, integration by parts gives the functional equation $\Phi_{F'}(s+1) = -s\Phi_F(s)$; this functional equation is a generalization of the functional equation for $\Gamma(s)$. Thus, the function $\Phi(s)/\Gamma(s)$ has no poles. Formula (1) contains a function of precisely this form. Unfortunately, in the case of interest to us, the function $F(x) = \frac{1}{e^x - 1}$ has a singularity at zero, so this argument does not apply. We must change the function; for instance, we can set $F(x) = \frac{x}{e^x - 1}$ and use the decomposition $x^{s-1} = x \cdot x^{s-2}$. As a result, we obtain the integral representation

$$\Gamma(s)\zeta(s) = \int_0^{\infty} \frac{x}{e^x - 1} x^{s-2} \, dx.$$

We have shifted the poles by 1: the poles of the function $\Gamma(s)\,\zeta(s)$ are at the points $1, 0, -1, -2, \ldots$, while the poles of $\Gamma(s)$ are at the points $0, -1, -2, \ldots$. Therefore, the function $\zeta(s)$ has a unique pole, at the point 1.

The general result is $\left(\frac{\Phi}{\Gamma}\right)(-k) = (-1)^k F^{(k)}(0)$. This can be expressed differently by the formula $\left.\frac{x^{s-1}}{\Gamma(s)}\right|_{s=-k} = \delta^{(k)}(x)$, where $\delta^{(k)}$ is the kth derivative of the Dirac function. This point of view is well explained in the book of Gel'fand and Shilov mentioned above. Applying this equality to the function $\zeta(s)$, we obtain

$$-k\zeta(1 - k) = (-1)^k \frac{d^k}{dx^k}\left(\frac{x}{e^x - 1}\right)\Big|_{x=0}; \tag{2}$$

here we have taken into account the shift by 1.

Consider the Bernoulli numbers B_k, which are defined by the identity

$$\frac{x}{e^x - 1} = \sum_{k=0}^{\infty} B_k \frac{x^k}{k!}.$$

Formula (2) gives

$$\zeta(1-k) = (-1)^{k+1} B_k/k, \quad k = 1, 2, 3, \ldots.$$

For instance, $\zeta(0) = B_1 = -1/2$, $\zeta(-1) = -B_2/2 = -1/12$, and $\zeta(-2) = 0$. Formally, $\zeta(-1) = 1 + 2 + 3 + \ldots$ and $\zeta(-2) = 1^2 + 2^2 + 3^2 + \ldots$. These are precisely the sums which we wanted to calculate.

Now, it is time for yet another portion of mathemagic. Let us formally set $B_k = B^k = B \times \cdots \times B$ and consider e^{Bx}. The equality $\frac{x}{e^x-1} = e^{Bx}$ implies $x = (e^x - 1)e^{Bx} = e^{(B+1)x} - e^{Bx}$. Therefore, for $n \neq 1$, we have $(B+1)^n = B^n$. For instance,

$$0 = (B+1)^2 - B^2 = 2B^1 + B^0 = 2B_1 + B_0.$$

We know that $B_0 = 1$; hence $B_1 = -1/2$. In a similar way, we obtain the equality $3B_2 + 3B_1 + B_0 = 0$ and calculate B_2, etc.

This construction can be formally described as follows. Consider the ring of polynomials over $\mathbb{C}$ in one variable B. Consider the linear mapping $ev \colon \mathbb{C}[B] \to \mathbb{C}$ defined at the basis elements by $ev(B^k) = B_k$ and extended over the entire $\mathbb{C}[B]$ by linearity. The mapping ev takes formal series in two variables B and x to formal series in one variable x. Now, we can repeat the above calculations, applying the mapping ev when needed. This is explained in Bourbaki's textbook on elementary analysis; it contains a chapter about the Bernoulli numbers.

It can also be proved (see the end of Section 2) that

$$\zeta(2k) = (-1)^{k+1} \frac{(2\pi)^{2k}}{2 \cdot (2k)!} B_{2k}.$$

Note that $B_3 = 0$, $B_5 = 0$, $\ldots$. An attempt to evaluate the zeta function at odd points by the same method leads to the equality $0 = 0$, which gives nothing. The reason why this happens is explained at the end of the second part.

2   Polylogarithmic functions

The polylogarithmic functions are specified by the equalities

$$\mathrm{Li}_k(z) = \sum_{n=1}^{\infty} \frac{z^n}{n^k}.$$

This definition is related to the equality $\mathrm{Li}_k(1) = \zeta(k)$. It is our hope that we might obtain some information about the zeta function by using polylogarithmic functions.

In the complex domain, the series for $\mathrm{Li}_k(z)$ converges at $|z| < 1$. The first problem is to construct an analytic continuation. The beginning is very simple:

$$\mathrm{Li}_0(z) = \sum_{n=1}^{\infty} z^n = \frac{z}{1-z}.$$

This is a rational function with pole $z = 1$.

My teacher Henri Cartan forbade me make cuts on the complex plane, because such an approach is not invariant. He required to always consider only Riemann surfaces. If you open his (very good) textbook on complex analysis, you will find definitions of analytic continuation in terms of sheaves or whatever, but none of them involves cuts. Nevertheless, I see no contradiction in cuts of the plane and, therefore, make a cut.

Consider the open simply connected set $U = \mathbb{C} \setminus [1, +\infty[$. If Φ is holomorphic on such a simply connected set, then it has a primitive function Ψ on this set. This function satisfies the normalizing condition $\Psi(0) = 0$, and $\frac{d\Psi}{dz} = \Phi$. The function Ψ is also holomorphic in the domain U.

It is easy to verify that $z\frac{d}{dz}\mathrm{Li}_k(z) = \mathrm{Li}_{k-1}(z)$. The function Li_0 is holomorphic in the domain U and $\mathrm{Li}_0(0) = 0$; therefore, the function Li_1 is also holomorphic in the domain U. Proceeding, we see that the same is true of Li_2, Li_3, ... All these functions can be analytically continued over U.

To go further, we need to investigate the limit behavior of these functions when they approach the cut from different sides, from above and from below. I consider only the case

$$\mathrm{Li}_1(z) = z + \frac{z^2}{2} + \frac{z^3}{3} + \cdots = \ln \frac{1}{1-z}.$$

The values of the logarithm above and under the cut differ by a constant. If γ_1 is the monodromy around 1, then $\gamma_1(\mathrm{Li}_1(z)) = -2\pi i$. A monodromy is the difference between two branches.[1] A difference between two branches can again be analytically continued to the entire plane. This monodromy was studied in detail by many algebraic geometers, such as Bloch, Deligne, Drinfeld, and others.

Let us refer to mathemagic again and ask Euler the question: How can we calculate the sum of the series $\sum\limits_{n=-\infty}^{\infty} z^n$? Euler answers: this sum equals 0. He argues as follows. Consider the sum $z + z^2 + \cdots + z^n + \cdots = \frac{z}{1-z} = \mathrm{Li}_0(z)$. It

---

[1] More precisely,

$$\gamma_1(\mathrm{Li}_1(z)) = \mathrm{Li}_1(z - i0) - \mathrm{Li}_1(z + i0)$$

for z in $]1, +\infty[$.

converges at $|z| < 1$. Now, consider the sum of reciprocals $z^{-1} + z^{-2} + \cdots + z^{-n} + \cdots = \frac{-1}{1-z} = \mathrm{Li}_0\left(\frac{1}{z}\right)$. It converges at $|z| > 1$. Recall that we have performed an analytic continuation. Both functions $\frac{z}{1-z}$ and $\frac{-1}{1-z}$ are rational; they are defined everywhere except at the pole at 1. To calculate the required sum, we must add 1 to these two functions:

$$1 + \mathrm{Li}_0(z) + \mathrm{Li}_0\left(\frac{1}{z}\right) = 1 + \frac{z-1}{1-z} = 0.$$

We can step-by-step generalize this argument and obtain the following result. The function $\mathrm{Li}_k(z)$ is holomorphic outside $[1, +\infty[$, and the function $\mathrm{Li}_k\left(\frac{1}{z}\right)$ is holomorphic outside $[0, 1]$. Therefore, the function $\mathrm{Li}_k(z) + (-1)^k \mathrm{Li}_k\left(\frac{1}{z}\right)$ is holomorphic outside $[1, +\infty[\cup [0, 1] = [0, \infty[$.

The logarithm is usually considered for the cut $]-\infty, 0]$, but it can be defined for the cut $[0, \infty[$ too. We shall assume that the function $\ln z$ is defined with the use of the cut $[0, \infty[$ and choose a branch of this logarithm in such a way that, approaching the cut from above, we obtain the usual real logarithm.[2]

For a logarithm so defined, we obtain the following formula:

$$\mathrm{Li}_k(z) + (-1)^k \mathrm{Li}_k\left(\frac{1}{z}\right) = -\frac{(2\pi i)^k}{k!} B_k\left(\frac{\ln z}{2\pi i}\right). \tag{3}$$

Here $B_k(t)$ is the Bernoulli polynomial. Informally, it is defined by $B_k(t) = (B + t)^k$. To obtain a formal definition, we must apply the mapping ev: first we consider a polynomial in the variables B and t and then replace each monomial B^k with B_k.

The Bernoulli polynomials can also be defined by the following properties, which completely characterize them:

- $\frac{d}{dt} B_k(t) = k B_{k-1}(t)$;

- $B_k(t+1) - B_k(t) = k t^{k-1}$;

- $B_0(t) = 1$;

- $B_k(0) = B_k$.

It is an easy exercise in algebra to prove that these properties uniquely determine some sequence of polynomials.

---

[2] Hence $\ln(re^{i\theta}) = \ln r + i\theta$ for $0 < \theta < 2\pi$.

We can write the corresponding generating function:

$$\sum_{k=0}^{\infty} B_k(t) \frac{x^k}{k!} = \frac{xe^{xt}}{e^x - 1}.$$

In particular, for $t = 0$, we obtain $B_k(0) = B_k$.

Setting $z = 1 + i\epsilon$, where $\epsilon > 0$, in (3) and letting ϵ tend to zero, we obtain (for $k \geqslant 2$)

$$\zeta(k) + (-1)^k \zeta(k) = -\frac{(2\pi i)^k}{k!} B_k.$$

For odd k, this formula gives no information: we obtain the identity $0 = 0$. For even k, we obtain the very formula mentioned at the end of Section 1.

3 Generalizations of polylogarithmic functions

We have seen that $\mathrm{Li}_1(z) = \ln \frac{1}{1-z}$. For this reason, Li_2 is called a *dilogarithm*, Li_3 is called a *trilogarithm*, etc.

We want to generalize this class of functions. And we want that the new class of functions include the usual logarithm, which is defined on the complex plane cut from $-\infty$ to 0. From now on, we shall use this standard definition of logarithm, rather than that used above.

All functions which we shall define will be holomorphic in the complex plane cut from 1 to $+\infty$ and from $-\infty$ to 0. The simplest way to define these functions is to use a differential equation. Recall that, for a polylogarithm, we have proved the formula

$$\partial_z \mathrm{Li}_k(z) = \frac{1}{z} \mathrm{Li}_{k-1}(z),$$

where ∂_z denotes differentiation with respect to z. The function to be defined is parameterized by indices $k_1, \ldots, k_t$; it is denoted by $\mathrm{Li}_{k_1,\ldots,k_t}(z)$.

Before proceeding to expand the function $\mathrm{Li}_{k_1,\ldots,k_t}(z)$ in a series, I shall write a differential equation for it. Let us introduce two noncommuting variables X_0 and X_1. Consider the sequence $\epsilon = (\epsilon_1, \ldots, \epsilon_\phi)$, where $\epsilon_i \in \{0, 1\}$. Every such sequence can be assigned the product $X_\epsilon = X_{\epsilon_1} \ldots X_{\epsilon_\phi}$. For example, $X_{010} = X_0 X_1 X_0$.

Consider the differential equation

$$\partial_z \Lambda(z) = \left(\frac{X_0}{z} + \frac{X_1}{1-z} \right) \Lambda(z).$$

In particular, if X_0 and X_1 are square matrices of order p, then this equation is a system of ordinary differential equations. This system is holomorphic, but it has

singularities at the points 0 and 1. Solutions to this system should be considered
in a simply connected domain not containing the points 0 and 1. For instance,
we can take the complex plane cut as above. According to the general theory
of holomorphic differential equations, the system has a holomorphic solution in
such a domain. Note that the Gauss equation for a hypergeometric function
can be written in such a form with $p = 2$; X_0 and X_1 are then some particular
matrices of order 2. A solution to the equation is a vector-function, i.e., a set of
two functions, each of which is a variant of the Gauss hypergeometric function
$_2F_1(a, b; c; z)$.

Let us return to general noncommuting variables X_0 and X_1. We seek
solutions of the form

$$\Lambda(z) = \sum_\epsilon \Lambda_\epsilon(z) X_\epsilon =$$

$$= \Lambda_\varnothing(z) + \Lambda_0(z) X_0 + \Lambda_1(z) X_1 +$$

$$+ \Lambda_{00}(z) X_0^2 + \Lambda_{01}(z) X_0 X_1 + \Lambda_{10}(z) X_1 X_0 + \Lambda_{11}(z) X_1^2 + \dots.$$

If X_0 and X_1 are matrices, then, as can be proved, this infinite series converges.
But now, I am not interested in its convergence; I consider a formal series.
Solving the differential equation for a formal series, we obtain the recursive
relations

$$\partial_z \Lambda_{0\epsilon}(z) = \frac{1}{z} \Lambda_\epsilon(z), \quad \partial_z \Lambda_{1\epsilon}(z) = \frac{1}{1-z} \Lambda_\epsilon(z).$$

For simplicity, we set $\Lambda_\varnothing(z) = 1$. Then

$$\partial_z \Lambda_0(z) = \frac{1}{z}, \quad \partial_z \Lambda_{00}(z) = \frac{1}{z} \Lambda_0(z)$$

$$\partial_z \Lambda_1(z) = \frac{1}{1-z}, \quad \partial_z \Lambda_{10}(z) = \frac{1}{1-z} \Lambda_0(z), \dots.$$

Therefore, $\Lambda_0(z) = \ln z + C_0$; this expression is defined on the complex plane
with two cuts. Next,

$$\Lambda_{00}(z) = \frac{1}{2} \ln^2 z + C_0 \ln z + C_1.$$

We normalize the solutions by $\Lambda_{0\dots0}(z) = \frac{1}{p!} \ln^p z$.

I apply the well-known fact that the integral $\int_0 x^\lambda \ln^p x \, dx$ converges for
$\lambda \geqslant 0$ (such an expression means that the integral converges at zero). The only
singularity at zero is due to the logarithm.

The other equation gives

$$\Lambda_1(z) = \ln \frac{1}{1-z}.$$

Here a constant could appear, but I impose the normalizing condition that the function must vanish at $z = 0$.

If we want to completely write out the asymptotic condition normalizing the function Λ, we can do this as follows. Let us write

$$\Lambda(z) = \tilde{\Lambda}(z) \exp(X_0 \ln z), \tag{4}$$

where

$$\exp(X_0 \ln z) = \sum_{p=0}^{\infty} \frac{1}{p!} \ln^p z \underbrace{X_0 \dots X_0}_{p}.$$

Here $\tilde{\Lambda}$ is holomorphic in a neighborhood of zero and $\tilde{\Lambda}(0) = 1$. This is the asymptotic condition on the function Λ.

Let us write the initial differential equation in the form

$$d\Lambda = \left(X_0 \frac{dz}{z} + X_1 \frac{dz}{1-z} \right) \Lambda.$$

In a neighborhood of zero, the term $X_1 \frac{dz}{1-z}$ is regular. If this regular term were absent, then the solution to the equation would be precisely the above exponential. Formula (4) is a special case of the Fuchsian form of a solution to a differential equation with singular points.

Certainly, if we replace X_0 and X_1 with matrices, we must give meaning to everything written above; in particular, we must define the exponentials. For instance, if $X_0 = \operatorname{diag}(\lambda_1, \dots, \lambda_n)$, then the corresponding function is $\operatorname{diag}(z^{\lambda_1}, \dots, z^{\lambda_n})$. This is what arises usually in the Fuchs theory. One of the difficulties about the classical Fuchs theory is that the theory does not work in the case where one of the differences $\lambda_i - \lambda_j$ is an integer. But we do not encounter this difficulty, because we consider formal series.

It turns out that our conditions uniquely determine the function $\tilde{\Lambda}$: a differential equation with such initial conditions has a unique solution.

The situation for $z = 0$ is symmetric to that for $z = 1$. Indeed, the mapping $z \mapsto 1 - z$ interchanges the two cuts, and the situation remains the same. (Although, the asymptotic expansion of the function $\Lambda(z)$ at the point $z = 1$ is, certainly, different.)

Let us again write a solution as an infinite series

$$\Lambda(z) = \sum_{\epsilon} \Lambda_\epsilon(z) X_\epsilon,$$

where ϵ is a finite sequence of zeros and ones. I have to alter the notation a little. First, I shall assume that $\epsilon = (\epsilon_1, \dots, \epsilon_p)$, where $\epsilon_p = 1$. It is easy to show that, in this case, Λ_ϵ has no singularities at zero: for the terms whose right-hand sides do not contain X_0, the multiplication kills the negative powers of z. These are the polylogarithmic functions which we wanted to define.

Let us introduce the following notation: $Y_1 = X_1$, $Y_2 = X_0 X_1$, $Y_3 = X_0 X_0 X_1$, ... Then we can represent the product $X_{\epsilon_1} \ldots X_{\epsilon_{p-1}} X_1$ in the form $Y_{k_1} \ldots Y_{k_t}$, where $k_1 \geqslant 1, \ldots, k_t \geqslant 1$. For instance,

$$X_1 X_0 X_0 X_1 X_0 X_1 = Y_1 Y_3 Y_2.$$

The polylogarithmic function is defined by

$$\mathrm{Li}_{k_1,\ldots,k_t}(z) = \Lambda_\epsilon(z).$$

For the remaining ϵ (with $\epsilon_p = 0$), the functions Λ_ϵ are represented as finite sums of the form

$$\sum \mathrm{Li}_{\underline{k}}(z) \ln^? z,$$

where $\underline{k} = (k_1, \ldots, k_t)$ and $\ln^? z$ denotes some power of the logarithm.

The expansion of the function $\mathrm{Li}_{\underline{k}}$ in a power series has the form

$$\mathrm{Li}_{\underline{k}}(z) = \sum_{n_1 > \cdots > n_t} \frac{z^{n_1}}{n_1^{k_1} \ldots n_t^{k_t}}.$$

The radius of convergence of this series equals 1.

We set $\mathrm{Li}_{\underline{k}}(1) = \zeta(\underline{k})$. For example, $\zeta(3,2) = \sum_{m>n} \frac{1}{m^3 n^2}$. We refer to the numbers $\zeta(\underline{k})$ so defined as *multiple zeta values* (MZVs). It is easy to prove that the series under consideration converges for $k_1 \geqslant 2$. But for, e.g., $\underline{k} = (1,1)$, we obtain the series $\sum_{m>n} \frac{1}{mn}$, which diverges. Indeed, the sum $\sum_{n=1}^{m-1} \frac{1}{n}$ approximately equals $\ln m$, and the series $\sum \frac{\ln m}{m}$ diverges. A similar argument shows that the series $\sum_{m>n} \frac{1}{m^2 n}$ converges, because the series $\sum \frac{\ln m}{m^2}$ converges.

This means the following. A polylogarithmic function is regular at zero. But for regularity near the point $z = 1$, some additional assumptions are required. For example, at $k_1 \geqslant 2$, the series absolutely converges on the unit circle. Thus, the function has a limit at the point $z = 1$, but it cannot be extended to a holomorphic function in a neighborhood of this point.

We can also consider polylogarithmic functions of many variables:

$$\mathrm{Li}_{\underline{k}}(z_1, \ldots, z_s) = \sum_{n_1 > \cdots > n_t} \frac{z_1^{n_1} \ldots z_s^{n_s}}{n_1^{k_1} \ldots n_t^{k_t}}, \qquad s \leqslant t.$$

We have defined the multiple zeta values by setting $z = 1$. Goncharov and some physicists discovered that very interesting numbers are obtained by taking a root of unity as z, i.e., by considering the series $\sum\limits_{n=1}^{\infty} \frac{\alpha^n}{n^k}$, where $\alpha^p = 1$. The sum of such a series can also be written in the form $\sum\limits_{j=0}^{p-1} \alpha^j \sum\limits_{n \equiv j \pmod{p}} \frac{1}{n^k}$.

4   Conjectures on the nature of some numbers

We shall consider the numbers $\zeta(k_1, \ldots, k_t)$, where $k_1 \geqslant 2$, $t \geqslant 1$, $k_2 \geqslant 1$, $\ldots$, $k_t \geqslant 1$. All these numbers are real. Among them are the familiar numbers $\zeta(2) = \frac{\pi^2}{6}$ and $\zeta(4) = \zeta(2)^2 \times \frac{2}{5}$. We are not interested[3] in the numbers $\zeta(2k)$. But the numbers $\zeta(3)$, $\zeta(5)$, $\ldots$ are of great interest to us.

The number $p = p(\underline{k}) = k_1 + \cdots + k_t$ is the *weight* of $\zeta(k_1, \ldots, k_t)$, and the number t is its *depth*. Let us denote the set of all rational linear combinations of all MZVs with weight p by $\mathcal{Z}_p$ (in honor of Zagier). We conventionally set $\mathcal{Z}_0 = \mathbb{Q}$ and $\mathcal{Z}_1 = (0)$, because $\mathcal{Z}_1$ can contain only the number $\zeta(1) = \sum \frac{1}{n} = +\infty$. (Zagier suggested to replace $\zeta(1)$ with the sum of the series $\sum \frac{1}{n}$ "in the sense of Euler," i.e., with the Euler constant

$$C = \lim_{n \to \infty} \left(1 + \frac{1}{2} + \cdots + \frac{1}{n} - \ln n\right) = \lim_{s \to 1} \left[\zeta(s) - \frac{1}{s-1}\right],$$

but this solution is not obviously right.) We have already found that $\mathcal{Z}_2 = \mathbb{Q}\pi^2$. It can also be proved that $\mathcal{Z}_3 = \mathbb{Q}\zeta(3)$, since $\zeta(3) = \zeta(2,1)$ (see below).

There are several conjectures on the spaces $\mathcal{Z}_p$, which are not completely independent of each other.

1. The sum of the vector spaces $\mathcal{Z}_0$, $\mathcal{Z}_1$, $\ldots$, $\mathcal{Z}_p$ is direct, i.e., if $z_0 + z_1 + \cdots + z_p = 0$, where $z_j \in \mathcal{Z}_j$ for $j = 0, 1, \ldots, p$, then $z_j = 0$ for all j.

Concerning this conjecture, it has been verified that, if $p(\underline{k}_j) = j$ for $0 \leqslant j \leqslant 17$ and $|m_j| \leqslant 10^{10}$ for $m_j \in \mathbb{Z}$, then $|\sum_j m_j \zeta(\underline{k}_j)| \geqslant 10^{-50}$ (except for $m_0 = \cdots = m_{17}$).

2. Let $d_p = [\mathcal{Z}_p : \mathbb{Q}]$ (the dimension of $\mathcal{Z}_p$ over $\mathbb{Q}$). Then, as Zagier conjectured,

$$\sum_{p=0}^{\infty} d_p t^p = \frac{1}{1 - t^2 - t^3}.$$

More detailed conjectures are based on the following observation. Consider, for example, the product $\zeta(2)\,\zeta(3) = \sum \frac{1}{n^2} \sum \frac{1}{m^3}$. Here m and n range over the

---

[3] By a previous formula, $\zeta(2k)/\pi^{2k}$ is a rational number, and we know that π is a transcendental number.

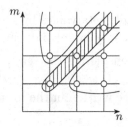

Figure 2

part of the integer lattice contained in the positive quadrant. Let us divide this set into three subsets: the diagonal elements, the superdiagonal elements, and the subdiagonal elements (Fig. 2).

It is easy to verify that the summation over the diagonal elements gives $\zeta(5)$, the summation over the subdiagonal elements gives $\zeta(2,3)$, and the summation over the superdiagonal elements gives $\zeta(3,2)$. Therefore,

$$\zeta(2)\,\zeta(3) = \zeta(5) + \zeta(2,3) + \zeta(3,2).$$

Generalizing this argument, we can obtain a relation of the form

$$\zeta(a_1,\ldots,a_s)\,\zeta(b_1,\ldots,b_t) = \sum \zeta(c_1,\ldots).$$

Here the number $\zeta(a_1,\ldots,a_s)$ is obtained by summation over the sequences $n_1 > \cdots > n_s$, and the number $\zeta(b_1,\ldots,b_t)$ is obtained by summation over the sequences $m_1 > \cdots > m_t$. We must shuffle these sequences, i.e., take their union and arrange the numbers in decreasing order (taking account of possible repetitions).

This is the first set of multiplicative relations. There is one more set of multiplicative relations. It is obtained as follows. We have proved above that

$$\zeta(3) = \frac{1}{2!} \int_0^\infty \frac{x^2}{e^x - 1}\, dx.$$

Making the change $x = \ln t$, we obtain

$$\zeta(3) = \frac{1}{2} \int_1^\infty \frac{(\ln t)^2}{t - 1}\frac{dt}{t}.$$

After some transformations, we come to the expression

$$\zeta(3) = \iiint_{1 > x_1 > x_2 > x_3 > 0} \frac{dx_1}{x_1}\frac{dx_2}{x_2}\frac{dx_3}{1 - x_3}.$$

Recall that, in combinatorial notation, $\zeta(3)$ corresponds to the product $Y_3 = X_0 X_0 X_1$, and the denominators of the three fractions in the integral are $x_1 - 0$, $x_2 - 0$, and $1 - x_3$. In the general case, the formula looks as follows. Put $\omega_0(x) = \frac{dx}{x}$ and $\omega_1(x) = \frac{dx}{1-x}$. Then

$$\zeta_{\epsilon_1,\ldots,\epsilon_p} = \int \cdots \int_{1 > x_1 > \cdots > x_p > 0} \omega_{\epsilon_1}(x_1) \ldots \omega_{\epsilon_p}(x_p).$$

To calculate $\zeta(k_1, \ldots, k_t)$, we have to consider the product

$$Y_{k_1} \ldots Y_{k_t} = X_{\epsilon_1} \ldots X_{\epsilon_p}$$

(here $p = k_1 + \cdots + k_t$). By definition, $\zeta(k_1, \ldots, k_t) = \zeta_{\epsilon_1,\ldots,\epsilon_p}$.
The number $\zeta(2)$ is expressed similarly in the form of a double integral as

$$\zeta(2) = \iint_{1 > x_1 > x_2 > 0} \frac{dx_1}{x_1} \frac{dx_2}{1 - x_2}.$$

Therefore, $\zeta(2)\zeta(3)$ is represented in the form of a quintuple integral. The variables must be shuffled (and ordered). The same argument as above gives a sum of expressions. But, in this case, the situation is simpler, because the diagonal is of measure zero and we can disregard it.

These considerations give the second set of multiplicative relations.
For $\zeta(2)\zeta(3)$, we obtain two expressions. One of them has the form

$$\zeta(2)\zeta(3) = \zeta(5) + \zeta(2,3) + \zeta(3,2).$$

(We could write out an explicit form of the other relation too, but we shall not do this.) As a result, we obtain one linear relation between MZVs of given weights $p = 5$. The integral representation gives also the equality $\zeta(3) = \zeta(2,1)$.

The main conjecture is that all independent linear relations with rational coefficients between MZVs of given weights can be obtained by this method. Maybe, it is possible to derive the two conjectures formulated above from this one in a purely algebraic way. My students are working on this reduction, but the work is not finished yet. It is not easy.

This conjecture is very strong. In particular, it implies that the numbers $\zeta(3)$, $\zeta(5)$, ... are transcendental and algebraically independent over the field of rationals. For instance, it is not known whether the number $\zeta(3)$ is transcendental and whether the number $\zeta(5)$ is irrational.

Pierre Cartier

Combinatorics of trees

Lecture on May 24, 1999

I shall start with the definition of the Catalan numbers, which are often encountered in combinatorics. The Catalan numbers are 1, 1, 2, 5, 14, 42, 132, 429, 1430, 4862, ... This sequence of numbers is defined by the first term $c_1 = 1$ and the recursive relation $c_n = \sum_{p+q=n} c_p c_q = \sum_{p=1}^{n-1} c_p c_{n-p}$. (It is convenient to set $c_0 = 0$.)

To obtain an explicit formula for the Catalan numbers, consider the generating function $c(t) = c_1 t + c_2 t^2 + \dots$. The recursive relation implies $c(t) = t + c(t)^2$. Solving this quadratic equation and taking into account that $c_0 = 0$, we obtain $c(t) = \frac{1}{2}(1 - \sqrt{1 - 4t})$. Therefore, $c_n = \frac{1}{n}\binom{2n-2}{n-1}$.

The Catalan numbers admit plenty of different combinatorial interpretations. For me, most important is the interpretation suggested by Cayley (1860). Namely, c_n is the number of different triangulations of a regular convex $(n+1)$-gon. Thus, the square has precisely two triangulations (Fig. 1); accordingly, $c_3 = 2$. The hexagon admits three different types of triangulations (Fig. 2); under each of these triangulations, the number of different triangulations of this type is written. The total number of different triangulations of the hexagon equals $6 + 2 + 6 = 14 = c_5$.

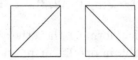

Figure 1. *Triangulations of the square*

6   2   6

Figure 2. *Triangulations of the hexagon*

Let c'_n be the number of different triangulations of the $(n+1)$-gon. To prove

Pierre Cartier, Institut de Mathématiques de Paris–Jussieu, CCNRS, Paris, France.

the equality $c'_n = c_n$, it is sufficient to verify that $c'_n = \sum\limits_{p+q=n} c'_p c'_q$. (Obviously, $c'_2 = 1$, and by convention $c'_1 = 1$.) I shall give the proof for the example of the hexagon. Let us mark its sides by the numbers 0, 1, 2, 3, 4, and 5. Consider a triangle from some triangulation that contains the side marked by 0 (Fig. 3). We mark the diagonals being the sides of this triangle also by 0. As a result (in the situation shown on Fig. 3), we obtain a $(3+1)$-gon and a $(2+1)$-gon, which are to be triangulated somehow. In the general case, we obtain a $(p+1)$-gon and a $(q+1)$-gon, where $p + q = n$.

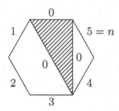

Figure 3. Marks on sides and diagonals

Exercise. Prove that a triangulation of a convex $(n+1)$-gon uses precisely $n-2$ diagonals, and precisely $n-1$ triangles are obtained.

Another interpretation of the Catalan numbers is related to rooted planar binary trees with n leaves. To a triangulation of the $(n+1)$-gon we can assign a tree with $n+1$ free (i.e., belonging to only one edge) vertices, as shown in Fig. 4. The unique free vertex corresponding to the side marked by 0 is distinguished. We call the marked vertex of the tree *root* and the remaining free vertices *leaves*.

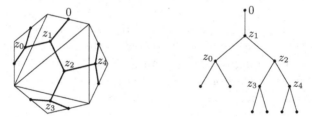

Figure 4. Construction of a tree from a triangulation

Now, let me explain what a tree and a planar binary tree are. A reduced graph is a subset $E \subset \binom{V}{2}$, where $\binom{V}{2}$ are the two-element subsets in V. The elements of the set V are the vertices of the graph, and the elements of the set E are its edges. The reducedness of a graph means that it has no loops (edges which begin and end in the same vertex) and no double edges. A graph is said to be disconnected if the vertex set V can be divided into two (nonempty) disjoint sets V' and V'' in such a way that both ends of any edge lie in V' or in V''. A tree is a connected graph without cycles. (A cycle is a sequence of edges forming a polygon.) A disconnected graph without cycles is called a forest.

A *planar tree* is a tree with a fixed embedding in the plane. From the combinatorial point of view, this means that, for each vertex, an order of traversing the edges incident to this vertex is fixed.

Now, let us define a *rooted binary tree*. First, this is a tree each of whose vertices is incident to either one or three edges. (The vertices incident to only one edge are the leaves and the root.) In addition, the edges of the tree must be oriented in such a way that each interior vertex is left by precisely two (hence the qualification "binary tree") edges and entered by precisely one edge. The root is left by one edge and each leaf is entered by one edge.

Let us denote the set of rooted planar binary trees with n leaves by $\mathcal{T}_n$. It can be proved that there is a one-to-one correspondence between the rooted planar binary trees with n leaves and the triangulations of the $(n+1)$-gon; the triangles in a triangulation correspond to interior vertices, and its diagonals correspond to interior edges. Thus, $c_n = |\mathcal{T}_n|$.

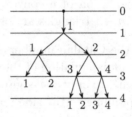

Figure 5. Distance from the roots

Any two vertices of a tree are joined by precisely one path. Therefore, we can determine the distance from each vertex to the root. Let us arrange the vertices in levels according to their distances from the root (Fig. 5). Suppose that G_k is the set of vertices of level k. To each vertex we assign the beginning of the edge entering it; this gives the sequence of mappings

$$G_0 \leftarrow G_1 \leftarrow G_2 \leftarrow G_3 \leftarrow G_4 \leftarrow \dots .$$

To encode this sequence, we enumerate the vertices of each level from left to right (for a planar tree, this order is determined uniquely). Let us introduce the notation $[n] = \{1, 2, \dots, n\}$. We have $G_j \simeq [\gamma_j]$, where γ_j is the number of elements in the set G_j. The sequence of mappings takes the form

$$[\gamma_0] \leftarrow [\gamma_1] \leftarrow [\gamma_2] \leftarrow [\gamma_3] \leftarrow [\gamma_4] \leftarrow \dots .$$

These mappings are nondecreasing: if numbers p and q are mapped to p' and q', then $p > q$ implies $p' \geqslant q'$.

1   The universal algebra

We call an arbitrary mapping $\mu\colon X \times X \to X$ a *magma*; this term is chosen because we impose no conditions on the binary composition law $(ab) = \mu(a,b)$ (magma has no structure).

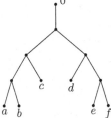

Figure 6. *Parenthesization with the use of a tree*

A rooted planar binary tree with n leaves determines a parenthesization of an n-tuple, that is, it determines an order in which the binary operation μ is applied to n elements. For example, the tree shown in Fig. 6 corresponds to the following arrangement of parentheses: $(((ab)c)(d(ef)))$. This parenthesization (i.e., the order in which the operations are performed) is obtained as follows. Take a pair of free edges with a common vertex, e.g., the edges with end-vertices a and b. We cut off this pair of edges and assign the element $\mu(a,b)$ to the new free vertex. As a result, we obtain a tree with a smaller number of leaves. For some pair of its free edges with a common vertex, we repeat the same procedure, and so on.

Thus, each tree $T \in \mathcal{T}_n$ determines a mapping $\mu_T\colon X^n \to X$. This mapping can be expressed as follows. Suppose that the rooted tree T has $l(T)$ leaves. Let us denote $\bigsqcup_T \{T\} \times X^{l(T)}$ by $M(X)$. The magma μ induces the mapping $\tilde{\mu}\colon M(X) \to X$ that takes $(T, x_1, \ldots, x_{l(T)})$ to $\mu_T(x_1, \ldots, x_{l(T)})$.

The universal property

(1) For any set X, the tree product $(T', T'') \mapsto T' * T''$ (Fig. 7) induces a multiplication on the set $M(X)$.

(2) If Y is a set with multiplication $\nu\colon Y \times Y \to Y$, then any mapping $\phi\colon X \to Y$ can be uniquely extended to a mapping $\Phi\colon M(X) \to Y$ consistent

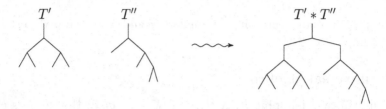

Figure 7. Tree product

with multiplication:

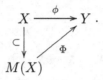

The embedding $X \subset M(X)$ is as follows. The set $\mathcal{T}_1$ comprises one tree τ; to each element $x \in X$ we assign (τ, x).

The universal property means that $M(X)$ is the free magma over X. Thus, we have obtained an explicit construction of the free magma.

Now, let us give a combinatorial interpretation of the identity $c(t) = t + c(t)^2$. Suppose that $|X| = t$. Let $M(X)_p$ denote the set of all trees T with $l(T) = p$. Then $M(X)_p = \mathcal{T}_p \times X^p$, and the set $M(X)$ decomposes as $M(X) = \coprod\limits_{p=1}^{\infty} M(X)_p$. Consider the Poincaré series $\sum c_p t^p$ for $M(X)$. The identity $c(t) = t + c(t)^2$ follows from $M(X) = X \sqcup M(X)^2$. Indeed, any element of the set $M(X)$ either lies in $M(X)_1$ or is uniquely represented in the form of a product of two elements of $M(X)$ (Fig. 8).

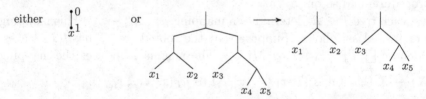

Figure 8. Tree decomposition

2   The Stasheff polyhedra

The polyhedron $\mathcal{P}_2$ is the singleton; $\dim \mathcal{P}_2 = 0$. The number of vertices in this polyhedron equals $1 = c_2$.

The polyhedron $\mathcal{P}_3$ is the closed interval; $\dim \mathcal{P}_3 = 1$. The number of vertices in this polyhedron equals $2 = c_3$.

The polyhedron $\mathcal{P}_4$ is the pentagon; $\dim \mathcal{P}_4 = 2$. The number of vertices in this polyhedron equals $5 = c_4$.

The polyhedron $\mathcal{P}_5$ is harder to describe. The number of its vertices must equal $c_5 = 14$. Its dimension must be 3. In addition, all Stasheff polyhedra are simple (in dimension 3, this means that the section near each vertex is a triangle). The Stasheff polyhedron $\mathcal{P}_5$ is obtained as follows. We glue together two tetrahedra by a common triangular face. The obtained polyhedron has three vertices, near which the sections are squares. But if we cut off each of these three vertices by a plane, we shall obtain a simple polyhedron (Fig. 9). This is precisely the polyhedron $\mathcal{P}_5$.

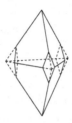

Figure 9. The Stasheff polyhedron $\mathcal{P}_5$

The Stasheff polyhedra are related to triangulations of polygons; moreover, there is a one-to-one correspondence between the vertices of the polyhedron $\mathcal{P}_n$ and the triangulations of the $(n + 1)$-gon. The simplest case is that of $\mathcal{P}_3$ (Fig. 10); the vertices of the interval $\mathcal{P}_3$ correspond to the two triangulations of the square, and the interior of the interval corresponds to the square itself.

Figure 10. The Stasheff polyhedron $\mathcal{P}_3$

Next in complexity is the polyhedron $\mathcal{P}_4$ (a pentagon); its vertices are in one-to-one correspondence with the triangulations of the pentagon (Fig. 11). The interior of $\mathcal{P}_4$ corresponds to the pentagon without diagonals; each side of $\mathcal{P}_4$ corresponds to the pentagon with one diagonal; each vertex of $\mathcal{P}_4$ corresponds to the pentagon with two diagonals. The two diagonals corresponding to a vertex of $\mathcal{P}_4$ are precisely the two diagonals determined by the two sides of $\mathcal{P}_4$ containing this vertex.

Similarly, for the cells of $\mathcal{P}_5$, we can establish the following correspondence: each vertex (0-cell) of $\mathcal{P}_5$ corresponds to a hexagon with three disjoint diagonals; each 1-cell of $\mathcal{P}_5$ corresponds to this hexagon with two disjoint diagonals; each 2-cell of $\mathcal{P}_5$ corresponds to the hexagon with one diagonal; and the 3-cell of $\mathcal{P}_5$ corresponds to the hexagon itself.

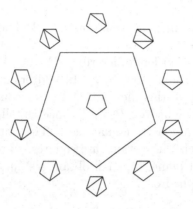

Figure 11. The Stasheff polyhedron $\mathcal{P}_4$

The boundary $\partial \mathcal{P}_5$ consists of six pentagons (i.e., polyhedra $\mathcal{P}_4$) and three squares (i.e., polyhedra $\mathcal{P}_3 \times \mathcal{P}_3$). The polyhedron $\mathcal{P}_2$ is the singleton; therefore, $\mathcal{P}_4 = \mathcal{P}_2 \times \mathcal{P}_4 = \mathcal{P}_4 \times \mathcal{P}_2$. Thus, $\partial \mathcal{P}_5$ consists of $2\mathcal{P}_2 \times \mathcal{P}_4$, $4\mathcal{P}_4 \times \mathcal{P}_2$, and $3\mathcal{P}_3 \times \mathcal{P}_3$. In the general case, it is natural to expect that $\partial \mathcal{P}_n = \bigcup_{p+q=n+1} p \mathcal{P}_p \times \mathcal{P}_q$. In this case, we can define mappings $\partial_k \colon \mathcal{P}_p \times \mathcal{P}_q \to \partial \mathcal{P}_n$ for $1 \leqslant k \leqslant p$.

The polyhedron $\mathcal{P}_n$ can be realized in the space $\mathbb{R}^{n-2}$ with coordinates $t_1, \ldots, t_{n-2}$ by the following model. Consider the set specified by the system of inequalities $t_1 \geqslant 0$, $t_2 \geqslant 0$, $\ldots$, $t_{n-2} \geqslant 0$ and $t_1 \leqslant 1$, $t_1 + t_2 \leqslant 2$, $t_1 + t_2 + t_3 \leqslant 3$, $\ldots$, $t_1 + \cdots + t_{n-2} \leqslant n - 2$ (Fig. 12 shows this set for $n = 2$). The polyhedron $\mathcal{P}_n$ can be obtained by additionally partitioning some faces of this polyhedron. I leave this to the reader as an exercise.

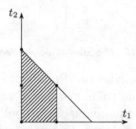

Figure 12. The model of the polyhedron $\mathcal{P}_4$

Now, let us describe the combinatorial construction of $\mathcal{P}_n$ as an abstract CW-complex. Let $n \geqslant 2$. Take a convex $(n+1)$-gon and a number $0 \leqslant k \leqslant n-2$. We define Γ_k (the of set k-cells) as follows: each element $\gamma \in \Gamma_k$ corresponds to a set of $n-2-k$ disjoint diagonals in the $(n+1)$-gon. Thus, for $k = n-2$, there is precisely one cell Γ_{n-2}, and the 0-cells Γ_0 correspond to the triangulations of the $(n+1)$-gon.

It remains to define an incidence relation. Suppose that $\gamma \in \Gamma_k$ and $\delta \in \Gamma_l$, where $k > l$. The cell δ is incident to the cell γ (geometrically, $\delta \subset \partial\gamma$) if $\gamma \subset \delta$ (as sets of diagonals).

The polyhedron $\mathcal{P}_n$ is defined as a geometric realization of this complex. Under such an approach, the proof that the geometric realization of this complex is homeomorphic to the $(n - 2)$-cell is nontrivial.

3 The space of (real) configurations

Consider the real projective line. Topologists denote it by $\mathbb{RP}^1$, and algebraic geometers use the notation $\mathbb{P}^1(\mathbb{R})$, or even simply $\mathbb{P}^1$. The standard definition is as follows: a point of $\mathbb{P}^1$ is a straight line passing through 0 in the vector space $V = \mathbb{R}^2$ with $\dim V = 2$. Suppose that $u_0, \ldots, u_n$ are different points in $\mathbb{P}^1$ considered up to projective transformation. In other words, sets $u_0, \ldots, u_n$ and $v_0, \ldots, v_n$ are equivalent if there exists an element $g \in G = \mathrm{GL}_2(\mathbb{R})$ such that $gu_i = v_i$.

In addition to the group G, there is another important group, $G_+ \subset G$, which consists of the matrices with positive determinants. The index $[G : G_+]$ equals 2. The transformations from the group G_+ preserve the orientation of $\mathbb{P}^1$.

We are interested in sets of points $u_0, \ldots, u_n$ considered up to a transformation from the group G_+. Let us introduce the normalization $u_0 = \infty$. Then $u_1, \ldots, u_n \in \mathbb{R}$ are different points considered up to a transformation $u_i \mapsto au_i + b$, where $a > 0$ and $b \in \mathbb{R}$.

On this set, the symmetric group S_n acts: an element $\sigma \in S_n$ takes a set $u_1, \ldots, u_n$ to the set $u_{\sigma(1)}, \ldots, u_{\sigma(n)}$.

The space of sets of points has $n!$ connected components. Each element of the group S_n permutes these connected components. Let us choose the component for which $u_1 < \cdots < u_n$. Such sets of points are called *real configurations*. The space of real configurations is denoted by conf_{n+1}^+.

The transformation group under consideration admits a normalization of points such that $u_1 = 0$ and $u_n = 1$. The configuration space for $n = 4$ is shown in Fig. 13. Indeed, in this case, $0 = u_1 < u_2 < u_3 < u_4 = 1$.

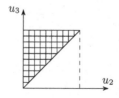

Figure 13. The configuration space

It is it is easy to obtain the pentagon $\mathcal{P}_5$ from the configuration space conf_5^+ (Fig. 14). The intersection points of triples of lines correspond to unstable configurations, where three points collide. At these points, blowing-ups are required.

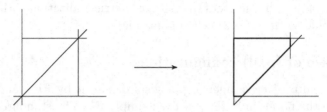

Figure 14. The compactification of the configuration space

Kapranov proved that, in the general case, the polyhedron $\mathcal{P}_n$ is a natural compactification of the space conf_{n+1}^+. The interior of the polyhedron $\mathcal{P}_n$ is identified with the space conf_{n+1}^+; the dimensions of these spaces are equal to $n-2$. The interior of the polyhedron $\mathcal{P}_n$ parameterizes the configurations.

Now, consider the interior points of the $(n-3)$-faces of the polyhedron $\mathcal{P}_n$. Such points belong to one of the sets $\mathrm{int}\,\mathcal{P}_p \times \mathrm{int}\,\mathcal{P}_q$, where $p+q = n+1$. To a point of such a set we assign a composition of two configurations (Fig. 15). For this purpose, we glue together two projective lines at one point. The configurations on one projective line are parametrized by $\mathrm{int}\,\mathcal{P}_p$, and the configurations on the other are parametrized by $\mathrm{int}\,\mathcal{P}_q$. For the first line, we label the point of tangency by the number k, and for the second line, we label this point by 0. We have $1 \leqslant k \leqslant p$, as it must be (p. 280).

In the general case, gluing together projective lines, we obtain a binary tree (Fig. 16).

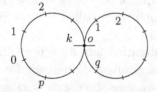

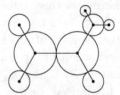

Figure 15. A composition of
configurations

Figure 16. A tree of
configurations

Pierre Cartier

What is an operad?

Lecture on May 27, 1999

The lecture consists of the following parts:

1. What is an operad?
2. (Co)homological operations;
3. Returning to the configuration space;
4. Conclusion: physics.

1   What is an operad?

The theory of operads is the theory of combining operations. Suppose given a magma $X \times X \to X$, $a, b \mapsto (ab)$. From this binary operation, we can construct more complicated operations.

Consider a function $f(x_1, \ldots, x_n) \in X$, where $x_i \in X$. In the language of computation theory, we can say that $x_1, \ldots, x_n$ are input variables and $f(x_1, \ldots, x_n)$ is an output (Fig. 1).

This operation can be assigned an $(n+1)$-gon; the side number 0 corresponds to the output, and the sides with numbers $1, \ldots, n$ correspond to the input variables (the sides are enumerated clockwise; see Fig. 2).

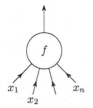

Figure 2. The polygon corresponding to the operation

Figure 1. Operation

Now, let us define an operation $f \circ_i g$. Suppose given a function $f(x_1, \ldots, x_n)$, a number $1 \leqslant i \leqslant n$, and a function $g(y_1, \ldots, y_p)$. Then we can make the substitution $x_i = g(y_1, \ldots, y_p)$. We treat the function

$$f(x_1, \ldots, x_{i-1}, g(y_1, \ldots, y_p), x_{i+1}, \ldots, x_n)$$

Pierre Cartier, Institut de Mathématiques de Paris–Jussieu, CCNRS, Paris, France.

as a function of $n + p - 1$ variables $x_1, \ldots, x_{n+p-1}$; for this purpose, we write it in the form

$$f(x_1, \ldots, x_{i-1}, g(x_i, \ldots, x_{i+p-1}), x_{i+p}, \ldots, x_{n+p-1}) = h(x_1, \ldots, x_{n+p-1}).$$

The operations $f \circ_i g$ can be interpreted both in terms of graphs (Fig. 3) and in terms of polygons (Fig. 4). Graphs are convenient because their vertices need not be enumerated: it is sufficient to define an orientation in the horizontal direction (e.g., from left to right, as in Fig. 3). When polyhedra are used, the 0th side of the polygon g is attached to the ith side of the polygon f.

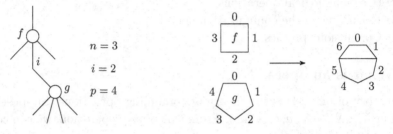

Figure 3. Gluing of graphs   Figure 4. Gluing of polygons

These interpretations show that combining operations is related to both trees and polygons with distinguished disjoint diagonals. The binary trees correspond to binary operations (Fig. 5).

Figure 5. A binary operation

Let X be a set with a binary operation $\mu \colon X \times X \to X$ (a magma!). We put $\mathcal{P}_n = \mathrm{Map}(X^{\times n}, X)$. In particular, $\mathcal{P}_0 = X$ and $\mathcal{P}_1 = \mathrm{Map}(X, X)$. If $f \in \mathcal{P}_n$ and $g \in \mathcal{P}_p$, then $f \circ_i g \in \mathcal{P}_{n+p-1}$. Thus, a mapping

$$\circ_i \colon \mathcal{P}_n \times \mathcal{P}_p \to \mathcal{P}_{n+p-1}$$

arises. On $\mathcal{P}_n$, the symmetric group S_n acts by permutations of variables:

$$\sigma f(x_1, \ldots, x_n) = f(x_{\sigma(1)}, \ldots, x_{\sigma(n)}).$$

For $x = (x_1, \ldots, x_n)$, we set $x \cdot f = f(x_1, \ldots, x_n)$. We assume that $x\sigma \cdot f = x \cdot \sigma f$. Notice that each rooted planar binary tree T defines an operation μ_T in $\mathcal{P}_n$

and, hence, a map $\mathcal{T}_n \to \mathcal{P}_n$, namely, $T \mapsto \mu_n$. More generally, an operad is a family $\mathcal{P}_n$ of operations, where $n = 0, 1, \ldots$, together with actions of S_n on $\mathcal{P}_n$ and with compositions $\circ_i$. We do not spell out explicitly the axioms, of associativity and permutation, respectively. The preceding discussion shows that the collection $\mathcal{T} = \bigsqcup_n \mathcal{T}_n$ of all rooted binary planar trees can be considered as an operad corresponding to the category of magmas.

An operad is linear if $\mathcal{P}_n$ are vector spaces (over a field K), the actions of the groups S_n are linear, and the compositions are bilinear.

For example, let

$$\mathcal{P}_n = \mathrm{Hom}_K(V^{\otimes n}, V), \quad \text{where} \quad V^{\otimes n} = \underbrace{V \otimes \cdots \otimes V}_{n}.$$

Here K is a field of characteristic zero and V is a vector space of dimension $d < +\infty$. On the space $V^{\otimes n}$, the group $G = \mathrm{GL}(V)$ acts as

$$g(v_1 \otimes \cdots \otimes v_n) = gv_1 \otimes \cdots \otimes gv_n$$

and the group S_n acts as

$$\sigma(v_1 \otimes \cdots \otimes v_n) = v_{\sigma^{-1}(1)} \otimes \cdots \otimes v_{\sigma^{-1}(n)}.$$

These actions commute. In addition, $V^{\otimes n}$ is a semisimple S_n-module, i.e.,

$$V^{\otimes n} = P_1 \oplus \cdots \oplus P_l,$$

where P_i are irreducible (simple) S_n-modules. In this decomposition, we can combine like terms, applying the relations $P \oplus P = P \otimes K^2$, $P \oplus P \oplus P = P \otimes K^3$, $\ldots$. As a result, we obtain the decomposition

$$V^{\otimes n} = \bigoplus_D P_D \otimes F_D.$$

We have $\sigma(p_D \otimes f_D) = \sigma p_D \otimes f_D$ and $g(p_D \otimes f_D) = p_D \otimes gf_D$. In the stable domain, where $d \geqslant n$, the irreducible representations of the groups S_n and $\mathrm{GL}_d(K)$ are parametrized by the same Young diagrams D. This is the Schur–Weyl duality.

Consider one more example (due to Macdonald and Milnor). Suppose that Vect_K^f is the category of finite-dimensional vector spaces over the field K and Vect_K is the category of vector spaces (not necessarily finite-dimensional) over the field K. Let T be a functor from the category Vect_K^f to the same category Vect_K^f. The functor T maps a vector space V to a vector space $T(V)$ and a linear mapping $\phi \colon V \to W$ to a linear mapping $T(\phi) \colon T(V) \to T(W)$. The linear mappings ϕ and $T(\phi)$ are represented by matrices (of different sizes). We require that the elements of the matrix $T(\phi)$ be expressed in terms of elements of the matrix ϕ as homogeneous polynomials of degree t. In this case, we say that T is a homogeneous functor of degree t. For $t = 0, 1, 2, \ldots$, let J_t be a vector space over K with a given action of the group S_t. We let S_t act on $V^{\otimes t}$ as above and on $V^{\otimes t} \otimes J_t$, by the diagonal action $\sigma(w \otimes j) = \sigma w \otimes \sigma j$. We set

$$T_t(V) = \left(\left(\underbrace{V \otimes \cdots \otimes V}_{t} \right) \otimes J_t \right)^{S_t}$$

(the fixed points of S_t) and $T_{\mathcal{T}} = \bigoplus_{t=0}^{\infty} T_t$. As a result, we obtain a functor $T_{\mathcal{T}}$ from Vect_K^f to Vect_K. Any functor which decomposes as a direct sum of homogeneous functors acts uniquely in this way.

The next examples are from universal algebra. Let V be a vector space (over the field K) with basis $e_1, \ldots, e_d$, and let $\mathrm{Sym}(V) = K[e_1, \ldots, e_d]$. Then

$$\mathrm{Sym}(V) = \bigoplus_{t=0}^{\infty} \mathrm{Sym}^t(V),$$

where $\mathrm{Sym}^t(V) = \left(\underbrace{V \otimes \cdots \otimes V}_{t} \right)^{S_t}$ is the symmetric part. This corresponds to the preceding example in which $J_t = K$ and the action of the group S_t is trivial.

Let Com_K be the category of commutative associative algebras with unity over the field K. Taking into account that $\mathrm{Sym}^1(V) = V$, we obtain the inclusion $V \subset \mathrm{Sym}(V)$; note that $\mathrm{Sym}(V)$ is an object in the category Com_K. Let A be an object in the category Com_K. Then any linear mapping $\lambda \colon V \to A$ is uniquely extended to a homomorphism $\Lambda \colon \mathrm{Sym}(V) \to A$. We assume that J_t is the space K with trivial action of the group S_t. We denote the multiplication operation in A by $\mu(a, b) = a \cdot b$. Let

$$\mathcal{P}(A)_n = \mathrm{Hom}_K(A^{\otimes n}, A).$$

Clearly, $\mu \in \mathcal{P}(A)_2$. Consider the minimal suboperad $\mathcal{C}(A)$ in $\mathcal{P}(A)$ containing μ. The associativity of multiplication implies that all trees of a given size determine the same operation. Hence $\mathcal{C}(A)_n$ is equal to $K \cdot \mu_n$ with

$$\mu_n(a_1 \otimes \cdots \otimes a_n) = a_1 \ldots a_n.$$

Moreover, $\sigma\mu_n = \mu_n$ for σ in S_n and $\mu_p \oplus_i \mu_q = \mu_{p+q-1}$ for $1 \leqslant i \leqslant p$. This describes the operad $\mathrm{Com}_K = \mathcal{A}(A)$ associated to the category Com_K.

If the multiplication is not only associative but also commutative, then invariance with respect to the action of S_n holds. In the absence of commutativity, we obtain the universal algebra

$$T(V) = \bigoplus_{t=0}^{\infty} \underbrace{V \otimes \cdots \otimes V}_{t}.$$

If $J_t = KS_t$ is the regular representation ($\dim J_t = t!$), we proceed as above and obtain the operad $\mathcal{A}ss_K$ associated to the category Ass_K of associative algebras with unity.

Let V be a vector space. The free Lie algebra $\mathrm{Lie}(V)$ over V is defined as follows. Consider the linear mapping $\Delta \colon T(V) \to T(V) \otimes T(V)$ such that $\Delta(tt') = \Delta(t) \otimes \Delta(t')$ and, for a $v \in V$, it has the form $\Delta(v) = v \otimes 1 + 1 \otimes v$. We have

$$\mathrm{Lie}(V) = \{u \in T(V) \colon \Delta(u) = u \otimes 1 + 1 \otimes u\}.$$

In particular, $V \subset \mathrm{Lie}(V)$. In $\mathrm{Lie}(V)$, we consider the commutator $[u, u'] = uu' - u'u$.

The Lie functor is a direct sum of homogeneous functors. Hence it is of the form

$$\mathrm{Lie}_t(V) = \left(\underbrace{V \otimes \cdots \otimes V}_{t} \otimes \mathcal{L}_t \right)^{S_t},$$

where $\mathcal{L}_0 = 0$, $\mathcal{L}_1 = 0$, the space $\mathcal{L}_2$ is generated by $[x, y] = -[y, x]$, and the space $\mathcal{L}_3$ is generated by the elements $[x, [y, z]]$, $[y, [z, x]]$, and $[z, [x, y]]$ related by the Jacobi identity $[x, [y, z]] + [y, [z, x]] + [z, [x, y]] = 0$. The action of the symmetric group is *via* permutation of the variables $x, y, \ldots$. To the category Lie_K of Lie algebras over K is therefore associated an operad $\mathcal{L})\!\!\upharpoonright_K$, which describes all "natural" operations that can be defined on a Lie algebra.

Now, let us define the Hochschild cohomology. Let A be an associative algebra with unity over the field K. We set

$$C^p = \mathrm{Hom}_K(A^{\otimes p}, A).$$

An element $c(a_1, \ldots, a_p) \in C^p$ is called a cochain of degree $p = |c| = \deg(c)$. Clearly, $\mu \in C^2$, where $\mu(a_1, a_2) = a_1 a_2$. We define the product of two cochains by

$$c \circ c' = c \circ_1 c' \pm c \circ_2 c' \pm c \circ_3 c' \pm \ldots$$

with suitable signs and put

$$[c, c'] = c \circ c' + (-1)^{|c| + |c'| + |c||c'|} c' \circ c.$$

As Gerstenhaber noticed (in the 1960s), the Jacobi identity

$$[[c, c'], c''] = [c, [c', c'']] + (-1)^{|c|+|c'|+|c||c'|}[c', [c, c'']]$$

holds (up to sign). In addition,

$$[c', c] = (-1)^{|c|+|c'|+|c||c'|}[c, c'].$$

Moreover, $\deg[c, c'] = \deg c + \deg c' - 1$. In particular, $\mu_2 = \mu \circ \mu = \frac{1}{2}[\mu, \mu]$ has degree 3. By associativity, $\mu_2(a, b, c) = (ab)c - a(bc) = 0$.

Let $\mathfrak{b}c = [\mu, c]$. Then $\mathfrak{b} \colon C^p \to C^{p+1}$, and the Jacobi identity implies $\mathfrak{b}\mathfrak{b} = 0$. The homology of the complex C with differential $\mathfrak{b}$ is called the Hochschild cohomology and denoted by $HH^*(A, A)$.

On the Hochschild cohomology, there are two operations, the Gerstenhaber commutator $[\, , \,]_G$ (its degree equals -1) and the $\cup$-product, which is defined by

$$c \cup c'(a_1, \ldots, a_p, a_{p+1}, \ldots, a_{p+q}) = c(a_1, \ldots, a_p)c'(a_{p+1}, \ldots, a_{p+q});$$

the degree of the $\cup$-product equals 0.

These operations have the following properties: the $\cup$-product is associative, and at the cohomology level it is even commutative up to sign; the operation $[\, , \,]_G$ is commutative up to sign and satisfies the Jacobi identity up to sign. At the cohomology level, the following Leibniz rule holds:

$$[c, c' \cup c'']_G = [c, c']_G \cup c'' \pm c' \cup [c, c'']_G.$$

2   (Co)homological operations

Now, we shall describe the construction of topological operads suggested by Stasheff in the 1960s. We start with defining the Pontryagin multiplication in homology. Let X be a topological space, and let $\mu \colon X \times X \to X$ be a continuous mapping. Consider the rational homology. By the Künneth formula, $H_*(X \times X) = H_*(X) \otimes H_*(X)$; therefore, the mapping μ induces a mapping $H_*(\mu) \colon H_*(X) \otimes H_*(X) = H_*(X \times X) \to H_*(X)$ (the Pontryagin multiplication). Thus, a multiplication in the space determines a multiplication in the homology. If the multiplication μ is associative or commutative, then so is the multiplication in homology.

The two trees shown in Fig. 6 correspond to the two different parenthesizations of the product of three elements. For a topological space with multiplication $\mu \colon X \times X \to X$, these two parenthesizations determine two mappings $X \times X \times X \to X$. If X is an H-space, that is, if these mappings are homotopic, then the multiplication in H_* is associative.

The two trees considered above correspond to the two triangulations of the square (Fig. 7), and these two triangulations correspond to the Stasheff polyhedron $\mathcal{P}_3$ (Fig. 8).

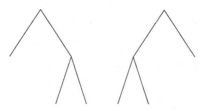

Figure 6. The two trees

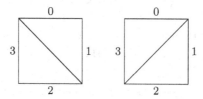

Figure 7. The two triangulations

Figure 8. The polyhedron $\mathcal{P}_3$

If μ is homotopy associative, a homotopy between the two triple products corresponds to a continuous mapping $\mathcal{P}_3 \times X^3 \to X$.

Hence, for n equal to 2 or 3, we hap a mapping

$$\phi_n \colon \mathcal{P}_n \times X^n \to X.$$

The mapping ϕ_2 corresponds to the definition of the multiplication μ in X, since $\mathcal{P}_2$ is reduced to a point, and ϕ_3 corresponds to the homotopy associativity of μ. Going over to the rational homology, a mapping ϕ_n as above defines defines a mapping

$$H_*(\phi_n) \colon H_*(\mathcal{P}_n) \otimes H_*(X) \otimes \cdots \otimes H_*(X) \to H_*(X).$$

Here $\mathcal{P}_n$ is the topological $(n-2)$-cell; therefore, $H_0(\mathcal{P}_n) = \mathbb{Q}$ and $H_i(\mathcal{P}_n) = 0$ for $i \geqslant 1$. As a result, we obtain a multilinear operation of order n on $H_*(X)$.

The fact that the multiplication $H_*(\mu)$ in $H_*(X)$ is associative is a consequence of the two embeddings of the point $\mathcal{P}_2 \times \mathcal{P}_2$ into $\partial\mathcal{P}_3$. In order to express the finer homotopy properties of an H-space, Stasheff introduces a collection of mappings ϕ_2, ϕ_3, ϕ_4, ... as above and expresses the compatibility with respect to the p mappings $\partial_i \colon \mathcal{P}_p \times \mathcal{P}_{n+1-p} \to \mathcal{P}_n$ corresponding to the decomposition $\partial\mathcal{P}_n = \bigcup\limits_{p+q=n+1} p\mathcal{P}_p \times \mathcal{P}_q$. But this is nothing else than an action on X of the topological operad $\mathcal{X}$ defined in the next section.

3   Returning to the configuration space

Since the space $\mathcal{M}_{0,n+1}(\mathbb{R})$ is homeomorphic to $S_n \times \operatorname{int}\mathcal{P}_n$, we have

$$\overline{\mathcal{M}_{0,n+1}(\mathbb{R})} \approx S_n \times \mathcal{P}_n.$$

In positive dimensions, the homologies $H_*(\overline{\mathcal{M}_{0,n+1}(\mathbb{R})})$ are trivial.

Let $\Pi_n = H_0(\overline{\mathcal{M}_{0,n+1}(\mathbb{R})}, \mathbb{Q})$. Then $\Pi_n = \mathbb{Q}S_n$ is a regular representation of the group S_n.

Consider the family of sets $\mathcal{X}_n = \overline{\mathcal{M}_{0,n+1}(\mathbb{R})}$, where $n = 0, 1, \dots$. The group S_n acts on $\mathcal{X}_n$. For $1 \leqslant i \leqslant p$, we can form a composition $\circ_i$ as shown in Fig. 9. As a result, we obtain a degenerate configuration; all such degenerate configurations form a space homeomorphic to $\operatorname{int}(\mathcal{P}_p \times \mathcal{P}_q)$. Compactifying it, we obtain a mapping $\circ_i$ from $\mathcal{X}_p \times \mathcal{X}_q$ to $\mathcal{X}_{p+q-1}$. Thus, the Stasheff polyhedra (or configuration spaces) can be organized in a topological operad $\mathcal{X}$.

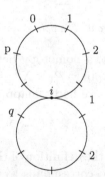

Figure 9. Composition

From a topological operad we can obtain an algebraic operad by employing homologies. The Stasheff topological operad $\mathcal{X}$ gives the trivial algebraic operad $\mathcal{A}ss_\mathbb{Q}$. But in the complex case, the theory is nontrivial.

The complex theory is constructed as follows.   Consider the space

$\mathcal{M}_{0,n+1}(\mathbb{C})$ obtained by factoring the space

$$\mathbb{C}_*^n = \{z = (z_0, \ldots, z_n) \in \mathbb{C}^n : z_j \neq z_k \text{ if } j \neq k\}$$

modulo the action of the affine group $z \mapsto az + b$. The construction of the compactification $\overline{\mathcal{M}_{0,n+1}(\mathbb{C})}$ is fairly complicated; it was described by Deligne and Mumford. The spaces $\overline{\mathcal{M}_{0,n+1}(\mathbb{C})}$ can be organized into a topological operad. We have already explained how to do this in the real case. In the real case, the compositions are represented by real polynomials with rational coefficients; they can be complexified by replacing real variables with complex ones.

In the real case, the homologies obtained are trivial, while in the complex case, the homologies $H_*(\overline{\mathcal{M}_{0,n+1}(\mathbb{C})}, \mathbb{Q})$ are very interesting. They form an algebraic operad $\Pi_{\mathbb{C}}$. This algebraic operad has two special elements, which generate it in a certain sense; these are the cup-product $\cup$ and the Gerstenhaber product $[\ ,\ ]_G$. Thus, the operad $\Pi_{\mathbb{C}}$ acts on $HH^*(A)$. This is proved by calculations. The question is why this is so.

In 1970, Arnold calculated the homologies $H_*(\mathbb{C}_*^n)$. This makes it possible to calculate the homology of a noncompactified space. The question is why it is the homologies $H_*(\overline{\mathcal{M}_{0,n+1}(\mathbb{C})}, \mathbb{Q})$ that control the operations in the Hochschild cohomologies. It is desirable to construct an action at the level of cochains.

4   Conclusion: physics

Now, we shall briefly outline the relation to physics. One of the most important problems at present is to obtain exact solutions to the Yang–Baxter equations. Their solutions can be obtained by solving certain differential equations, known as the Knizhnik–Zamolodchikov equations, which are constructed on $\mathbb{C}_*^n$. In solving these equations, Drinfeld obtained power series related to the MZVs considered in the first lecture.

There is also a relation to the Feynman diagrams and to the scattering matrix.

A. A. Kirillov

The orbit method beyond Lie groups.
Infinite-dimensional groups

Lecture on September 2, 1999

My today's and tomorrow's lectures are in some sense a continuation of the lectures delivered here in the winter of 1997/98. Those two lectures were entitled "The Orbit Method and Finite Groups," and these two lectures are entitled "The Orbit Method beyond Lie Groups." I shall not dwell on the orbit method. I only mention that it is applied to Lie groups. Thus, the main object under consideration is Lie groups. But the orbit method applies also to other groups, which are not Lie groups. I have prepared three series of such examples:

(1) infinite-dimensional groups;
(2) finite groups;
(3) quantum groups.

My last-year lectures were concerned with the second series, finite groups; so I shall not talk about them, although an interesting progress has been made in this direction. Today I shall talk about infinite-dimensional groups, and tomorrow, about quantum groups. Quantum groups is a very fashionable direction in modern mathematics. Their success is largely due to the sonorous name. The fine point is that the quantum groups are not groups; this is an object of a different nature. But they still have some group features, and we could try to apply the orbit method to them. I shall talk about these attempts tomorrow. Today I shall talk about infinite-dimensional groups, which are not Lie groups either.

The usual Lie groups are (finite-dimensional) manifolds endowed with a group structure which is compatible in a certain sense with the structure of a manifold. Infinite-dimensional groups are almost the same thing, but the manifolds are infinite-dimensional, i.e., local coordinate systems are infinite-dimensional. But, whereas all finite-dimensional spaces of the same dimension over the real number field are isomorphic, the infinite-dimensional spaces are not. There are many different infinite-dimensional spaces, and it is not always clear which of them should be considered. A typical example of infinite-dimensional spaces is spaces of functions. For instance, we can consider functions on the real line. But there are various functions on the line. First, we might consider all functions whatever, but this makes no sense; usually, functions with certain properties are considered. We can consider continuous functions; this makes more sense. It is even better to consider smooth function, which have one, two, three... or infinitely many derivatives. We can consider analytic functions. We can also impose some conditions at infinity, e.g., consider rapidly decreasing functions or compactly supported functions, which do not vanish only in a finite domain. As you see, there are plenty of infinite-dimensional spaces. In this respect, infinite-dimensional groups are not specific sets, as well as algebraic manifolds. They usually consist of functions and become sets only after these functions are specified. Before that, we only have a mere rule determining a group law.

The problems which arise in the attempt to apply the orbit method to infinite-dimensional groups are very interesting. Sometimes, they coincide with well-known classical problems, solved or not. Sometimes, new problems emerge. A part of these new problems I want to discuss today.

Now, let me digress for a moment. People are often interested in news in mathematics, and I usually try to satisfy audience's curiosity, as far as I can. At present, mathematics manifests a special interest in exceptions, apart from general theories, which cover many special examples. There have always been lovers of exceptions, but now the role of exceptions is growing more and more important. All sciences have exceptional objects. For example, among the complex simple Lie groups are four infinite series of groups and five special groups which fit into none of these series. Similar things occur in other sciences too. I shall tell about one such example in more detail. There is an object related to the theory of Lie groups, to geometry, and to many physical applications; I mean *lattices*. A lattice is a discrete subgroup in Euclidean n-space such that the quotient group by this subgroup is compact (such subgroups are sometimes called *cocompact*). For any lattice on the line, we can choose a scale in such a way that the lattice consist of all integers numbers. In this case, the problem of classification of lattices is trivial.

From the point of view of group theory, all lattices in n-space are the same: they are all isomorphic to $\mathbb{Z}^n$. But geometrically, these lattices may be different. For example, there is the standard integer lattice in the plane. Another lattice can be constructed as follows. Consider three coordinate axes in the plane that make angles of $120°$ with each other (Fig. 1). To each point we assign three coordinates (x, y, z) (signed projections) rather than two. These coordinates are related by $x + y + z = 0$. The points for which all the three coordinates x, y, and z are integer form a lattice (Fig. 2).

We can compare the densities of these two lattices. The density of a lattice is defined as follows. At all points of the lattice, we place balls of maximal size in such a way that these balls be disjoint. The ratio of the area covered by the balls to the entire area is the density of the lattice. We see that the second

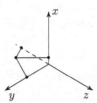

Figure 1. Coordinates on the plane

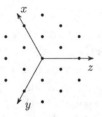

Figure 2. The lattice on the plane

lattice is denser than the rectangular one. In reality, the rectangular lattice is one of the sparsest lattices, and the second lattice is the densest one.

The theory of lattices was initiated long ago, in the nineteenth century. First, it was a very popular direction; then it had been forgotten a little, but currently, this theory is attracting attention of mathematicians again. In the theory of lattices, in addition to series of lattices, there are singular, exotic, lattices, which do not fit into these series. One of such lattices is related to the existence in 4-space of much more regular polyhedra than in the other spaces. In 3-space, there are 5 regular convex polyhedra: the regular tetrahedron, cube, octahedron, icosahedron, and dodecahedron. In 4-space, there are 6 regular polyhedra (there is a polyhedron bounded by 24 octahedra, which has no analogues in spaces of other dimensions). All the other spaces have only 3 regular polyhedra (the regular simplex, cube, and the polyhedron dual to cube). Thus, 4-space is exceptional, and it has an exceptional lattice related to the regular 24-hedron. This lattice consists of points with integer coordinates x_1, x_2, x_3, and x_4 such that all of them are even or odd simultaneously. It can be verified that each point has precisely 24 nearest neighborhoods; the distance between neighboring points equals 2. For example, the origin has eight neighboring points of the form $(\pm 2, 0, 0, 0)$ and 16 neighboring points of the form $(\pm 1, \pm 1, \pm 1, \pm 1)$. All points in the lattice are equivalent: any point can be mapped to any other point by a symmetry of the lattice.

These examples of lattices are fairly simple. A more interesting example is the *Leech lattice* in space $\mathbb{R}^{24}$. This lattice was discovered by geometers; in recent years, it has often been used by physicists in relation to string theories. I shall give the "physical" definition of the Leech lattice rather than the original one. Many physicists believe that we live in Minkowski 26-space $\mathbb{R}^{1,25}$, in which one coordinate is time and 25 coordinates are spatial: $(t, x_0, \ldots, x_{24})$. Our senses say to us that we live in the 4-dimensional world with one time and three spatial coordinates. But our senses are very imperfect, they cannot distinguish between a very small 10-manifold and a point. Thus, the space may be of very small size in 22 directions, on the order of 10^{-33} cm say; this is quite sufficient for preventing any instrument from detecting these directions. The remaining three spatial directions are visible. But this is a digression. Let us return to mathematics. The Minkowski space $\mathbb{R}^{1,25}$ is endowed with the quadratic form $t^2 - x_0^2 - \cdots - x_{24}^2$, and the light cone contains the remarkable integer vector $(70, 0, 1, 2, \ldots, 24) = \rho$. Consider an integer lattice M in $\mathbb{R}^{1,25}$. Take the orthogonal complement $\rho^\perp$ in $\mathbb{R}^{1,25}$; we have $\rho^\perp \ni \rho$. Consider the quotient space $\rho^\perp / \mathbb{R}\rho$, which has dimension 24. Taking the integer points in $\rho^\perp$ and factoring them modulo $\mathbb{R}\rho$, we obtain an integer lattice in 24-space, which is remarkable in many respects. This lattice has no analogues in spaces of other dimensions.

Now, I return to the main topic – the orbit method for infinite-dimensional groups. I want to consider two infinite-dimensional groups, the first of which is well studied and the second is hardly studied at all. As a rule, abstract groups arise from transformation groups, and transformation groups are usually formed of transformations which preserve something; i.e., usually, we take a set with some structure and consider all one-to-one transformations preserving this structure. For example, we can take the plane as the set and the distance between points as the structure. Then we obtain the group of motions and reflections, i.e., the isometry group. We could also consider the larger group of the transformations preserving the projective structure on the plane, or of the transformations preserving the topology on the plane (that is, continuous transformations). The group of continuous transformations of the plane is infinite-dimensional.

One of the best-studied infinite-dimensional groups is the group of one-to-one smooth transformations of the circle; it is denoted by $\mathrm{Diff}(S^1)$. We can represent the circle as the quotient space $S^1 = \mathbb{R}/\mathbb{Z}$, i.e., introduce a coordinate $t \pmod 1$ (two points on the straight line are identified if the difference between them is integer). A diffeomorphism of the circle is specified by a function $s = f(t)$; here s is a new coordinate. The difference between $f(t+1)$ and $f(t)$ must be integer. The function f can be either monotonically increasing or monotonically decreasing. Clearly, this determines two connected components of the group $\mathrm{Diff}(S^1)$. We shall consider the connected component that contains the identity element; we denote it by $\mathrm{Diff}_+(S^1)$. The condition on this group is $f(t+1) = f(t) + 1$. The function f is smooth and satisfies the condition $f'(t) > 0$. The condition $f(t+1) = f(t)+1$ in terms of the derivative is written as

$$\int_0^1 f'(t)\, dt = 1.$$

In addition, $f'(t+1) = f'(t)$.

This group has a fairly evident description. If the role of parameter is played by f' rather than by f, then the group consists of positive periodic functions (with period 1) for which the integral over the period is equal to 1. Topologically, this set is trivial: it is contractible, being a convex set in a linear space. Both conditions (that the derivative is positive and its integral equals 1) are convex. But to obtain the group itself, we must pass from derivatives to functions. For this purpose, it is sufficient to specify an initial condition belonging to the circle. Thus, we see that the group $\mathrm{Diff}_+(S^1)$ of interest to us is homotopy equivalent to the circle. Sometimes, it is convenient to consider simply connected groups. We can obtain a simply connected group by considering monotonic functions instead of factorizing modulo integer-valued functions. The obtained group $\widetilde{\mathrm{Diff}}_+(S^1)$ is the universal covering of the group $\mathrm{Diff}_+(S^1)$.

Now, let us discuss how the orbit method can be applied to the group $G = \widetilde{\mathrm{Diff}}_+(S^1)$. For this purpose, we must first pass from the Lie group G to the Lie algebra $\mathfrak{g} = \mathrm{Lie}(G)$. The Lie algebra is the tangent space at the identity, i.e., it consists of the infinitesimal transformations of the circle. As a geometric object, the Lie algebra consists of vector fields on the circle. Indeed, we consider transformations $t \mapsto t + \epsilon v(t)$. In the group under consideration, the group law corresponds to superposition of functions $f_1 \circ f_2(t) = f_1(f_2(t))$. At the level of the Lie algebra, the group law transforms into the commutator $[v, w] = vw' - v'w$; here $v' = dv(t)/dt$. As is known, every Lie group acts on its Lie algebra by linear transformations (this is the adjoint representation of the Lie group). In the case under consideration, the group consists of changes of variables $t = t(s)$ (coordinate systems) on the circle. For vector fields, we can make changes of coordinate systems too; that is, a vector field can be written in various coordinate systems. It turns out that the adjoint representation of a group is precisely the same changes of variables, but for vector fields.

We can specify a vector field by a function $v(t)$. But the geometric meaning of this function is that it determines a vector field. Thus, it is more correct to write $v(t)\frac{d}{dt}$ rather than $v(t)$. For vector fields, a change of variables $t = t(s)$ gives

$$v(t)\frac{d}{dt} \mapsto v(t(s))\frac{d}{dt(s)} = v(t(s))\frac{d}{t'_s\, ds}.$$

If we trace only the coefficient, then, introducing the notation $t = f(s)$, we obtain

$$v \mapsto \frac{v \circ t}{t'} = \frac{v \circ f}{f'}.$$

The orbit method deals with the dual space $\mathfrak{g}^*$ rather than with the Lie algebra $\mathfrak{g}$ itself; it consists of linear functionals on $\mathfrak{g}$. What is the geometric meaning of the objects dual to vector fields? The dual object together with the vector field gives a number. It turns out that the dual object is a quadratic differential, i.e., an expression of the form $p(t)(dt)^2$. Under a change of variables $t = f(s)$, the quadratic differential changes as $p(t)(dt)^2 \mapsto p \circ f \, (f')^2$. It can be verified that, if we have a vector field $v(t)\frac{d}{dt}$ and a quadratic differential $p(t)(dt)^2$, then the expression $p(t)v(t)\, dt$ composed of them is a differential form, i.e., under a change of variables, it is multiplied by the first power of the derivative. Being a differential form, it can be integrated. The integral $\int_{S^1} p(t)v(t)\, dt$ does not depend on the choice of parameter. Thus, the quadratic differential is a linear functional on the space of vector fields. We chose the notation in such a way that the action of the group on the space $\mathfrak{g}^*$ dual to the natural action on the space $\mathfrak{g}$ is natural as well: this is the usual change of variables for quadratic differentials according to the formula written above.

Generally, the function p is a distribution. But we are interested in the case where p is a usual smooth function, because most interesting examples of coadjoint orbits are related to such functions p.

I have to make yet another remark. The orbit method is used to construct and study infinite-dimensional unitary representations. But for the main applications of infinite-dimensional unitary representations (quantum field theory and quantum mechanics in general), usual representations are not so important as projective representations. The set of symmetries of a quantum system is the projective space corresponding to the set of unitary operators rather than the unitary operators themselves. Two proportional unitary operators are not distinguished from each other, because two wave functions differing by a multiplier with absolute value 1 coincide. Thus, the usual representations, which are functions on the group that satisfy the relation

$$\pi(g_1 g_2) = \pi(g_1)\,\pi(g_2),$$

are replaced by the projective representations, which are functions on the group that satisfy the relation

$$\pi(g_1 g_2) = c(g_1, g_2)\,\pi(g_1)\,\pi(g_2),$$

where $c(g_1, g_2) \in \mathbb{C}$ and $|c(g_1, g_2)| = 1$.

The projective representations of a group G reduce to usual representations of a slightly larger group, namely, of a central extension of the group G. If the group has no nontrivial central extensions for some reason, then each projective representation is actually a usual representation. But many interesting groups have nontrivial central extensions and, hence, admit nontrivial projective representations that do not reduce to usual representations. For example, the diffeomorphism group $\mathrm{Vect}(S^1)$ of the circle does have a nontrivial central extension.

The orbit method must respond. We replace a group by its central extension; the Lie algebra is then also replaced by a central extension, and the coadjoint action is replaced by a slightly more complicated action. The coadjoint action of a central extension differs from the initial action in that a linear action becomes affine.

The group G and its central extension $\widetilde{G}$ are included in the exact sequence

$$0 \to \mathbb{R} \to \widetilde{G} \to G \to 1.$$

I started with zero, because $\mathbb{R}$ is an additive group, and I ended with unity, because the group G is usually represented in the multiplicative form; each of the groups 0 and 1 comprises precisely one element.

For Lie algebras, the exact sequence looks as

$$0 \to \mathbb{R} \to \tilde{\mathfrak{g}} \to \mathfrak{g} \to 0.$$

There is also the dual sequence

$$0 \leftarrow \mathbb{R} \leftarrow \tilde{\mathfrak{g}}^* \leftarrow \mathfrak{g}^* \leftarrow 0.$$

At the level of linear spaces, this exact sequence splits; therefore, we can assume that $\tilde{\mathfrak{g}}^*$ and $\mathfrak{g}^* \oplus \mathbb{R}$ coincide as linear spaces. As we know, the space $\mathfrak{g}^*$ consists of quadratic differentials; the elements of this space are called *momenta*. An extended momentum consists of a quadratic differential and a number. We must define an action of the group on the set of pairs of quadratic differentials and numbers. This action must look as

$$K(f)(p, c) = (p \circ f\,(f')^2 + cS(f), c). \tag{1}$$

That c remains the same is a general fact for all central extensions; the additional parameter does not change under the coadjoint action. Because of this, it is sometimes called a *charge*. An affine transformation of p consists of two parts. We already know the linear part $p \circ f\,(f')^2$. The additional term does not depend on p, and it depends linearly on c; therefore, it has the form $cS(f)$. For (1) to be an action, the quadratic differential $S(f)$ must have the following property:

$$S(f_1 \circ f_2) = S(f_1) \circ f_2\,(f_2')^2 + S(f_2).$$

The lovers of cohomologies might say that this expression is the equation of a cocycle.

There exists precisely one S with the required property, namely,

$$S(f) = \frac{f'''}{f'} - \frac{3}{2}\left(\frac{f''}{f'}\right)^2. \tag{2}$$

Now, we can forget everything said above and consider that as a mere motivation. The mathematical setting of the problem is as follows. It is required to classify the coadjoint orbits, i.e., periodic functions of an argument t, with respect to the transformations of form (1).

Expression (2) has been well known to mathematicians since the nineteenth century. It was discovered by the German mathematician Schwarz; for this reason, $S(f)$ was called the *Schwarz derivative*. It is quite natural to ask what the functions for which $S(f) = 0$ are. The functions for which the Schwarz derivative vanishes have the remarkable property that they form a group with respect to composition. These functions are solutions to a differential equation of order 3, so the corresponding group has dimension 3. There are not that many three-dimensional Lie groups; the group in question is isomorphic to the group PSL_2 and can be represented as a group of linear-fractional transformations. Thus, the general solution to the equation $S(f) = 0$ looks like $f(x) = \frac{ax+b}{cx+d}$.

The even more remarkable property of the Schwarz derivative is that the value of the Schwarz derivative of some good function is usually a number of a simple form. For example, if $f(x) = e^{\lambda x}$, then

$$S(f) = \frac{\lambda^3 e^{\lambda x}}{\lambda e^{\lambda x}} - \frac{3}{2} \left(\frac{\lambda^2 e^{\lambda x}}{\lambda e^{\lambda x}} \right)^2 = -\frac{\lambda^2}{2}.$$

For the function $\tan x$, the result is also a number, but I shall not say what number.

We have to solve the following difficult problem: reduce a function to a simplest form by using transformations that act by the complicated rule described above. This problem seems to be very artificial. But if I state it differently, it will look quite natural. It turns out that this problem is equivalent the following geometric problem. Let us define a notion of projective manifold structure for the circle S^1. To obtain a usual smooth manifold, we cover a topological space by local coordinate systems and require that the transitions between the coordinate systems be smooth functions. Now, we impose an additional condition. Suppose that two local coordinate systems with coordinates t and s intersect. On the intersection, we have two coordinates, and one of them is a function of the other, say $t = f(s)$. The usual definition of a manifold requires only that the function f be smooth. We narrow the class of functions and require that this function be linear-fractional, i.e., that $t = \frac{as+b}{cs+d}$. The result is called a projective structure on the circle. The question is: Can we introduce any projective structure whatever on the circle, and if we can, then in how many ways?

As is known, it is impossible to cover the circle by one chart. But using the stereographic projection, we can introduce two coordinates on the circle

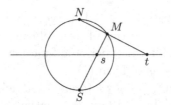

Figure 3. The projective structure on the circle

(Fig. 3). Suppose that N is the north pole, S is the south pole, and M is a point on the circle. We can project the point M onto the x-axis from the point N or from the point S. As a result, we obtain two coordinates, t and s. The coordinate t is defined everywhere except at the north pole N; the coordinate s is defined everywhere except at the south pole S. In the domain where both coordinates are defined, they are related by $st = 1$. This relation is linear-fractional: $t = 1/s$. Thus, we have defined a projective structure on the circle. It turns out that the circle admits other projective structures.

Theorem. *A classification of the momenta for a central extension of the circle diffeomorphism group is equivalent to a description of the projective structures on the circle.*

The circle in not the simplest one-dimensional manifold. The simplest 1-manifold is the straight line. Let us try to solve the problem of describing the projective structures on the line. The line has one trivial projective structure: we can introduce one local coordinate x (it is also a global coordinate). Another example of a projective structure can be obtained as follows. Let us map the line on the circle by the rule $x \mapsto e^{ix}$. There is a projective structure on the circle; we can transfer it to the line. We obtain a projective structure different from that determined by the coordinate x. These projective structures are not equivalent, because there exists no function which expresses x in terms of the coordinates on the circle and is linear-fractional.

Now, I shall give one more example of a projective structure on the straight line, which makes it possible to describe all projective structures on the straight line. First, note that the interval $(0, 1)$ with a coordinate y varying from 0 to 1 is not projectively equivalent to the straight line. As smooth manifolds, the line and the interval coincide: there exists a smooth mapping of the line onto the interval for which the inverse mapping is smooth too. But this function cannot be linear-fractional. It is also clear that the half-line with a coordinate z varying from 0 to $+\infty$ is projectively equivalent to the interval. Indeed, as the transition function we can take $y = \frac{z}{z+1}$.

Let us denote the line with trivial projective structure by $\mathbb{R}$ and the line

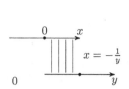

Figure 4. The projective structure $\frac{3}{2}\mathbb{R}$

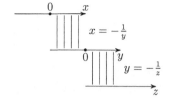

Figure 5. The projective structure $2\mathbb{R}$

with the projective structure of the half-line by $\frac{1}{2}\mathbb{R}$. Now, we shall manufacture a projective manifold, which we shall call $\frac{3}{2}\mathbb{R}$ (Fig. 4). For this purpose, we take two straight lines with coordinates x and y and attach the positive ray $(0, +\infty)$ of the first line to the negative ray $(-\infty, 0)$ of the second line by the rule $x = -\frac{1}{y}$. The projective manifold $\frac{3}{2}\mathbb{R}$ contains projective submanifolds $\frac{1}{2}\mathbb{R}$ and $\mathbb{R}$, but it differs from them.

Developing this construction, we obtain a projective manifold $2\mathbb{R}$. To this end, we take three lines with coordinates x, y, and z and additionally attach the positive ray of the second line to the negative ray of the third line by the rule $y = -\frac{1}{z}$ (Fig. 5). The projective manifold $\frac{3}{2}\mathbb{R}$ does not contain $2\mathbb{R}$ as a projective submanifold.

Similarly, we can define projective structures $\frac{m}{2}\mathbb{R}$ on the straight line, where m is an arbitrary positive integer or ∞. In reality, there are only three possibilities for an infinite straight line: infinite to the right, infinite to the left, and infinite in both directions. (The last case corresponds to the projective structure transferred to the line from the circle.) It can be proved that these exhaust all projective structures on the straight line. The invariant which distinguishes between these projective structures is the minimum number of charts required to cover the entire line.

Such is the answer for the straight line. For the circle, things are much more interesting. The answer for the circle is equivalent to a description of the orbits of coadjoint representation for the diffeomorphism group of the circle.

I spent a lot of time stating solved problems; now, I shall state one unsolved problem. Consider the group $G = \mathrm{Diff}(D, \partial D, \sigma)$, where

$$D = \{(x, y) \in \mathbb{R}^2 \colon x^2 + y^2 \leqslant 1\}, \quad \partial D = \{(x, y) \in \mathbb{R}^2 \colon x^2 + y^2 = 1\},$$

and $\sigma = dx\, dy$ is the area form. This means that G is the group of the diffeomorphisms of the disk D that are smoothly extended to the boundary ∂D and preserve the form σ.

An example of a transformation from the group G is a rotation of the disk. It is not that easy to write an explicit formula for a different transformation from the group G, except for

$$r \mapsto r, \quad \phi \mapsto \phi + f(r).$$

For this group, I want to state the problem of classifying the coadjoint orbits and discuss it. First, we must figure out what the Lie algebra of this group is. If we took all the diffeomorphisms, we would obtain the algebra of vector fields tangent to the boundary on the boundary. But we want that vector fields be such that the small shifts along them preserve area. As is known, such vector fields are called divergence-free. A vector field

$$v = a(x, y) \frac{\partial}{\partial x} + b(x, y) \frac{\partial}{\partial y}$$

is divergence-free if

$$\operatorname{div} v = \frac{\partial a}{\partial x} + \frac{\partial b}{\partial y} = 0.$$

Under this condition, a small shift along the vector field preserves area.

Now, let us discuss the dimension of the group under consideration. For infinite-dimensional groups, the notion of dimension does make sense, but this dimension is functional. In topology, the topological invariance of dimension for finite-dimensional spaces is proved: spaces of different dimensions are not homeomorphic. For infinite-dimensional manifolds, the situation is not so simple; nevertheless, those who deal with the so-called global analysis know that there is the notion of functional dimension. For example, the functions of one variable form a space of functional dimension 1, and the functions of two variables form a space of functional dimension 2. Therefore, functions of two variables cannot be expressed through functions of one variable. You might argue that there is Hilbert's 13th problem, which was solved by Kolmogorov and Arnold, who managed to express an arbitrary function of n variables as a superposition of functions of one variable. But there is something fishy, because they consider continuous functions. Continuous functions are something incomprehensible; there is no way of describing them. As to smooth or analytic functions, a function of two variables contains much more information than functions of one variable; no finite set of functions of one variable is sufficient to replace one function of two variables. So, the diffeomorphism group of the plane has functional dimension corresponding to two functions of two variables, i.e., $2\infty^2$. Indeed, a diffeomorphism of the plane is determined by two functions of two variables. The Lie algebra in this case has the same dimension, because a vector field on the plane is also determined by a pair of functions. But we have the additional condition $\operatorname{div} v = 0$. Hence there exists a function h for which $a = -\frac{\partial h}{\partial y}$ and $b = \frac{\partial h}{\partial x}$. (The vector field v is called the skew gradient of the

Figure 6. The tree of the components of level lines

function h and denoted by $v = $ s-grad h; the skew gradient is the usual gradient rotated through 90°.) Thus, the group under consideration has dimension ∞^2.

The condition that a vector field is tangent to the boundary has a very simple expression, namely, $h \upharpoonright \partial D = $ const. Indeed, the skew gradient of the function h is tangent to the boundary; therefore, the usual gradient is perpendicular to the boundary. Hence the boundary is a level line, i.e., the function is constant on the boundary. The skew gradient, as well as the usual gradient, does not change under the addition of a constant to the function. Therefore, we can assume that $h \upharpoonright \partial D = 0$. Then the Lie algebra $\mathfrak{g} = C^\infty(D, \partial D)$ is the space of smooth functions on the disk vanishing on the boundary. The commutator in this Lie algebra is the usual Poisson bracket $[f, g] = f'_x g'_y - f'_y g'_x$. The dual space $\mathfrak{g}^*$ consists of the generalized functions F for which $\langle F, f \rangle = \iint_D F f \sigma$. We consider only the smooth part of the space $\mathfrak{g}^*$, i.e., assume that F is a smooth function. Then the coadjoint action is the usual action of diffeomorphisms on functions.

We have arrived at the following problem. Suppose given a smooth function on the disk. Two functions are considered equivalent if one of them can be transformed into the other by an area-preserving change of variables. What possibilities do we have?

This problem is far from being trivial. Apparently, it has a comprehensible solution. I shall outline it. Each function $f \in C^\infty(D, \partial D)$ is related to the interesting topological invariant Y_f – the tree of the components of the level lines of the function f (Fig. 6). It is defined as follows. Consider a level line. It may consist of several connected components. We treat each connected component as a separate point. The set of such points has a natural topology. The corresponding topological space is the tree of level components.

Clearly, the tree of the components of level lines is invariant with respect to changes of variables, including the area-preserving changes of variables. The functions for which the trees of level line components have the simplest form (two vertices joined by an edge) admit precisely one invariant with respect to area-preserving changes of variables. This is an unusual, functional, invariant.

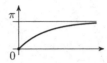

Figure 7. The graph of the function S

Namely, consider the function

$$S(c) = \text{area}\{x \colon f(x) \leqslant c\}.$$

If f is nonnegative, then $S(0) = 0$, $S(\infty) = \pi$, and the graph of this function has the form shown in Fig. 7. A theorem, which is neither very simple nor very difficult, asserts that, for functions with simplest trees of components, this invariant is unique. This means that, if we know the function S, then we know the function f up to an area-preserving change of variables. To be more precise, we can change variables in such a way that the function f depend only on the radius (knowing the function S, we can easily determine the precise form of this dependence).

Conjecturally, a similar theorem is valid for any tree of components, but this is not proved. We must modify the function S and specify its values for each edge of the tree separately. Such a function is still invariant, but it is not known whether there are other invariants.

The infinite Lie group $\text{Diff}(D, \partial D, \sigma)$ and its Lie algebra are, apparently, a rich source for new studies. The point is that this algebra is dual to itself, because it has the form $\langle F, f \rangle = \iint_D F f \sigma$, which is invariant with respect to the coadjoint representation. Thus, the adjoint representation is equivalent to the coadjoint representation. This allows us to treat such a Lie algebra as a usual compact Lie algebra, and transferring the basic finite-dimensional constructions to this infinite-dimensional case is a very promising enterprise. In particular, describing the flag manifold for this group is a very challenging problem, to which I draw the attention of young mathematicians. Although, there are several flag manifolds rather than one; they correspond to different types of trees.

I had not enough time to tell about the symplectic structure of orbits and about the complex structure, which is also encountered sometimes.

A. A. Kirillov

The orbit method beyond Lie groups.
Quantum groups

Lecture on September 3, 1999

Today I shall talk about quantum groups. First, I shall tell about my understanding of what quantum groups are. A usual Lie group is simultaneously a smooth manifold and a group. I shall not discuss the group structure at the moment, but I shall talk a little about the structure of a smooth manifold. There exist various definitions of smooth manifolds. One of them is algebraic; it frequently turns out to be most useful from the computational point of view. The general principle of computations in mathematics is that everything must be reduced to algebraic problems, which can be solved algorithmically. How can we replace a construction as geometric as a smooth manifold by a purely algebraic notion? For this purpose, instead of a smooth manifold M, we consider the algebra $A(M)$ of smooth (real-valued) compactly supported functions on M. "Compactly supported" means that each function vanishes outside some compact set. If the manifold is compact, then this requirement is not needed. For compact manifolds, the entire approach looks simpler; the theorems have shorter formulations and simpler proofs. But for the result to be general, I state it for all manifolds.

The algebra $A(M)$ is topological; in this algebra, the notion of limit is defined. Convergence on compact manifolds means the convergence of functions together with all their derivatives. The algebra $A(M)$ completely describes the manifold M. Thereby, all geometry is banished and algebra alone remains.

How can we reconstruct the manifold M? If there is another manifold N and a smooth mapping $\phi \colon M \to N$ is given, then we can construct a dual mapping of function algebras $\phi^* \colon A(N) \to A(M)$. Namely, to a function $f \in A(N)$ we assign $\phi^*(f) = f \circ \phi$. The question arises as to whether the compactly supported functions remain compactly supported. For a compact manifold, the answer is clear, but if the manifold is noncompact, a compactly supported function may be mapped to a noncompactly supported one. Therefore, for noncompact manifolds, we must restrict the class of mappings and consider only *proper* mappings, under which the preimages of all compact sets are compact. They map compactly supported functions to compactly supported ones. This is not very convenient; for example, such a mapping never takes a noncompact manifold to a compact manifold, because the preimage of the compact manifold itself must be compact.

The mapping ϕ^* is a homomorphism of algebras, so the geometric notion of a smooth mapping of manifolds is replaced by an algebraic notion. Most remarkably, any homomorphism of function algebras is generated by a smooth mapping of manifolds. This assertion also requires certain reservations, but I shall not discuss them now.

First, consider the simplest case where the manifold M_0 consists of one point. Clearly, $A(M_0) = \mathbb{R}$; therefore, a smooth mapping $\phi \colon M_0 \to M$ induces a homomorphism $\phi^* \colon A(M) \to \mathbb{R}$. In this special case, the above theorem is stated as follows.

Theorem 1. *Any nonzero continuous homomorphism*

$$\chi \colon A(M) \to \mathbb{R}$$

has the form χ_m, where $m \in M$ and $\chi_m(f) = f(m)$.

Proof. The kernel $\operatorname{Ker} \chi = \{f \in A(M) : \chi(f) = 0\}$ of the homomorphism χ is an ideal; we denote it by I_χ.

Lemma. *For any nontrivial ideal in the algebra of functions on a manifold, there exists a point in the manifold at which all functions from the ideal vanish.*

Proof. Suppose that there are no such points. Then, for any point $m \in M$, there exists a function $f_m \in I_\chi$ such that $f_m(m) \neq 0$. Choose a neighborhood U_m of the point m in which the function f_m does not vanish.

Take any function $g \in A(M)$. We want to prove that this function also belongs to the ideal I_χ. This implies $I_\chi = A(M)$, which contradicts the nontriviality of the ideal.

By definition, the function g is compactly supported, i.e., it vanishes outside some compact set K. The set K can be covered by a finite family $U_1, \ldots, U_N$ of the neighborhoods defined above. They correspond to functions $f_1, \ldots, f_N$ from the ideal. Consider the function $f = \sum_{i=1}^{N} f_i^2$. Clearly, $f > 0$ on K. Therefore, $g = fh$, where h is a smooth function. But $f \in I_\chi$; hence $g \in I_\chi$. $\qquad\square$

In the case under consideration, the ideal has codimension 1, because the entire algebra is mapped to a one-dimensional space. For such an ideal, there can exist only one point at which all functions from the ideal vanish. Therefore, the homomorphism evaluates the functions at this point. $\qquad\square$

Thus, the entire theory of smooth manifolds can be set in purely algebraic terms; the role of points of manifolds is played by algebras of functions. The next step is most important. Consider the question: Can we replace the algebra of smooth functions by some more general algebras, say noncommutative? It turns out that we can. The result of such a replacement is now called a *noncommutative manifold*. This is not a manifold in the classical sense; we cannot consider points of this manifold. We can only consider the algebra of functions on this manifold, and this algebra of functions can be arbitrary, in particular, noncommutative. The benefit is that the intuition of smooth manifolds is carried over to noncommutative algebras. An effort to carry over constructions from the theory of smooth manifolds to noncommutative algebras brings forth many interesting things. In particular, we can define a noncommutative Lie group; its noncommutativity means that it is noncommutative as a manifold rather than as a group.

We have come to the following definition: a quantum group is a noncommutative manifold which is simultaneously a group. We must only explain what "a noncommutative manifold which is simultaneously a group" means, because noncommutative manifolds are not sets. We must formulate the condition that some set is a group in the language of algebras of functions on sets, i.e., we must abandon points and only deal with functions.

The usual definition of a group is as follows. A group is a set G together with a mapping $G \times G \xrightarrow{\Pi} G$ possessing certain properties. For a Lie group, this mapping must be smooth. It is also required that taking the inverse element is smooth, but I leave formulating the corresponding assertion as an exercise. Now, I shall translate the associativity property of multiplication into the language of functions. Under the usual definition, associativity is the commutativity of the diagram

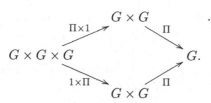

For algebras of functions, we have the mapping $A(G) \xrightarrow{\Pi^*} A(G \times G)$. If the group is finite, then $A(G \times G)$ algebraically coincide with the usual tensor product $A(G) \otimes A(G)$. For infinite groups, this is not so. In this case, we must take the completed (in the topological sense) tensor product $A(G) \widehat{\otimes} A(G)$. The associativity of the group then corresponds to the commutativity of the diagram

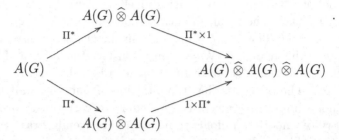

This definition makes sense for noncommutative manifolds as well.

Now, we can define a *quantum group* as an algebra A with a mapping $A \xrightarrow{\Pi^*} A \widehat{\otimes} A$ such that the corresponding diagram is commutative. In addition, the condition corresponding to the existence of an inverse for each element must hold, but I leave formulating it as an exercise.

A special case of quantum groups is the usual Lie groups.

In reality, the term *quantum groups* is used for a very special class of non-commutative manifolds, for which the function algebras, while noncommutative, are small deformations of commutative algebras. The definition given above is suitable for any noncommutative algebras, but little can be said in such generality. The most interesting examples require introducing additional structures. So, let us restrict the class of noncommutative algebras.

Why are quantum groups called quantum? The point is that one of the ways to attach a mathematical meaning to the notion of quantization consists in replacing an appropriate commutative operation by a noncommutative one. There are also other approaches. For example, some people believe that quantization consists in replacing continuous parameters by discrete parameters. Or in replacing some simple construction by complicated ones. The term "quantization" has no adequate translation into the mathematical language. It has several translations, but they are not equivalent. This is quite natural, because physics and mathematics are two different languages, and their scopes of notions must not coincide. The point is simply that there is no word in mathematics that has the same scope of notions as the word "quantization" in physics.

We shall deal with a special class of noncommutative algebras. Namely, consider the space of formal power series in a parameter h over some commutative (and associative) algebra A. In this space, we can introduce a noncommutative multiplication as follows. Take an $a = a_0 + h a_1 + h^2 a_2 + \ldots$ and a $b = b_0 + h b_1 + h^2 b_2 + \ldots$. It is required to multiply these elements. They can be multiplied as formal power series, but this multiplication is commutative. Let us try to define a noncommutative multiplication which acts as the multiplication of series in the zeroth approximation and may differ from it for terms containing higher powers of h. Under the assumption that h commutes with the other elements, it is sufficient to learn to multiply the elements of the algebra A itself. Then we will be able to multiply power series. Now, we define multiplication as follows:

$$a\,\widehat{h}\,b = a_0 b_0 + h\{a, b\}_1 + h^2\{a, b\}_2 + \dots$$

The sign $\widehat{h}$ symbolizes the dependence of multiplication on h. The braces $\{a, b\}_i$ denote bilinear operations in the algebra. The associativity of multiplication implies plenty of relations between braces. These relations can be resolved in a certain sense. I shall not describe the overall science, which is called the theory of deformations of associative algebras; I shall only state its results.

The theory of deformations of associative algebras, as well as all the other sciences, strongly depends on the notion of homology. Its results are represented by some cohomology classes, i.e., by some cocycles modulo coboundaries. In the simple language, this means that we might make some change of variables in the algebra A itself, i.e., replace each power series by a new power series all of whose terms are uniquely determined by the zeroth term: $a = a_0 + hA_1(a_0) + h^2 A_2(a_0) + \dots$, $b = b_0 + hA_1(b_0) + h^2 A_2(b_0) + \dots$. This gives the new multiplication law

$$\{a, b\}_1^{\text{new}} = \{a, b\}_1^{\text{old}} + aA_1(b) - A_1(ab) + A_1(a)b.$$

This change should be considered trivial: we have written the same multiplication in different coordinates. It turns out that the equation to which the brace satisfies says that the general brace is the sum of such a trivial added brace and a skew-symmetric expression. Therefore, up to a change of variables, we can assume that multiplication is anticommutative for the first term; the symmetric part can be removed. For the zeroth term, multiplication is always commutative, and for the first term, it can be made anticommutative. Next, it turns out that the associativity of multiplication for the new brace is equivalent to the Jacobi identity. We might foresee this, because there is a general theorem which says that the Jacobi identity is an odd analogue of associativity.

For the algebra of smooth functions on a manifold, this looks as follows. Let $f_1, f_2 \in C^\infty(M)$. Then

$$f_1 \,\widehat{h}\, f_2(m) = f_1(m)f_2(m) + h\{f_1, f_2\}(m).$$

This series can be extended, but the first terms are interesting enough. Under the additional physically sensible requirement that the support of the product must be contained in the intersection of the supports of the factors, any bilinear operation not increasing the supports must be determined by a differential operator. If, moreover, the Jacobi identity holds, then the differential operator contains no derivatives of order higher than 1. Therefore, in local coordinates, the brace has the form

$$c^{ij}(m)\,\partial_i f_1\,\partial_j f_2.$$

The constants c^{ij} are skew-symmetric with respect to i and j and satisfy a quadratic relation implied by the Jacobi identity. Such a geometric structure on a manifold is called a *Poisson structure*, and a manifold with such a structure is a *Poisson manifold*.

Geometrically, a Poisson structure is determined by a bivector c^{ij}. The abbreviated notation is $c^{ij}\,\partial_i \wedge \partial_j$. The polyvector fields on a manifold can be defined as expressions which have the following form in local coordinate systems:

$$c^{i_1\cdots i_k}\,\partial_{i_1} \wedge \cdots \wedge \partial_{i_k}.$$

Under changes of coordinate system, these expressions are transformed in a standard way, as partial derivatives; the exterior product is also transformed in a standard way, as a skew-symmetric associative multiplication. It turns out that there exists a natural bilinear operation

$$\mathrm{Vect}^k(M) \times \mathrm{Vect}^l(M) \to \mathrm{Vect}^{k+l-1}(M),$$

where $\mathrm{Vect}^k(M)$ is the set of all k-vector fields on the manifold M. In particular, a pair of vector fields is assigned a vector field, namely, the usual commutator of these vector fields. A pair of bivector fields is assigned a trivector field. This operation is natural in the sense that it does not depend on the choice of local coordinates. The operation is anticommutative on vector fields and commutative on the bivector fields. I is always either commutative or anticommutative.

On bivectors, the operation acts as follows:

$$[c,c]^{ijk} = c^{s[i}\partial_s c^{jk]};$$

the brackets on the right-hand side denote antisymmetrization with respect to the superscripts. The Jacobi identity is equivalent to the relation $[c,c] = 0$. To see this, it is sufficient to recall that

$$\{f_1, f_2\} = c^{ij}(m)\,\partial_i f_1\,\partial_j f_2$$

and write out the Jacobi identity for the left-hand side. We obtain a skew-symmetric relation involving three functions, which determine a trivector. This trivector is precisely $[c,c]$.

We return to quantum groups. I remind you that a quantum group is related to a noncommutative manifold. We already have the first approximation to this noncommutative manifold; we have considered two terms of the series, the zeroth (commutative) and the first (anticommutative). This quasi-classical approximation to a quantum group is determined by a completely classical object, a Lie group with Poisson structure. Such objects are called *Poisson–Lie groups*.

A Poisson–Lie group consists of a Lie group G and a bivector field c on G satisfying two conditions:

- the Jacobi identity $[c, c] = 0$ holds;

- if L_x is the left translation by x and R_y is the right translation by y, then

$$c(xy) = L_x c(y) + R_y c(x). \tag{1}$$

The first condition can be formulated not only for a Lie group but also for an arbitrary manifold. The second condition is the translation of the associativity of multiplication into the language of homomorphisms of function algebras.

The second condition is computationally inconvenient, because c is a section of a bundle rather than a function. The value of the bivector c at a given point x belongs to the space of bivectors tangent to G at the point x. Therefore, its values at different points belong to different spaces. But the Lie groups are parallelizable; hence we can introduce a new function, namely,

$$\gamma(x) = L_x^{-1} c(x).$$

This γ is a usual function on G taking values in $\mathfrak{g} \wedge \mathfrak{g}$, because $c(x)$ takes values in the exterior square of the tangent space at the point x and $L^{-1}(x)$ maps the tangent space at the point x to the tangent space at the identity, i.e., to $\mathfrak{g}$. For the function γ, equality (1) takes the form

$$\gamma(xy) = \gamma(y) + \operatorname{Ad} y^{-1} \gamma(x). \tag{2}$$

This is a functional equation. We can easily reduce it to a differential equation by setting $y = \exp(tY)$ and differentiating with respect to t:

$$Y\gamma(x) = Y\gamma(1) - e^{\operatorname{ad} Y} Y(x). \tag{3}$$

We have obtained a system of ordinary differential equations with respect to the function $\gamma(x)$; the right-hand side contains its first derivative at the initial point. *A propos*, setting $x = y = 1$ in (2) gives $\gamma(1) = 0$. The right-hand side of (3) contains only the first derivatives of γ at the point 1; therefore, the bivector γ is completely determined by this differential equations, provided that its first derivatives at 1 are known.

The first derivative of a bivector at the point 1 is a three-index tensor; indeed, a bivector acts on functions, but simultaneously, it linearly depends on the tangent vector. This gives an element of the space $\mathfrak{g}^* \otimes \mathfrak{g} \wedge \mathfrak{g}$, which is isomorphic to $\mathrm{Hom}(\mathfrak{g}, \mathfrak{g} \wedge \mathfrak{g})$. This element completely determines the Poisson–Lie group. It must satisfy certain conditions, and yet it contains all the information about the Poisson–Lie group. This situation strongly resembles the very notion of Lie group. As is known, Lie groups are largely determined by their Lie algebras, and Lie algebras are determined by sets of structural constants. The structural constants are a third-rank tensor. The set of structural constants determines a mapping $\mathfrak{g} \wedge \mathfrak{g} \to \mathfrak{g}$, and for a Poisson–Lie algebra, we obtain a mapping in the reverse direction, $\mathfrak{g} \to \mathfrak{g} \wedge \mathfrak{g}$. But we can easily pass from one kind of mappings to the other by considering dual mappings. For a Poisson–Lie algebra, the dual mapping has the form $\mathfrak{g}^* \wedge \mathfrak{g}^* \to \mathfrak{g}^*$. Thus, there is a complete analogy. A Lie group is determined by the structural constants c_{ij}^k, and in a Poisson–Lie group, a bivector c_k^{ij} defining multiplication on the dual space is added. It turns out that the conditions imposed on this bivector ensure that the second set of c_k^{ij} is a set of structural constants in the dual space, i.e., the multiplication defined by this set also satisfies the Jacobi identity.

Thus, a very beautiful algebraic construction arises. There is a space $\mathfrak{g}$, which is a Lie algebra with commutator $[\ ,\ ]$, and a dual space $\mathfrak{g}^*$, which is also a Lie algebra with some commutator $[\ ,\ ]_*$. Certainly, to yield a Poisson–Lie group, these structures must be related. Algebraists have considered this question. The final solution is as follows. The mapping $\mathfrak{g} \to \mathfrak{g} \wedge \mathfrak{g}$ treated as a 1-chain on the Lie algebra $\mathfrak{g}$ with values in $\mathfrak{g} \wedge \mathfrak{g}$ must be a 1-cycle. The dual condition on the mapping $\mathfrak{g}^* \to \mathfrak{g}^* \wedge \mathfrak{g}^*$ is equivalent to the first condition.

Ten years ago, a remarkable interpretation of both these equivalent conditions was discovered; it is easy to memorize even for people not dealing with cohomology. What is the simplest way to make the commutators in a space consistent with the commutators in the dual space? We start with coding the notion of dual space. For this purpose, consider the direct sum $\mathfrak{g} \oplus \mathfrak{g}^* = \mathfrak{D}$. The space $\mathfrak{D}$ has an additional structure; namely, we can define a bilinear form on $\mathfrak{D}$ such that it vanishes on $\mathfrak{g}$ and on $\mathfrak{g}^*$, and if one argument belongs to $\mathfrak{g}$ and the other belongs to $\mathfrak{g}^*$, then the usual value of a functional at a vector is taken. This form is nondegenerate on $\mathfrak{D}$, and the subspaces $\mathfrak{g}$ and $\mathfrak{g}^*$ are isotropic with respect to it. In addition, $\mathfrak{g}$ and (separately) $\mathfrak{g}^*$ are endowed with Lie algebra structures. It turns out that all the additional conditions involving cocycles are equivalent to the requirement that, on $\mathfrak{D}$, there exists a Lie algebra structure such that it preserves this bilinear form (i.e., all the ad operators are skew-symmetric with respect to the corresponding inner product) and the restrictions of this Lie algebra structure to $\mathfrak{g}$ and to $\mathfrak{g}^*$ coincide with the Lie algebra structures given on these spaces.

The invariance of the bilinear form corresponds to the condition

$$([x, y], z) = ([y, z], x).$$

The resulting object assigned to a Poisson–Lie group is what is called a *Manin triple*. It consists of a Lie algebra $\mathfrak{D}$ with a nondegenerate invariant bilinear form and of two subalgebras $\mathfrak{g}$ and $\mathfrak{g}^*$ isotropic with respect to this bilinear form. These algebraic data completely code a Poisson–Lie group, which is a geometric object.

Now, I shall tell about what this has to do with the orbit method. Consider a trivial Poisson–Lie group, i.e., a Lie group G with bivector $c = 0$. The corresponding Manin triple is $\mathfrak{g} \oplus \mathfrak{g}^*$, where the commutator operation in $\mathfrak{g}$ coincides with that in the usual Lie algebra and the commutator in $\mathfrak{g}^*$ is zero. The bilinear form is as it should be on the direct sum of a space and its dual. For this reason alone, the commutators of elements from $\mathfrak{g}$ and from $\mathfrak{g}^*$ are uniquely determined by the condition of form invariance. Namely, a commutator is the coadjoint action of an element from $\mathfrak{g}$ on a functional on the Lie algebra. Thus, the Lie algebra $\mathfrak{D}$ corresponds to the Lie group $G \ltimes \mathfrak{g}^*$ being the semidirect product of the Lie group G and the space $\mathfrak{g}^*$ dual to the Lie algebra. The group action is as follows: on G, it is as it should be, $\mathfrak{g}^*$ is an Abelian group under addition, and G acts on $\mathfrak{g}^*$ coadjointly. We obtain the main object of the orbit method – the coadjoint action of a Lie group G on the space $\mathfrak{g}^*$. This object arises from the consideration of the trivial Poisson–Lie group with zero bivector c. In the general case, we obtain a Lie group D with two subgroups G and G^*. It should be mentioned that, in this construction, G and G^* play fully symmetric roles. The quantum groups related to deformations occur in pairs. To each group G, the group G^* is associated. If G is a usual Lie group, then G^* is the space dual to its Lie algebra.

The Lie algebra of the group D is the direct sum of the Lie algebras of the groups G and G^*; therefore, in the first approximation, D is the product of G and G^*; but in the general case, this is not so. The situation resembles decomposing a group of matrices into the product of upper triangular and lower triangular matrices: such a decomposition is possible for general matrices, but it is not possible for special matrices. Thus, not every element of D is a product of an element from G and an element from G^*.

I find it useful to study the action of G not only on the space $\mathfrak{g}^*$ but also on the group G^*. This action is no longer linear.

If $G = \mathrm{SO}(3)$ or $\mathrm{SL}(2)$, then G^* is a solvable group. In general, the group dual to a simple Lie group is always solvable. The group G^* depends on the choice of braces. But for a simple group G, all groups G^* can be described.

As I said, they are related to certain cocycles, and all cocycles for a simple group are trivial, i.e., all of them are coboundaries. Therefore, the required cocycle can be written out explicitly. If this cocycle (which is a coboundary) is sufficiently nondegenerate, then a fairly explicit description is obtained. It is due to Belavin and Drinfel'd.

For $G = SO(3)$, one dual group is obtained, and for $SL(2)$, there are three dual groups, because $SL(2)$ has elements of three types – elliptic, parabolic, and hyperbolic.

A problem studied not at all is to describe the Manin triple for an infinite-dimensional Lie algebra, namely, for the algebra discussed in the last lecture. As the Lie algebra $\mathcal{A}$ I suggest to take the algebra of smooth complex functions on the disk D that vanishes on the boundary, and as the invariant inner product,

$$(f_1, f_2) = \operatorname{Im} \int_D f_1 f_2 \, d\sigma.$$

In the finite-dimensional situation, any Lie algebra is associated with some Lie group. But this infinite-dimensional Lie algebra is associated with no Lie group; in the infinite-dimensional case, there are many examples of this kind. In return, $\mathcal{A}$ can be represented as the direct sum of subspaces $\mathfrak{g}$ and $\mathfrak{g}^*$ that are isotropic with respect to the inner product defined above and are Lie algebras associated with Lie groups. Namely, as $\mathfrak{g}$ we take the real-valued functions $C_{\mathbb{R}}^{\infty}(D, \partial D)$ (this space is isotropic because we take the imaginary part of the integral, and for real-valued functions the imaginary part of the integral vanishes). The space $\mathfrak{g}^*$ is constructed as follows. Let us write a function on the disk in the form

$$f = f(r, \phi) = \sum_{k \in \mathbb{Z}} c_k(r) e^{ik\phi}.$$

It is convenient to replace r by a different coordinate. In mechanics, there are canonical coordinates called action–angle coordinates. In these coordinates, the form is represented as a product of differentials of coordinates. In the case under consideration,

$$\sigma = r \, dr \wedge d\phi = d(\pi r^2) \wedge d\left(\frac{\phi}{2\pi}\right) = dS \wedge d\theta;$$

here S is the area (action in mechanics) and θ is the normal angle. We set

$$\mathfrak{g}^* = \left\{ f = \sum_{k \geqslant 0} c_k(S) e^{2\pi i k \theta} \right\}.$$

The operation in the Lie algebra is the Poisson bracket. The Poisson bracket can be taken in arbitrary coordinates satisfying the only condition that the area has the standard expression $dx\,dy$. In particular, as canonical coordinates S and θ can serve. Then

$$\{f_1(S)e^{2\pi ik\theta},\ f_2(S)e^{2\pi il\theta}\} = 2\pi i\big(lf_1'(S)f_2(S) - kf_1(S)f_2'(S)\big)e^{2\pi i(k+l)\theta}.$$

The numbers k and l are added together, hence the functions for which $k \geqslant 0$ form a subalgebra (the positive harmonics form a subalgebra).

Thus, it is possible to manufacture a Manin triple from a Lie algebra $\mathcal{A}$. The space is infinite-dimensional, so all familiar finite-dimensional constructions must be studied all over again. This problem has been untouched so far.

It often happens in science that, when some problem is known to be very difficult, nobody tackles it. And, then, a young man who does not know that it is difficult comes across this problem and solves it. People ask him: "How did you manage to solve such a difficult problem?" And he replies: "I did not know that it is difficult." So, I appeal to everybody who does not yet recognize all the difficulties involved to tackle this problem.

I. M. Krichever

Conformal mappings and the Whitham equations

Lecture on December 23, 1999

The topic named in the first part of the title of this lecture is familiar to every student. My ultimate goal is to show how the theory of integrable equations, which has been extensively developed during the past twenty years, and the Whitham theory, which already has a ten-year history, are related to the classical problem of complex analysis. The Riemann theorem asserts that, if a domain in the complex plane has a boundary containing more than two points, then there exists a conformal mapping of this domain onto the unit disk. This is an existence theorem. Many applied sciences are engaged in constructing such conformal mappings in particular situations; moreover, these problems are related to applications in hydrodynamics, in the theory of oil-fields, and in aerodynamics. The necessity of constructing conformal mappings of special domains emerges very often.

I want to present a recent remarkable observation of Zabrodin and Wiegmann, who discovered a relation between the classical problem on conformal mappings of domains and the dispersionless Toda lattice a couple of months ago. I shall tell about the development of this observation in our joint paper (not yet published), namely, about its generalization to nonsimply connected domains and about the role which the methods of algebraic geometry play in it.

Before proceeding to the problem proper, I want to give a brief overview of the entire context in which it has arisen, in order to clarify what the Whitham equations are. Surprisingly, the same structures related to the Whitham equations arise in various fields of mathematics, not only in the theory of conformal mappings. For example, they arise in the problem of constructing n-orthogonal curvilinear coordinates, which was the central problem of differential geometry in the nineteenth century. Let $x^i(u)$ be a curvilinear coordinate system in $\mathbb{R}^n$, where x^i are the Cartesian coordinates expressed in terms of curvilinear coordinates u. Such a coordinate system is called *n-orthogonal* if all the level hypersurfaces $u_i = \text{const}$ intersect at a right angle. An example of such a coordinate system is polar coordinates. In the two-dimensional case, the problem is trivial, but starting with dimension 3, it becomes very rich. Theoretically, this problem was solved by Darboux, again at the level of an existence theorem. He proved that the local problem of constructing an n-orthogonal curvilinear coordinate system depends on $n(n-1)/2$ functions of two variables. There are fairly many particular examples of n-orthogonal coordinates system. One of such examples is elliptic coordinates. In essence, solving a given system of differential equations reduces to constructing apt coordinates, in which the system becomes trivial. This is why good coordinate systems are so important: they increase the chances of solving the equation.

The problem about n-orthogonal coordinate systems can be set in intrinsic terms as the problem of finding flat diagonal metrics $ds^2 = \sum H_i^2(u)(du)^2$. Egorov considered such metrics satisfying the additional condition $H_i^2 = \partial_i \Phi$. This is the metric symmetry condition. Such metrics are called *Darboux–Egorov metrics*. They have many special features.

This problem, the Whitham equations, and the problem about conformal mappings belong to one complex of ideas and methods. A little later, I shall tell how the problem about n-orthogonal curvilinear coordinates is related to topological quantum models of field theory.

Another theme, which has eventually united all these diverse problem, is the theory of integrable equations, commonly referred to nowadays as soliton equations. This theory emerged about 30 years ago. The best-known (and oldest) soliton equation is the Korteweg–de Vries (KdV) equation

$$u_t - \frac{3}{2}uu_x + \frac{1}{4}u_{xxx} = 0.$$

There are many other soliton equations (fortunately, having important applications) which can be integrated by the methods of soliton theory. I shall not talk about these methods; they were largely developed 10–20 years ago and continue being developed at present.

Korteweg and de Vries found the simplest solution to the KdV equation very shortly after they wrote it. This is a stationary solution $u(x,t)$ not depending on t. In this case, $u_t = 0$, and we can integrate the equation:

$$\frac{3}{4}u^2 = \frac{1}{4}u_{xx} + g_2.$$

Then, we can multiply it by u_x and integrate again:

$$\frac{1}{2}(u_x)^2 = u^3 + g_2 u + g_3.$$

The solutions to such an equation are expressed in terms of the Weierstrass function as

$$u = 2\wp(x + \text{const}; \omega_1, \omega_2);$$

the Weierstrass function $\wp$ is a doubly periodic function with periods $2\omega_1$ and $2\omega_2$ having a second-order pole at zero; i.e., $\wp = \frac{1}{x^2} + O(x)$ at zero. This solution depends on three constants; thus, we have obtained the complete set of solutions, because the equation is of the third order.

A stationary solution is constructed from an elliptic curve, i.e., a curve of genus 1. In the 1970s and in the early 1980s, in a cycle of papers by Dubrovin, Novikov, Matveev, myself, and a number of other authors, the so-called algebro-geometric methods for constructing solutions to soliton equations were developed; given a set of algebro-geometric data, they yield solutions to various nonlinear equations, including the KdV equation, the sine-Gordon equation, and other equations pertaining to this science. A solution is obtained by processing data by a machine called finite-zone integration. A solution is represented explicitly, but in terms of the Riemann theta-functions rather than in terms of elliptic functions. The algebro-geometric data set consists of a Riemann surface Γ_g of genus g with fixed points $P_1, \ldots, P_N$ and fixed local coordinates $z_1, \ldots, z_N$ in neighborhoods of these points; it also includes a fixed point on the complex multidimensional torus $J(\Gamma_g)$ being the Jacobian of this surface. A solution is constructed from such data. This makes it possible to solve very diverse equations, depending on the number of fixed points and on the classes of curves.

For the stationary solution to the KdV equation, the algebro-geometric data set includes an elliptic curve $y^2 = E^3 + g_2 E + g_3$ and a fixed point at infinity. These data play the role of integrals, for they do not change with time. But the point on the Jacobian moves. The phase space of the equation looks as follows. There is a space of integrals being curves with marked points and fixed local coordinates at these points; over each point of the space of integrals, a torus hangs. The motion on the torus is a rectilinear winding, in full accordance with the spirit of the theory of completely integrable finite-dimensional systems, i.e., with Liouville theory.

Such is the answer for soliton equations. The procedure for constructing solutions is another story. I shall not tell it now. Instead, I want to tell about what happened to this science thereafter, starting with the mid-1980s. At that time, a particular emphasis was placed on the theory of perturbations of integrable equations. Usually, we are interested not only in a specific equation but also in what happens in its neighborhood. The basic element of the perturbation theory of integrable equations is Whitham theory.

Before proceeding to Whitham theory, I want to write one formula; its various forms are encountered in all the sciences mentioned above. As I said, the description of motion for soliton equations in terms of systems of integrals and rectilinear windings of the torus is fully consistent with Liouville theory. The ultimate goal of Liouville theory is specification of action–angle variables. A Hamiltonian system is constructed from a manifold M^{2n} (phase space), a symplectic structure ω on it, and a Hamiltonian H. A Hamiltonian system is called completely integrable if, in addition to a Hamiltonian, it has n integrals in involution, for which $\{F_i, F_j\} = 0$. The compact surface levels of these integrals must be n-dimensional tori, and the motions on them must be rectilinear windings. The torus has natural coordinates, cycles. If Φ_i are the angular coordinates for the basis cycles, then the action variables are defined as the coordinates A_i canonically conjugate to the angular variables, i.e., such that the symplectic structure in these coordinates has the standard Darboux form $\omega = \sum dA_i \wedge d\Phi_i$. Selecting such coordinate systems among all coordinate systems is a separate nontrivial problem. The Liouville theorem in Arnold's setting says that we must integrate the primitive form over the basis cycles. But it is unclear how to explicitly describe this n-dimensional torus in the $2n$-dimensional manifold. Thus, this theorem also has the character of an existence theorem. All attempts to explicitly construct action–angle variables have failed. In the early 1980s, Novikov and Veselov made a remarkable observation. Analyzing the first integrable Hamiltonian equations known at that time, they discovered that the action–angle variables have the same form for all these systems. Namely, integration over a cycle on n-space is replaced by integration over a cycle on the corresponding Riemann surface, that is,

$$A_i = \oint_{a_i} Q \, dE. \tag{1}$$

Here Q is a meromorphic differential; to each Hamiltonian system, its own differential Q corresponds. These differentials may be multivalued. Nobody knew why this is so. Novikov and Veselov called these formulas analytic Poisson brackets. Their nature has been explained analytically only recently, three years ago, in my joint paper with Phong (in *Journal of Differential Geometry*). We analyzed the answers for the symplectic structures which arise in Seiberg–Witten theory for the supersymmetric Yang–Mills model and noticed that the same symplectic brackets as those describing the case of hyperelliptic curves (I should mention that everything considered by Novikov and Veselov referred to the case of hyperelliptic curves) were rediscovered by Seiberg and Witten.

Memorize formula (1), because precisely the same integral of a multivalued differential solves the problem about conformal mappings of domains.

What are the Whitham equations? Suppose that we have slightly changed (perturbed) the equation. Then the integrals of the initial equation cease to be integrals. They begin to slowly vary; as physicists say, they become adiabatic integrals. For the nonperturbed equation, a point of the phase space moves on a torus. As soon as we perturb the equation, a slow drift along the space of integrals begins. The system of equations on the moduli space of curves with

Figure 1. An overturning wave

marked points describing this motion is precisely the Whitham equations. It turns out that they are themselves integrable.

I shall not talk about algebro-geometric data any longer. I shall continue the discussion at a quite elementary level, where only curves of genus 0 are considered. The point is that the solution to the KdV equation that involves the Weierstrass function is not the simplest one; the simplest solution is a constant. In the theory of KdV equation, curves of genus 0 are trivial, and nobody was interested in this solution. But when we consider perturbations of a constant solution rather than this solution itself, the theory becomes interesting. In Whitham theory, genus 0 plays a nontrivial role. This case is called the dispersionless limit of soliton equations. It can be treated as a special case of a more general problem or considered separately.

Why "dispersionless"? The coefficients in a KdV equation are inessential, because it can be reduced to the form $u_t = uu_x + u_{xxx}$ by scale transformations. In what follows, I shall not trace the coefficients. If a solution is almost constant, we can forget about the third derivative. A good approximation is the equation $u_t = uu_x$. It is this equation that is called the dispersionless limit, because in the KdV equation, the term u_{xxx} is responsible for dispersion. The equation $u_t = uu_x$ is the simplest Whitham equation.

The KdV equation is an infinite-dimensional analogue of integrable (in the sense of Liouville) Hamiltonian systems; the equation $u_t = uu_x$ is also integrable, but in a completely different sense. Solving the equation $u_t = uu_x$ (it is called the Riemann–Hopf equation) is child's play. Indeed, take an arbitrary function $f(\xi)$ and consider the equation $u = f(x + ut)$. This equation implicitly defines a function $u(x, t)$. This function is a solution to the equation $u_t = uu_x$. Moreover, all the solutions are obtained in this way.

This solution is commonly used as a basis for explaining the role of non-linearity in hydrodynamics. Treating u as altitude (wave amplitude), we see that velocity of a point is proportional to its altitude. Therefore, if the function is not monotone, then the "hump" begins to outrun everything else; the wave becomes steeper and overturns (Fig. 1). At the overturning point, the third derivatives cannot be neglected, for they grow large. Hydrodynamics explains this as regularization of the behavior of the wave by dispersion (viscosity).

All Whitham equations are integrated by similar methods, which consist in writing some implicit expression for a solution.

What is the more general setting of dispersionless Lax equations? Let me remind you that constructing solutions to soliton equations is based on the Lax representation $\dot{L} = [L, A]$. For the KdV equation, we have $L = \partial^2 + u(x, t)$ (the Sturm–Liouville operator) and $A = \partial^3 + \frac{3}{2}u\partial + \frac{3}{4}u_4$. The generalizations of the KdV equation have arisen from consideration of operators with matrix coefficient and higher-order operators

$$L = \partial^n + u_{n-2}\partial^{n-2} + \cdots + u_0.$$

The Lax representation is a consistency condition for the overdetermined system of linear equations $L\psi = E\psi$, $L_2\psi = A\psi$. In general, the idea of the inverse problem method is not to start from the equation but go in the reverse direction, i.e., construct an operator and a solution from a given function ψ.

As we have agreed to begin with considering the simplest solutions to Lax equations (when u is a constant), we can solve the corresponding linear differential equation very easily. The solution is an exponential, and the eigenvalues are polynomials. Taking the eigenfunction $\psi = e^{px}$, we obtain $E(p) = p^n + u_{n-2}p^{n-2} + \cdots + u_0$ (the symbol of the corresponding differential operator). The Whitham equations are written as $\partial_i E = \{E_+^{i/n}, E\}$; here $\{f, g\} = f_p g_x - f_x g_p$ is the Poisson bracket. We shall express $u_d(X, T)$ in terms of the slow variables $X = \epsilon x$ and $T = \epsilon t$.

The subscript i is not a misprint. Each integrable equation arises as a part of the large hierarchy formed by a whole family of integrals commuting with this equation. This is in the spirit of Liouville integrability: if we have a set of integrals in involution, then each of these integrals regarded as a Hamiltonian generates its own Hamiltonian dynamics. That the integrals are in involution means that the corresponding dynamics commute.

Now, I shall explain what $E_+^{i/n}$ is. Let $E^{1/n}(p) = p + \sum v_i p^{-i}$ be the Laurent expansion. Then $E^{i/n}(p) = p^i + \cdots + O(p^{-1})$; $E_+^{i/n}$ means that only nonnegative powers of p are taken, i.e., $O(p^{-1})$ is crossed out. We obtain a polynomial whose coefficients are polynomials in u. Therefore, the result is a closed system of equations, which is the dispersionless limit of Lax equations. In the simplest case, where $E = p^2 + u$ and $i = 3$, we obtain the Riemann–Hopf equation mentioned above.

How does the general solution procedure for a dispersionless limit look like? Consider the space of pairs (Q, E), where E and Q are polynomials of forms $E = p^n + u_{n-2}p^{n-2} + \cdots + u_0$ and $Q = b_0 p + \cdots + b_{m-1}p^m$, respectively. On this space, we can introduce the Whitham coordinates $T_i = \frac{1}{i}\operatorname{res}_\infty(E^{-i/n}QdE)$. The T_i so defined are functions of u and b (they linearly depend on b and polynomially on u). These T_i vanish at large i; there are precisely as many nonzero T_i as required. We can locally invert the T_i as functions of u and b

and obtain functions $u(T)$ and $b(T)$. Substituting these values $u(T)$ into E, we obtain a function $E(T)$. It turns out that $E(T)$ is a solution to an equation of dispersionless hierarchy. The polynomial Q seems to play an auxiliary role. But $Q(T)$ is also a solution to the same equation with the same Hamiltonian; namely, $\partial_i Q = \{E_+^{i/n}, Q\}$. Moreover, we always have $\{Q, E\} = 1$. The equation $\{Q, E\} = 1$ is called the *string equation*.

In the dispersionless limit, as opposed to the usual hierarchy of Lax equations, only one solution survives, since all solutions are parametrized by different higher times; the general solution satisfies a suitable string equation.

The dispersionless science had been known for several years when Dijkgraaf, Verlinde, and Witten published a paper. They considered a quite different problem, namely, classification of the topological models of field theory. Solving this problem, they obtained the very same formula in a completely different context. It became clear that, behind the dispersionless science, a very important element was hidden; now, it is known as the tau-function. The whole structure related to the dispersionless limit of the KdV equation or of the general Lax equation is coded by only one function

$$F(t) = \frac{1}{2} \operatorname{res}_\infty \left(\sum_{i=1}^\infty T_i k^i \, dS \right).$$

Here $dS = Q \, dE = \sum_{i=1}^\infty T_i \, dk^i + O(k^{-1})$ and $k = E^{1/n}(p) = k(p) = p + O(p^{-1})$. I remind you that we deal with the case of a curve of genus 0; the marked point can be driven to infinity. The only surviving parameter is the local coordinate p. It can be verified, although this is far from being obvious, that the derivatives of the function F with respect to times T_i give all the remaining coefficients. For example, $\partial_i F = \operatorname{res}_\infty(k^i \, dS)$ and $\partial_{ij}^2 F = \operatorname{res}_\infty(k^i \, d\Omega_j)$, where $\Omega_j = E_+^{j/n}$. There is the remarkable formula

$$\partial_{ijk}^3 F = \sum_{q_s} \operatorname{res}_\infty \left(\frac{d\Omega_i \, d\Omega_j \, d\Omega_k}{dQ \, dt} \right),$$

where the summation is over the critical points of the polynomial E (such that $dE(q_s) = 0$).

Now, I return to the initial problem about conformal mappings. I shall consider only the case of domains bounded by analytic curves. Let us denote the interior domain by D and the exterior domain by $\overline{D}$. I shall be interested in schlicht conformal mappings of the exterior of the unit disk to $\overline{D}$. For reading, I recommend the book A. N. Varchenko and P. I. Etingof. *Why the Boundary of a Round Drop Becomes a Curve of Order Four* (Providence, RI: Amer. Math. Soc., 1992). It contains many beautiful particular examples of conformal mappings related to the following problem, which arises in the oil industry. Imagine that the domain under consideration is an oil-field. There are several oil wells through which the oil is pumped out. This somehow deforms the domain. The equation describing the dynamics of the domain boundary is as follows. Let Φ be a solution to the equation

$$\Delta\Phi = \sum q_i\,\delta(z - z_i)$$

with zero boundary condition $\Phi \restriction \partial D$. Then $\operatorname{grad}\Phi$ is the velocity of the boundary.

This problem is integrable in a certain sense. It turns out that the final shape of the drop does not depend on the oil pumping schedule, as it must be for commuting flows. The result depends only on the amount of oil pumped out through each oil well; the particular procedure of pumping does not matter.

The main contribution to this science was made by Richardson, who discovered an infinite set of integrals. It is these integrals that I am going to discuss next.

It is fairly easy to prove that any domain (simply connected or not) is completely determined by its harmonic moments. The harmonic moments of a domain D are defined as follows. Let $u(x, y)$ be a harmonic function. Then the corresponding harmonic moment is equal to

$$t_u = \iint_D u(x, y)\,dx\,dy.$$

When the domain changes, the harmonic moment of some function also changes. This is a local assertion. The harmonic moments are local coordinates.

It is not necessary to consider all harmonic moments; it is sufficient to take only some of the functions. For example, the set of functions

$$t_n = \iint_D z^{-n}\,dz\,d\bar{z}, \quad n \geqslant 1$$

together with the function

$$t_0 = \iint_{\bar{D}} dz\,d\bar{z},$$

where $\bar{D}$ is the exterior domain, is a local set of coordinates for a simply connected domain.

The fundamental observation made by Wiegmann and Zabrodin is a follows. Consider, in addition, the moments

$$v_n = \iint_{\bar{D}} z^n\,dz\,d\bar{z}$$

of the complement. Clearly, the functions v_n can be expressed in terms of $t_0, t_1, \ldots$. It turns out that

$$\frac{\partial v_n}{\partial t_m} = \frac{\partial v_m}{\partial t_n}.$$

This means that there exists a function $F(t)$ for which $\partial_n F(t) = v_n$. It turns out that $\partial_0 \, \partial_n F$ are the expansion coefficients of a schlicht function implementing a conformal mapping. We assume that this function is normalized as follows. In the complement to the unit disk, there is a coordinate w, and in $\overline{D}$, a coordinate z. We consider the mapping of the exteriors and suppose that infinity is mapped to infinity; moreover, we assume that $z = rw + O(w^{-1})$. In this case, $w(z) = r^{-1}z + \sum(\partial_0 \, \partial_n F)z^{-n}$. Again, it turns out that all the conformal mappings are coded by one function. This is precisely the function which I mentioned above.

First, I want to give a new proof that locally, the coordinates t_n form a complete coordinate system. From the proof, it will be seen how this all is related to the dispersionless science.

I need the notion of the Schwarz function. Locally, a smooth curve can be specified in the form $y = f(x)$. In the complex form, this can be written as $\bar{z} = S(z)$. The function S is called the Schwarz function of the curve. For example, for the unit circle, we obtain the equation $\bar{z} = z^{-1}$.

For a real-analytic curve (without corners), the function S can be extended to a complex-analytic function in a small neighborhood of the curve.

The first assertion which I want to prove is as follows. Suppose that a contour deforms, i.e., we have a family of Schwarz functions $S(z, t)$, where t is a deformation parameter. If none of the harmonic moments t_n changes under such a deformation, then the curve is fixed, i.e., the deformation is trivial.

Assertion 1. *The 1-differential $S_t(z, t)\, dz$ is purely imaginary on the contour ∂D, i.e., all of its values on the vectors tangent to the contour are purely imaginary.*

This follows easily from the definition of the Schwarz function.

The next assertion uses the specifics of the coordinates t_n under consideration.

Assertion 2. *If $\partial_t t_n = 0$, then the holomorphic differential $\partial_t S\, dz$ defined in a small neighborhood of the curve can be extended to a holomorphic differential on the entire exterior.*

Before proving the second assertion, I shall explain how to derive the required result from these two assertions. Any domain $D \subset \mathbb{C}$ with coordinate z determines a closed Riemann surface. To construct it, we take another copy of the same domain with coordinate $\bar{z}$ and attach it to the given domain along the boundary. The obtained Riemann surface is called the Schottky double. Let us apply the Schwarz symmetry principle: any function analytic in the upper half-plane and real on the real axis can be analytically continued to the lower half-plane. We have a holomorphic differential in D. It can be extended to the complex conjugate, because it is purely imaginary on the boundary. As a result, we obtain a holomorphic differential on the sphere. But there are no nonzero holomorphic differentials on the sphere.

We proceed to prove the assertion that the holomorphic differential $\partial_t S \, dz$ can be extended to the entire exterior. Using the Cauchy integral, we can represent an arbitrary function on a smooth contour as the difference of a function holomorphic in the exterior domain and a function holomorphic in the interior domain. Let

$$\widehat{S}(z) = \oint_{\partial D} \frac{\partial_t S(w) \, dw}{z - w}.$$

The function $\widehat{S}(z)$ is holomorphic outside the contour, it can be extended to the boundary, and $S^+ - S^- = \partial_t S$. If the origin lies inside the domain and $|z| < |w|$, then

$$\widehat{S}(z) = \sum z^n \oint_{\partial D} \partial_t S(w) w^{-n} \, dw = \sum z^n \partial_t t_n,$$

because

$$t_n = \iint_D z^{-n} \, dz \, d\bar{z} = \oint_{\partial D} z^{-n} \bar{z} \, dz$$

by the Stokes theorem.

If the moments do not vary, then the expansion coefficients of $\widehat{S}$ at $z = 0$ are identically zero. Therefore, the function S^- is identically zero in some neighborhood of $z = 0$. But this function is analytic; hence it vanishes identically. For $\partial_t S \, dz$ to be holomorphic, one more coefficient should be zero, because we have multiplied the function by dz, and the differential has a pole of the second order.

Now, it is clear what changes when we differentiate with respect to t_n. The first assertion is valid for an arbitrary variable. The expansion coefficients are no longer identically zero; one of the coefficients is nonzero. This means, in particular, that $\partial_{t_0} S \, dz$ is a meromorphic differential with a simple pole at infinity.

When we take the double, a second pole emerges according to the symmetry principle. We obtain a differential having residue ± 1 at two points. (There is only one such differential.) This is a global property, as the Liouville theorem. An analytic function on a compact surface is constant. These two facts allow us to use global properties. The first fact makes it possible to pass from a domain with boundary to a compact surface. And the second fact, which requires special assumptions, gives an analytic continuation to a meromorphic object.

Thus, we have proved that $\partial_0 \bar{z}\, dz = \frac{dw}{w}$. Here we differentiate $\bar{z}$ at a constant z. An equivalent expression is

$$\{z(w, t_0), \bar{z}(w, t_0)\} = 1.$$

This is an assertion about the zeroth moment. The assertion about all the remaining moments is $\partial_n \bar{z}\, dz = dz^n_+$. Here the following notation is used. Let $z(w) = w + \dots$. Then $z^n(w) = w^n + \dots$. The plus sign means that we take only the positive part (a polynomial on the sphere).

You may ask why the differential has a pole at only one point, although, by the symmetry principle, it must have another pole at the symmetric point. But the functions t_n are not analytic; these are functions of both the real and imaginary parts: $t_n = x_n + iy_n$. We have

$$\frac{\partial}{\partial t_n} = \frac{\partial}{\partial x_n} - i\frac{\partial}{\partial y_n}.$$

Therefore,

$$\frac{\partial}{\partial x_n} \bar{z}\, dz = dz^n_+ - d\bar{z}^n_+$$

and

$$\frac{\partial}{\partial y_n} \bar{z}\, dz = i(dz^n_+ + d\bar{z}^n_+).$$

The point is that we can write down hierarchies with respect to t_n and with respect to the complex conjugate variable $\bar{t}_n$. The result is a dispersionless Toda lattice.

The following remarkable formula holds:

$$F(t) = -\frac{t_0^2}{2} + \sum_{n \geqslant 0}(n-2)(t_n v_n + \bar{t}_n \bar{v}_n).$$

This formula contains a plenty of nontrivial identities. For example, the identity $\partial_n F = v_n$ looks almost naïve. But the v_n themselves depend on t_n in a puzzling way. Substituting and differentiating these dependences, we obtain precisely v_n.

For the ellipse, the function F can be calculated explicitly:

$$F = \frac{1}{2}t_0^2 \ln t_0 - \frac{3}{4}t_0^2 - \frac{1}{2}\ln(1 - 4|t_2|^2) + t_0\frac{|t_1|^2 + t_1^2\bar{t}_2 + \bar{t}_1^2 t_2}{1 - 4t_2\bar{t}_2}.$$

This example shows how F depends on the first three moments. (The complement to the ellipse has only three nonzero moments.)

For nonsimply connected domains, the first assertion (about the derivative of the Schwarz function) remains valid. The second one relies on the summation of a geometric progression for a Cauchy integral. In the late 1980s, studying the quantization operator for boson strings, Novikov and I developed a Fourier–Laurent theory for arbitrary Riemann surfaces. The basis z^n is replaced by another basis.

The formula written above is symmetric with respect to t and v. This suggests that it makes sense to try to apply it to the old classical problem of constructing a mapping of domains from a schlicht conformal mapping of their complements. The relation between these mappings may be nontrivial. For example, the complement of the ellipse is mapped onto the complement of the disk by a simple algebraic function, while the mapping of the interior of the ellipse to the interior of the disk is an elliptic function.

V. Yu. Ovsienko

Projective differential geometry: old and new

Lecture on January 6, 2000

1   Symmetry groups and differential invariants

I shall talk about projective and differential geometry and some classical theorems of the theory of smooth curves, which is now known as Sturm theory.

Let M be a smooth manifold. Following Klein (or Thurston), by a differential geometry on M I understand the study of the invariants of the action of a Lie group G (this action is assumed to be defined at least locally). I shall begin with examples.

Example 1. The manifold M is $\mathbb{R}^n$ on which the Euclidean group $\mathcal{E}(n) = \mathrm{SO}(n) \ltimes \mathbb{R}^n$ (the semidirect product of the rotation group and the translation group) acts.

Euclidean geometry studies the invariants of the action of the group $\mathcal{E}(n)$. The basic invariant is the metric $g = \sum (dx^i)^2$. The metric can be used to calculate diverse curvatures (of curves, surfaces, etc.).

Example 2. The manifold M is $\mathbb{R}^n$ on which the affine group $\mathrm{Aff}(n) = \mathrm{GL}(n) \ltimes \mathbb{R}^n$ acts.

There exists no Aff-invariant metric. Nevertheless, the notion of curvature can be defined.

For simplicity, consider the case of $n = 2$. Let $\gamma(t)$ be a locally convex curve, i.e., a curve having no inflection points. Then the vectors $\dot{\gamma}$ and $\ddot{\gamma}$ are not collinear (Fig. 1). In Euclidean geometry, curvature can be defined as a function of curves distinguishing between curves up to Euclidean transformations. The same approach can be used in the affine case. We have assumed that the vectors

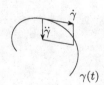

Figure 1. The curve γ

$\dot{\gamma}$ and $\ddot{\gamma}$ are not collinear. Let us write the vector $\dddot{\gamma}$ in the form of their linear combination as

$$\dddot{\gamma} = a\dot{\gamma} + b\ddot{\gamma}.$$

The curve $\gamma(t)$ is determined by the functions $a(t)$ and $b(t)$. Let us choose a parameter t so as to get rid of one of these functions. Namely, we choose t in such a way that the area of the parallelogram spanned by the vectors $\dot{\gamma}$ and $\ddot{\gamma}$ be constant: $[\dot{\gamma}, \ddot{\gamma}] = \text{const}$. Differentiating this equality, we obtain $[\dot{\gamma}, \dddot{\gamma}] = 0$, i.e., $\dddot{\gamma} = k(t)\dot{\gamma}$. The function $k(t)$ is called the *affine curvature*. It can be proved that two curves are affine-equivalent if and only if they have equal affine curvatures.

Note that the condition $[\dot{\gamma}, \ddot{\gamma}] = \text{const}$ is affine-invariant, because the area form is affine-invariant up to multiplication by a constant.

Exercise. Prove that

$$k(t) = \frac{[\ddot{\gamma}, \dddot{\gamma}]}{[\dot{\gamma}, \ddot{\gamma}]}.$$

Example 3. The manifold M is the projective space $\mathbb{RP}^n$, i.e., the set of straight lines in $\mathbb{R}^{n+1}$ passing through the origin.

This is our basic example.

It is convenient to choose affine coordinates on each affine hyperplane not passing through the origin. Almost all points of the projective space are parametrized by the intersection points of the lines with this hyperplane. The exception is the points belonging to a projective subspace of codimension 1.

On the linear space $\mathbb{R}^{n+1}$, the group $G = \text{SL}(n+1, \mathbb{R})$ acts (this is the group of volume-preserving linear transformations). This action carries over to $\mathbb{RP}^n$, because it transforms straight lines into straight lines.

According to our definition, projective geometry must study all invariants of geometric objects (curves, submanifolds, diffeomorphisms) in the projective space with respect to the action of this group.

Projective geometry used to be extremely popular. Why is it interesting? It is easy to see that the group $\text{SL}(n+1, \mathbb{R})$ contains both the Euclidean and affine groups as subgroups. It turns out that the projective symmetry group is maximal in a certain sense. Namely, no Lie group containing $\text{SL}(n+1, \mathbb{R})$ can act on an n-dimensional manifold even locally. Thus, if we define projective invariants (curvature and so on), these will be the strongest invariants, being invariants with respect to the maximal group; they are most difficult to find.

There is yet another example of a symmetry group maximal in the same sense. This is the group of conformal transformations. There are other maximal groups too; classification of maximal geometries was considered by many authors.

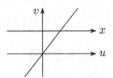

Figure 2. Coordinates on the projective line

In dimension 1, conformal geometry coincides with projective geometry. We shall start with this simplest case, namely, consider the geometry of the projective line.

2   The Schwarz derivative

Let us fix a coordinate $x = \frac{u}{v}$ on the projective line $M = \mathbb{RP}^1$ (Fig. 2). The group $G = \mathrm{SL}(2, \mathbb{R})$ acts on the projective line as

$$\begin{pmatrix} a & b \\ c & d \end{pmatrix} x = \frac{ax + b}{cx + d}.$$

The action of the center $\{\pm 1\}$ is trivial. The quotient group which effectively acts on $\mathbb{RP}^1$ is

$$\mathrm{PSL}(2, \mathbb{R}) = \mathrm{SL}(2, \mathbb{R})/\{\pm 1\}.$$

In this case, the only natural objects whose invariants can be studied are the diffeomorphisms $f \colon \mathbb{RP}^1 \to \mathbb{RP}^1$. For such a diffeomorphism, we define the *Schwarz derivative* by the formula

$$S(f) = \frac{f'''}{f'} - \frac{3}{2} \left(\frac{f''}{f'} \right)^2.$$

The meaning of this complicated expression is explained by the following classical theorem.

Theorem 1. *Diffeomorphisms f and g are projectively equivalent $\big($i.e., $g(x) = \frac{af(x)+b}{cf(x)+d}\big)$ if and only if $S(f) = S(g)$.*

The proof of the *only if* part is simple. But the proof of the *if* part is difficult.

Remark 1. In the one-dimensional case, the Euclidean transformations reduce to translations; therefore, two diffeomorphisms f and g of the straight line are equivalent with respect to the Euclidean group if and only if $f(x) = g(x) +$ const. The equivalent condition is $f' = g'$. Hence, in the Euclidean case, diffeomorphisms are distinguished from each other by the usual derivative.

Exercise. Prove that, on the affine line, the role of derivative is played by the logarithmic derivative f''/f'.

Now, let us find out where the Schwarz derivative arises from. I must say that this object is very universal; it emerges in tens or even hundreds of contexts. I have selected the two most classical and simple among them.

The usual derivative measures the degree to which a diffeomorphism f fails to preserve the distance between two close points x and $x + \epsilon$:

$$f(x + \epsilon) - f(x) = \epsilon f'(x) + (\epsilon^2).$$

In projective geometry, the distance between two points makes no sense, because any two (and even three) points can be mapped to any two other points by a projective transformation. But quadruples of points x_1, x_2, x_3, x_4 on the projective line do have a (unique) invariant; this is the cross ratio

$$[x_1, x_2, x_3, x_4] = \frac{(x_3 - x_1)(x_4 - x_2)}{(x_2 - x_1)(x_4 - x_3)}.$$

Traditionally, projective geometry begins with this invariant, which has been known since the time of the ancient Greeks. In projective geometry, the cross ratio plays the same role as the distance plays in Euclidean geometry. And the Schwarz derivative arises in the same way as the usual derivative. I found the following definition in a work of Elie Cartan, but probably, it had been known earlier. Consider points x, $x+\epsilon$, $x+2\epsilon$, and $x+3\epsilon$. Applying a diffeomorphism f and calculating the difference of the cross ratios, we obtain

$$[f(x), f(x + \epsilon), f(x + 2\epsilon), f(x + 3\epsilon)] - [x, x + \epsilon, x + 2\epsilon, x + 3\epsilon] = -2\epsilon^2 S(f) + (\epsilon^3).$$

However, historically, the Schwarz derivative was defined (in the nineteenth century) in a different way. Consider the simplest differential equation of the second order

$$y''(x) + u(x) \cdot y(x) = 0, \quad u(x) \in C^\infty(\mathbb{R}).$$

Take arbitrary linearly independent solutions $y_1(x)$ and $y_2(x)$ and consider the function $f(x) = \frac{y_1(x)}{y_2(x)}$ (in the domain where $y_2(x) \neq 0$). It turns out that $u(x) = \frac{1}{2}S(f)$. This is the second definition of the Schwarz derivative.

The space of solutions is two-dimensional. Therefore, if $\tilde{y}_1(x)$ and $\tilde{y}_2(x)$ is another pair of linearly independent solutions, then

$$\tilde{f}(x) = \frac{\tilde{y}_1(x)}{\tilde{y}_2(x)} = \frac{ay_1(x) + by_2(x)}{cy_1(x) + dy_2(x)} = \frac{af(x) + b}{cf(x) + d}.$$

Figure 3. An osculating circle

From this we can obtain a proof of the *if* part of Theorem 1. But it is difficult to prove that the space of solutions is two-dimensional; for this reason, the proof of the *if* part of the theorem is difficult.

These are only two of the many ways to obtain the Schwarz derivative. It has an enormous number of various properties and is used in many sciences, such as complex analysis, the theory of dynamical systems, and mathematical physics.

3 A remarkable property of curvatures

The classical four-vertex theorem is as follows. *Let γ be a closed convex curve on the plane. Then its Euclidean curvature has at least four critical points, i.e., the curvature function has at least four extrema.* The points of extremum of the curvature are called *vertices*.

This theorem was proved in 1909 by Indian mathematician Mukhopadhyaya.

Let us give two geometric interpretations for a vertex of a curve. Consider a generic curve γ. For each point on the curve, we can construct an osculating circle that approximates the curve with second-order accuracy. The radius of the osculating circle equals $1/k$, where k is the curvature. Since the circle approximates curve up to the second order, the curve passes from the exterior of the circle to its interior (Fig. 3). This happens at all points not being vertices. And only at the vertices, where the circle approximates the curve with third-order accuracy, the curve locally lies on one side of the circle.

Another way to geometrically define a vertex is related to the so-called *caustics* of curves. A caustic is the envelope of a family of normals to the curve (Fig. 4). The cusps of a caustic correspond to the vertices of the curve. The caustics of the ellipse have been known for a long time; even Apollonius and Jacobi knew about them.

The four-vertex theorem, in spite of its simplicity, attracts attention even now. It has tens of different proofs and numerous generalizations. One of the generalizations is the affine six-vertex theorem proved by Mukhopadhyaya in the same 1909 paper. The six-vertex theorem asserts that the affine curvature has at least six critical points.

Figure 4. A caustic

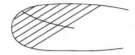

Figure 5. An affine caustic

The points of extremum of the affine curvature (affine vertices) have geometric descriptions similar to those given above for the points of extremum of the usual (Euclidean) curvature. First, an affine vertex is a point of an extremely tight contact between the curve and a conic. Secondly, the affine vertices can be defined with the use of affine caustics. An affine caustic is the envelope of a family of affine normals, and an affine normal is a tangent line to the curve swept out by the midpoints of the chords parallel to a tangent (Fig. 5). The affine vertices correspond to the cusps of an affine caustic.

All affine normals of the ellipse pass through its center. This is a degenerate case, as well as the circle in Euclidean geometry. For the ellipse, the osculating conic at any point is the ellipse itself, like a circle is its own osculating circle.

Recently, V. I. Arnold proved the four-vertex theorem by means of symplectic topology.

4 The Ghys theorem and the zeros of the Schwarz derivative

In 1995, Etienne Ghys from Lyons discovered the following amazing fact.

Theorem 2. *Let f be an arbitrary diffeomorphism of the projective line $\mathbb{RP}^2$. Then the Schwarz derivative $S(f)$ vanishes at least at four different points.*

At first sight, the Ghys theorem has nothing in common with the four-vertex theorem, except the number 4. But Ghys obtained his result as an analogue of the four-vertex theorem.

Instead of a diffeomorphism f, he considered its graph, which is a smooth closed curve on the two-dimensional torus $\mathbb{T}^2 = \mathbb{RP}^1 \times \mathbb{RP}^1$. This curve is nowhere vertical and nowhere horizontal (Fig. 6). If $f \in \mathrm{PSL}(2, \mathbb{R})$, then the

Figure 6. The graph of a diffeomorphism

graph is a hyperbola. It turns out that, in this situation, hyperbolas play the role of osculating circles. To be more precise, for every point x, there exists a unique projective transformation $g_x \in \mathrm{PSL}(2, \mathbb{R})$ approximating the diffeomorphism f up to a 2-jet. The Schwarz derivative corresponds to the deviation of the diffeomorphism at a given point from the corresponding projective transformation. The zeros of the Schwarz derivative are the points at which the projective transformation approximates f up to a 3-jet. Ghys used these considerations to obtain the first proof of his theorem, which follows classical Kneser's proof of the four-vertex theorem based on osculating circles.

The proof of the Ghys theorem was published by S. Tabachnikov and myself.

5   The relation to the Lorentzian geometry

In reality, the Ghys theorem is not an analogue of the four-vertex theorem; it is precisely the four-vertex theorem, but in Lorentzian geometry. Consider the flat Lorentzian metric $g = dx\,dy$ on the torus; the isotropic cone is formed by the vertical and horizontal directions. Let us calculate the Lorentzian curvature of the curve $\gamma = (x, f(x))$ (this curve is precisely the graph of the diffeomorphism f). We parametrize it by a parameter t. We have $\dot{\gamma} = (\dot{x}, f'\dot{x})$. Let us choose the parameter t in such a way that $\|\dot{\gamma}\| \equiv 1$, i.e., $\dot{x} = \frac{1}{\sqrt{f'}}$. The Lorentzian curvature is equal to the length of the vector of second derivative. Let us evaluate it:

$$\ddot{\gamma} = \left(\frac{1}{\sqrt{f'}}, \sqrt{f'} \right)^{\cdot} = \frac{1}{2}\left(-\frac{f''}{(f')^2}, -\frac{f''}{f'} \right).$$

Thus, the Lorentzian curvature equals

$$\|\ddot{\gamma}\| = \frac{1}{2}\frac{f''}{(f')^{3/2}} = k(\gamma).$$

This gives the quite unexpected expression

$$k' = \frac{1}{2\sqrt{f'}}S(f). \tag{1}$$

Figure 7. A strictly convex curve in $\mathbb{RP}^2$

It turns out that the Schwarz derivative is related to the Lorentzian curvature. The points of extremum of the curvature are the zeros of the Schwarz derivative.

Remark 2. Expression (1) can be written in the more elegant form $2dk\,dt = S(f)$ (as a quadratic differential). This expression is invariant. The question arises: What are the Lorentzian metrics for which such a relation holds? It can be proved that it holds precisely for the Lorentzian metrics of constant curvature

$$g = \frac{dx\,dy}{(axy + bx + cy + d)^2}.$$

6   The decisive intervention of projective geometry

All the results mentioned so far referred to different geometries, such as Euclidean, affine, and Lorentzian. How are they related to projective geometry? There is one general assertion which implies everything else.

Consider a closed smooth curve $\gamma \subset \mathbb{RP}^n$. In 1956, M. Barner introduced the notion of strictly convex curves and proved a remarkable theorem. Let us call a closed curve γ in the projective space *strictly convex* if, through any $n-1$ points of the curve γ, there passes a hyperplane that intersects γ only in these points. An example of a strictly convex curve in $\mathbb{RP}^2$ is shown in Fig. 7.

The theorem of Barner is as follows.

Theorem 3. *Any strictly convex curve in $\mathbb{RP}^n$ has at least $n + 1$ flat points.*

Recall that a flat point is a point at which the tangent hyperplane approximates the curve with an accuracy of order higher than usual, i.e., the vectors $\dot{\gamma}$, $\ddot{\gamma}$, ..., $\gamma^{(n)}$ are linearly dependent. In the 2-dimensional case, the flat points are inflection points.

Let us show that all theorems stated above reduce to the Barner theorem.

We start with the four-vertex theorem. Consider the Veronese mapping $v\colon \mathbb{R}^2 \to \mathbb{RP}^3$, which is defined by

$$(x, y) \mapsto (x^2 + y^2 : x : y : 1).$$

This mapping establishes a one-to-one correspondence between the circles in the plane $\mathbb{R}^2$ and the hyperplanes in $\mathbb{RP}^3$. The vertices of a convex curve γ in the plane $\mathbb{R}^2$ correspond to the flat points of the curve $v(\gamma)$ in $\mathbb{R}P^3$. Indeed, consider an osculating circle for the curve γ. The equation of the circle is precisely the equation of a hyperplane in $\mathbb{R}P^3$. The image of $\mathbb{R}^2$ in $\mathbb{R}P^3$ is a two-dimensional surface. The image of the circle is the intersection of this two-dimensional surface with a hyperplane. That the curve has a vertex at the point under consideration means (at least in the generic case) that the curve lies locally on one side of the osculating circle. In such a case, the image of the curve lies locally on one side of the hyperplane, because the image of $\mathbb{R}^2$ locally divides $\mathbb{R}P^3$ into two parts. This corresponds precisely to a flat point.

It is also easy to verify that the image of a convex curve under the Veronese mapping is a strictly convex curve. Indeed, for a convex curve in the plane, any straight line passing through its two points does not intersect this curve at other points. The straight line is a special case of the circle; therefore, its image also lies in some hyperplane. This hyperplane has the required property, namely, it intersects $v(\gamma)$ at precisely two points.

To prove the six-vertex theorem, we must consider the mapping $\mathbb{R}^2 \to \mathbb{R}P^5$ defined by

$$(x,y) \mapsto (x^2 : xy : y^2 : x : y : 1).$$

This mapping establishes a one-to-one correspondence between conics and hyperplanes. An affine vertex is a point at which the osculating conic approximates the curve better than usual. Such conics correspond to hyperplanes which approximate the image of the curve better than usual. In turn, such hyperplanes correspond to flat points.

To prove the Ghys theorem, we consider the Segre mapping $\mathbb{R}P^1 \times \mathbb{R}P^1 \to \mathbb{R}P^3$, which is defined by

$$((x : y), (y : z)) \mapsto (xz : xt : yz : yt).$$

It turns out that the zeros of the Schwarz derivative correspond to the flat points of the image. This follows readily from the interpretation of the zeros of the Schwarz derivative based on osculating hyperbolas.

A modern proof of the Barner theorem and its reduction to the four-vertex theorem are given in a recent paper of S. Tabachnikov.

7 Discretization

O. Musin and V. Sedykh (1996) proved a discrete analogue of the four-vertex theorem. Let P be a convex m-gon in the plane. Consider the circle passing through its three successive vertices v_i, v_{i+1}, and v_{i+2}. The vertices v_{i-1} and v_{i+3} can lie either on different sides or on one side of this circle. If they lie on one side, we say that this triple of consecutive vertices is *extremal*.

Theorem 4. *For any convex m-gon, where* $m \geqslant 4$, *there exist at least four extremal triples.*

Exercise. Formulate a discrete version of the theorem about six affine vertices.

Now, let us formulate the discrete version of the Ghys theorem. We replace a diffeomorphism f by a pair of ordered sets of points in $\mathbb{RP}^1$; we denote them by $X = (x_1, \ldots, x_m)$ and $Y = (y_1, \ldots, y_m)$. The graph of f is replaced by a polygonal line. The Schwarz derivative becomes the difference of cross ratios

$$[x_i, x_{i+1}, x_{i+2}, x_{i+3}] - [y_i, y_{i+1}, y_{i+2}, y_{i+3}].$$

The discrete version of the Ghys theorem asserts that this difference changes sign at least four times.

The Barner theorem can be discretized too. The definition of a strictly convex m-gon remains the same. The discrete version of the Barner theorem can be proved by induction on the number of polygon vertices.

Discretization is interesting because smooth theorems are weaker than their discrete versions, as they are obtained from the latter by passing to the limit. At the same time, the proofs of many discrete theorems are simpler; they can be obtained by induction on the number vertices.

The four-vertex theorem has two discrete versions. The second discrete version, which was suggested by Wegner, differs from the theorem of Musin and Sedykh in that the circumscribed circles are replaced by inscribed ones.

Most amazing is that there exists a discrete version of the four-vertex theorem which is almost a hundred years older than the theorem itself. This is the celebrated 1813 lemma of Cauchy, who invented it for the purpose of proving the Cauchy theorem on the rigidity of convex polyhedra.

Lemma (Cauchy). *Let P and P' be convex m-gons, where* $m \geqslant 4$. *Suppose that these polygons have equal respective sides. Then the difference between the respective angles of these polygons changes sign at least four times.*

Cauchy gave an incorrect proof of this lemma. It was corrected by Hadamard.

Acknowledgments

The author sincerely thanks V. Prasolov for carefully taking notes and preparing this lecture for publication.

S. V. Matveev

Haken's method of normal surfaces and its applications to classification problem for 3-dimensional manifolds – the life story of one theorem

Lecture on September 7, 2000

In mathematics' development, there are milestones, problems whose solutions take us to new discoveries and understanding of the intrinsic nature of mathematical objects. In topology, one such problem is that of classification of three-dimensional manifolds. In the course of investigation of this problem were the following most important stages.

- 1960 Haken: in order to understand the structure of a three-dimensional manifold, it is necessary to study two-dimensional surfaces $F^2 \subset M^3$ inside it. This method proved to be highly productive. Over half of all papers in three-dimensional topology which appeared since then are based on this method. The classification theorem for sufficiently large manifolds was obtained precisely with the help of this method.

- 1980 Gabai: it is necessary to study foliations, that gives information on manifolds.

- 1980 Thurston: it is necessary to study geometries on manifolds, hyperbolic manifolds, elliptic manifolds, Sol-manifolds, and so on, there are 8 homogeneous geometries in total.

- 1990 Witten: it is necessary to consider state sums and use them to construct invariants of three-dimensional manifolds.

- Recently, the interest in algorithmic topology has strengthened. There, Haken's method plays a key role.

What does Haken's method consist in? It is hopeless to attempt to consider all surfaces in a three-dimensional manifold – there are just too many of them. Even if surfaces are considered up to isotopy, there are still too many. Hence it is natural to single out a certain class $\mathcal{N}$ of surfaces in the manifold so that it possess two properties:

(1) class $\mathcal{N}$ is informative (for instance, in the following sense: the class contains all interesting surfaces in a given manifold);

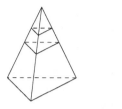
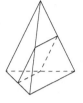

Figure 1. Admissible intersections

Figure 2. Prohibited intersections

(2) class $\mathcal{N}$ admits an explicit description.

The notation $\mathcal{N}$ refers to the notion of a "normal surface." Let me explain what a normal surface is. Further on it will always be assumed that our manifold M^3 is triangulated, i.e., decomposed into tetrahedra so that every two tetrahedra either do not intersect or intersect along a common vertex, edge, or face. Roughly speaking, a normal surface (with respect to a given triangulation) is defined in such a way: it intersects all tetrahedra in the triangulation nicely.

First I shall say how a normal surface is allowed to intersect tetrahedra and then will list prohibited intersections (see Figs. 1 and 2).

A surface may intersect a tetrahedron along a triangle (or several parallel copies of a triangle) or along a rectangle (Fig. 1). For every tetrahedron there are 4 types of triangles (each type corresponds to one of the faces) and 3 types of rectangles (each type corresponds to a pair of opposing edges).

Generally speaking, the intersection of the surface with a tetrahedron must consist of discs, and the boundary of each disc must intersect each edge no more than once. However, the surface itself might intersect the edge many times.

Now I can explain why the class of normal surfaces is informative, i.e., contains all interesting surfaces. First, we should find out which surfaces are uninteresting. Uninteresting are those surfaces that lie in $\mathbb{R}^3$, because such surfaces can be found in all three-dimensional manifolds.

All surfaces with "tubes" (Fig. 3) should be prohibited. A tube is characterized by having a disc which intersects the surface along its boundary such that the intersection curve does not bound a disc on the surface. Such discs are

Figure 3. A surface with a tube

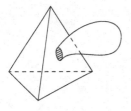
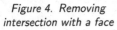

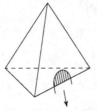

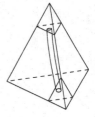

Figure 4. Removing intersection with a face

Figure 5. Removing intersection with an edge

Figure 6. Removing a "tube"

called *compressing*. The name is due to the fact that a compressing disc can be strongly compressed and thus the surface can be simplified (cut along the disc). An incompressible surface in M^3 is a surface which has no compressing discs.

A manifold M^3 is called *irreducible* if every embedded sphere $S^2 \subset M^3$ bounds a ball. Any three-dimensional manifold is a connected sum of irreducible ones, with the exception of the manifold $S^2 \times S^1$, which is irreducible but is not a nontrivial connected sum. Thus, if we understand the structure of irreducible three-dimensional manifolds, we will understand the structure of all three-dimensional manifolds.

Theorem 1. *Every closed incompressible surface in an irreducible manifold M^3 is isotopic to a normal surface.*

Proof. Consider an arbitrary closed incompressible surface and shift it in general position with respect to tetrahedra of the triangulation. If there is a prohibited situation (a tube intersecting a tetrahedron's face in a circle), then there is a compressing disc D. The surface is incompressible, therefore the boundary ∂D of that disc bounds a disc D' on the surface, see Fig. 4. Since the manifold is irreducible, the sphere $D \cup D'$ bounds a ball, which can be dragged in or out of the tetrahedron. After that, the prohibited intersection would be destroyed.

If there is a prohibited situation shown in Fig. 5, then the intersection of such a component of the surface with the tetrahedron is destroyed in an obvious way, by means of an isotopy of the surface.

Further, if inside a tetrahedron is a tube that connects two triangular sections (Fig. 6) then that tube can be destroyed by the incompressibility of the surface.

The remaining cases are treated in a similar fashion. $\qquad\square$

The first requirement is thus met: the class $\mathcal{N}$ is informative.

Now it is necessary to check whether the second requirement is met as well. For that, we need to describe the class of all normal surfaces explicitly. Each normal surface can be presented by an integer vector. This is done in the following way. Each tetrahedron contains 7 types of allowed intersections. Denote all possible types of allowed intersections (with all tetrahedra) by $E_1, \ldots, E_n$; here $n = 7t$, where t is the number of tetrahedra in the triangulation. Then we can assign to the surface F an integer vector $(x_1, x_2, \ldots, x_n)$, where x_i is the number of triangles or rectangles of type E_i in the surface's intersection with tetrahedra. It is easy to see that if all these numbers are known then the surface can be easily reconstructed from them: the pieces of the surface can only be put together in a unique way. However, not every collection of numbers gives a surface. For instance, the collection $(1, 0, \ldots, 0)$ cannot correspond to a (closed) surface.

Let us find out which vectors can be realized by surfaces. Consider a tetrahedron, choose a face of it, and single out one of the angles of that face. Draw a segment joining the sides of this angle. The segment may belong to a triangular section of type E_i or to a rectangular section of type E_j. The chosen face belongs to just one more tetrahedron. In that tetrahedron, the segment in question may belong to sections of type E_k or of type E_m. As a result, we obtain an equation $x_i + x_j = x_k + x_m$. In addition, if a surface is embedded then no tetrahedron can contain rectangular sections of two different types (such sections always intersect); a solution of the system of equations thus obtained is called admissible if it does not contain two, simultaneously nonzero x_i corresponding to distinct types of rectangular sections of the same tetrahedron.

Theorem 2. *The set of all normal surfaces is in one-to-one correspondence with the set of all admissible integer nonnegative solutions of the above system of equations.*

The proof of this theorem is really simple, and I will not dwell on it.

The number of equations in the system is $6t$, therefore the system is underdetermined. It is likely to have many solutions. An important observation by

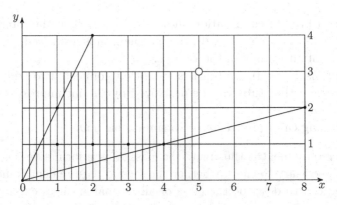

Figure 7. The system of inequalities

Haken is that the solution set for this system has a finite basis of fundamental solutions.

A solution $\bar{x}$ is called fundamental (indecomposable) if an equality $\bar{x} = \bar{y} + \bar{z}$, where $\bar{y}, \bar{z}$ are nonnegative integer solutions, implies that either $\bar{y} = 0$ or $\bar{z} = 0$.

Theorem 3 (Haken). *The set of fundamental solutions is finite.*

A similar theorem is well known in linear algebra, but there, all kinds of coefficients are allowed, integer and noninteger, positive and negative. Here the coefficients must be nonnegative integers.

The proof of Haken's theorem is rather easy; instead of a general proof, I will consider a simple example.

First of all, let me point out the following fact. We are looking for non-negative solutions; in doing so, we include inequalities into our system. We can assume that the system consists of homogeneous inequalities only. Indeed, equation $x = 0$ is equivalent to the system of two inequalities $x \geqslant 0$ and $x \leqslant 0$.

Consider the system of inequalities $-x + 4y \geqslant 0$ and $2x - y \geqslant 0$ (Fig. 7). Add up the coordinates of fundamental solutions which belong to the lines: $(5, 3) = (4, 1) + (1, 2)$. It is clear that all fundamental solutions are contained in the domain $x \leqslant 5$, $y \leqslant 3$. Indeed, if a solution is outside of that domain then one of the two fundamental solutions, either $(4, 1)$ or $(1, 2)$, can be subtracted from it. In general case the argument is similar.

Now I can tackle one of my goals, to describe the algorithm for recognition of the unknot. The unknot is characterized by its property to bound an embedded disc. This is a question about a surface. (Does there exist a surface of a given type?) In this case the surface is nonclosed, though. However, everything I have said relates to nonclosed surfaces as well, although the number of equations is less: the segments in triangles on the boundary of the manifold produce no equations. We only need to disregard triangles and rectangles that are adjacent to the surface's boundary.

For precision, the knot should be replaced by a solid torus, which is a tubular neighborhood of the knot. A knot is trivial if and only if there is a disc spanning some longitude of the solid torus.

It is easy to show that, if there is at least one disc spanning a longitude then such a disc can be found among fundamental surfaces. As soon as this is known, the recognition algorithm for the unknot is very simple, at least in the theoretical sense. Let us remove a tubular neighborhood of the knot and triangulate the manifold thus obtained. Write down the system of equations (inequalities) and find fundamental solutions. Then, for each fundamental solution we check whether it gives a disc whose boundary is a longitude. This is easy to do by calculating the Euler characteristic of the surface and checking whether its boundary is a circle intersecting the meridian just once.

Now I would like to tell about the classification of sufficiently large three-dimensional manifolds. An irreducible manifold M^3 is called *sufficiently large* if it contains an incompressible surface other than S^2, $\mathbb{R}P^2$, or D^2. For instance, every orientable irreducible manifold with a boundary is either sufficiently large or homeomorphic to a handlebody of some genus. Closed manifolds with infinite first homology groups are also sufficiently large.

Theorem 4 (Haken, Waldhausen, Johannson, Hemion). *There is an algorithm for recognition of sufficiently large three-dimensional manifolds. (For every two sufficiently large manifolds the algorithm says if they are homeomorphic.)*

It is easy to extract from this theorem the proof of existence of algorithmic classification for sufficiently large three-dimensional manifolds. Namely, we construct the algorithm for enumerating all three-manifolds first. For that, we need to choose among three-dimensional simplicial complexes those that are manifolds. The list thus obtained would contain duplicates. We can then get rid of duplicates by applying the recognition algorithm.

The story of the classification theorem for sufficiently large manifolds went as follows. It was firstly proven by Haken, after he'd constructed the theory of normal surfaces, in 1962. However, shortly afterwards a serious gap was found in his proof. For a long time, various mathematicians (Waldhausen, Johannson, Jaco, Shalen, and others) were attempting to close it. Incidentally, a great deal of work was done, in particular, this led to development of the theory of characteristic submanifolds.

At last, a crucial obstacle was singled out: for a very particular class of manifolds Haken's method does not work. This obstacle consists of so-called

Stallings manifolds. Later I will explain why Haken's method does not work for them.

The obstacle related to Stallings manifolds was overcome by Hemion in 1976. He managed to solve the recognition problem for Stallings manifolds by an independent method, which had nothing to do with Haken's theory.

After that, it was announced that classification theorem for sufficiently large manifolds had been proved. A few expository papers on that subject appeared, and Hemion's book was published in 1991. The theorem is important, a lot of papers refer to it.

All publications on this topic were structured according to the same pattern. First, the Haken's proof and the arising obstacle were described, then Hemion's way of overcoming that obstacle was exposed, and finally the conclusion was made that the theorem had been proved. Nonetheless, no text claiming to be comprehensive ever appeared. This gave rise to a natural question, why are there no other obstacles? I decided to understand that and write down a complete proof. It turned out that there indeed was another obstacle, which I called Stallings quasi-manifolds and which could not be overcome by Hemion's method. There were no more obstacles. Doing away with the second obstacle required very powerful methods of theory of surface homeomorphisms, which were developed by Thurston much later than 1976, the time when the theorem's proof had been announced. Thurston's work appeared in the eighties, and we need the algorithmic version of his methods, the construction of so-called train tracks suggested by Bestvina and Handel even later, in 1995.

Thus, until 1998 when my paper appeared in Russian Mathematical Surveys, the theorem in fact remained unproven. Working out the complete proof took 20 years.

The classification theorem for sufficiently large manifolds immediately implies the existence of algorithmic classification of knots. Or, rather, it follows not from the theorem itself, but from its method of the proof: it is necessary to repeat the proof of Haken's theorem all the while watching how the meridian of the tubular neighborhood of the knot behaves.

The classification theorem is proved by means of a construction called a *hierarchy*, or a *skeleton*. Let us consider a sufficiently large manifold. It contains an incompressible surface. Let us cut the manifold along that surface. We will obtain one or two manifolds. They will be called chambers. Chambers have boundaries, hence they are sufficiently large. Cut them along incompressible surfaces, and so on. We stop when all remaining chambers are balls. It is also necessary to make sure that the surfaces be in general position. For instance, no four surfaces should intersect in the same point. Technically it is more convenient to erect walls rather than cut the manifold. In the former case, inside the manifold we obtain a two-dimensional polyhedron P^2, which I call a skeleton. When making cuts, the result is called a hierarchy. However, in this case it is hard to follow what is actually going on. Considering a skeleton (an object) rather than a hierarchy (a process) substantially simplifies proving the classification theorem.

The proof that we finally stop is based on the notion of the *complexity* of a three-dimensional manifold. Cuttings always decrease the complexity (unless it is zero).

Consider another manifold and construct a skeleton for it. The first important observation is, if the skeletons are homeomorphic then the manifolds are homeomorphic as well. Indeed, a homeomorphism between the boundaries of two balls can be extended to a homeomorphism between the balls by means of a cone construction.

The second observation is as follows. Suppose that we can impose on the erection of admissible walls restrictions so strict that at every step it is possible to erect only a finite amount of walls (up to a homeomorphism of the manifold onto itself). Then only a finite collection of skeletons can be constructed for every manifold, and the manifolds are homeomorphic if and only if the collections of skeletons are pairwise homeomorphic. Indeed, if, on the one hand, some two skeletons are homeomorphic then, as it has already been shown, the manifolds themselves are homeomorphic. On the other hand, if two manifolds are homeomorphic then their associated collections of skeletons are also homeomorphic, because their construction is defined up to a homeomorphism.

The whole thing is reduced to ensuring the finite choice of the ways to erect an admissible wall. Let us insert surfaces of minimal complexity. The *complexity* of a surface F is the number $c(F) = -\chi(F) + N$, where N is the number of intersection points with singularities of the polyhedron obtained at the previous step. The finiteness theorem for the number of surfaces of minimal complexity is a version of the same theorem from the theory of normal surfaces which takes into account surface's complexity. The number of fundamental surfaces with regard to the complexity remains finite as well.

Haken noticed the following fact. Suppose that at each step all fundamental surfaces have positive complexity. Then surfaces of minimal complexity form a part of fundamental surfaces, in particular, there are only finitely many surfaces of minimal complexity. This statement is obvious, since the complexity is additive with respect to summation of surfaces. Therefore, should a surface F be the sum of two surfaces F_1 and F_2 of positive complexity, then $c(F) > c(F_2)$ and $c(F) > c(F_1)$. Hence any surface of minimal complexity is fundamental.

This is an easy case. Haken didn't go any further. At this point a question arises, what do we do in the case when there are fundamental surfaces of zero

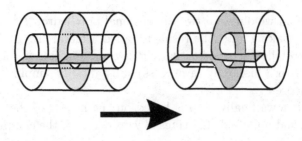

Figure 8

complexity? Haken thought that it would be easy to do away with this case, but it turned out it wasn't so.

What kind of surfaces have zero complexity? In other words, when do we have equality $-\chi(F) + N = 0$? The Euler characteristic can be positive only in cases of the sphere, the projective plane $\mathbb{R}\mathrm{P}^2$ and the disc. But due to incompressibility spheres cannot occur. For the same reason projective planes cannot be present either, because the boundaries of their regular neighborhoods are spheres. Discs can be disregarded by incompressibility: surfaces are inserted so as to be incompressible.

Thus, a surface of zero complexity does not contain singular points and has zero Euler characteristic. Surfaces with zero Euler characteristic are annulus, Möbius band, torus, and Klein bottle. Tori and Klein bottles are no problem for us, since any chamber contains only a finite number of such surfaces (up to homeomorphisms of the chamber fixed on its boundary).

The case of annuli and Möbius bands is much more complicated, because the number of such surfaces in a chamber can be infinite even up to homeomorphisms fixed on the boundary of the whole manifold.

For instance, by twisting several times one annulus along another (see Fig. 8), it is possible to construct an infinite number of nonequivalent annuli, since twists of a chamber along one annulus shift the boundary of the other.

The correct strategy consists in decomposing all annuli into two types: longitudinal and transverse ones (the cases of Möbius bands is treated in a similar fashion, so we can omit it). Here I have the following situation in mind. The whole annulus lies inside a chamber, and the boundary of the annulus belongs to the chamber's boundary. An annulus A is called *longitudinal* if any other annulus A_1 can be shifted so that for the new annulus A_1' the following condition hold: the intersection $A \cap A_1'$ either is empty or consists of circles parallel to the core circle of A (Fig. 9). Otherwise the annulus is called *transverse*.

A circle bounding a disc inside an annulus can be destroyed by incompressibility. Therefore the intersection of any annulus with a transverse annulus can

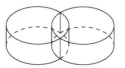

Figure 11.
Structure of
direct product

Figure 9.
Longitudinal annulus

Figure 10.
Transverse annulus

be transformed to the one shown in Fig. 10.

It turned out that longitudinal annuli are much nicer than transverse ones. The presence of annuli was a problem for us, because twists along an annulus could lead to chamber homeomorphisms which cannot be extended to homeomorphisms of the whole manifold. However, a twist along any other annulus leaves a longitudinal annulus invariant, since the intersection of those annuli consists of circles parallel to the core circle of the longitudinal annulus. Such circles are invariant, therefore a longitudinal annulus is not affected by twists. It implies that there are only a finite number of longitudinal annuli.

Now we can describe the final version of the method of inserting surfaces. If there are no annuli, the previous method works. If there is at least one longitudinal annulus, we insert it; there are only finitely many longitudinal annuli. We stop when only balls and chambers without longitudinal annuli but with transverse annuli are left. It turns out that such chambers are easy to describe: they have the form $F \times I$ or $F \mathbin{\widetilde{\times}} I$ (a nontrivial bundle with a fiber segment).

This is proven quite easily. Let us draw a transverse annulus as a cylinder. It follows from it being transverse that there exists another annulus which intersects it along two vertical segments (Fig. 11). Then in a neighborhood of the union of these two annuli there exists a direct product structure. The construction is extended, and each time the structure of a direct product, more precisely, of a bundle with fiber a segment, is preserved (in the process of extension some segment may flip over, i.e., we might obtain a skew product).

I will not pay attention to the balls. If there is a chamber $F \times I$, there is something glued to it, and so on. In the end we obtain a Stallings manifold. For it, Haken's algorithm doesn't work, because after the first cut we get a manifold $F \times I$, which contains lots of different annuli. The annuli correspond to curves on the surface, and the number of curves on a surface is infinite. Hence there is no hope to make the procedure of choosing an annulus finite.

For manifolds containing Stallings manifolds the classification problem must be solved separately. This was what Hemion did. After that, it was announced that the classification problem for sufficiently large three-dimensional manifolds had been solved. However, there are Stallings quasi-manifolds, which are glued

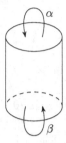

Figure 12. Two involutions

from $F \widetilde{\times} I$ in the following way. Let us take the manifold $F \times I$ and on each its boundary component consider some orientation reversing involution (homeomorphism of period 2) without fixed points. Perform the gluing along those involutions (Fig. 12). The manifold thus obtained is a Stallings quasi-manifold. For such manifolds Haken's method does not work for the same reasons as earlier, and the problem must be solved separately. A simple argument reduces the recognition problem for Stallings quasi-manifolds to the following problem on surface homeomorphisms. Suppose there are two homeomorphisms $f, g \colon F \to F$ of the surface onto itself. It is necessary to find out whether there exists a number n such that $f^n = g$ (the equality is up to isotopy). Surface homeomorphisms considered up to isotopy are described in terms of homomorphisms of fundamental groups. Therefore the problem is purely algebraic. It is difficult only because the number n is not bounded. If there is an upper bound for n, the problem immediately becomes easy.

A desired bound for n can be obtained with the help of Thurston's theory of so-called *stretching factors*. The theory is simplest in the case of a torus. Torus homeomorphisms are given by matrices of order 2 with determinant 1. Imaginary eigenvalues happen very rarely; this is not interesting. If the eigenvalues are real, they have the form λ and λ^{-1}. Thus, in one direction we have an expansion and in the other a contraction. In the torus case it had long been known, and Thurston proved that this was how all surface homeomorphisms work, not just those of a torus. For every surface homeomorphism h there is a stretching factor $\lambda(h) > 1$. Then there is the following bound for n: it does not exceed any N such that $\lambda(f)^N \geqslant \lambda(g)$. This completes the proof of the theorem on algorithmic recognition of sufficiently large manifolds and that of the theorem on their algorithmic classification.

Printed in the United States
by Baker & Taylor Publisher Services